Secondary Metabolites from Natural Products: Extraction, Isolation and Biological Activities

Secondary Metabolites from Natural Products: Extraction, Isolation and Biological Activities

Guest Editors

Radosław Kowalski
Tomasz Baj

MDPI

Basel • Beijing • Wuhan • Barcelona • Belgrade • Novi Sad • Cluj • Manchester

Guest Editors
Radosław Kowalski
Department of Analysis and Evaluation of Food Quality
University of Life Sciences in Lublin
Lublin
Poland

Tomasz Baj
Department of Pharmacognosy with Medicinal Plants Garden
Medical University of Lublin
Lublin
Poland

Editorial Office
MDPI AG
Grosspeteranlage 5
4052 Basel, Switzerland

This is a reprint of the Special Issue, published open access by the journal *Molecules* (ISSN 1420-3049), freely accessible at: https://www.mdpi.com/journal/molecules/special_issues/4N9X5907SM.

For citation purposes, cite each article independently as indicated on the article page online and as indicated below:

Lastname, A.A.; Lastname, B.B. Article Title. *Journal Name* **Year**, *Volume Number*, Page Range.

ISBN 978-3-7258-6129-3 (Hbk)
ISBN 978-3-7258-6130-9 (PDF)
https://doi.org/10.3390/books978-3-7258-6130-9

Cover image courtesy of Radosław Kowalski

Contents

About the Editors

Radosław Kowalski

Radosław Kowalski is a Full Professor in food technology and nutrition at the University of Life Sciences in Lublin, Department of Analysis and Food Quality Assessment. His research focuses on phytochemistry, extraction and the analysis of secondary metabolites (polyphenols, essential oils, alkaloids), bioactivity of plant extracts, dietary supplement technology, and instrumental techniques (GC, GC-MS, HPLC, LC-MS/MS, metabolomics). He also studies contaminants in food (heavy metals, pesticides, mycotoxins), the stability of fats and flavors, and agroecological applications of plant compounds in biocontrol. He is the author of over 260 publications, 2 patents, and implemented dietary supplement formulations.

According to the Stanford University ranking (2024), he is among the 2% most influential scientists worldwide. He has participated in national and international research projects and cooperates with centers in Algeria, Brazil, Pakistan, and Bangladesh. He serves on Editorial Boards of international journals and supervises Ph.D. students and graduates, while actively performing expert and reviewing functions.

Tomasz Baj

Tomasz Baj is an Associate Professor at the Medical University of Lublin, working in the Department of Pharmacognosy with the Medicinal Plant Garden. He is a Pharmacist and specialist in industrial pharmacy and herbal medicine. His scientific activity covers phytochemistry, pharmacognosy, essential oils, and innovative forms of herbal medicines, with particular emphasis on the application of statistical modeling methods in phytochemical research. He is the author and co-author of more than 150 scientific and popular science articles and a co-inventor of over 10 patents, including novel methods of obtaining plant extracts, compositions and formulations based on essential oils, as well as innovative solutions in health prevention.

At the university, he teaches students of pharmacy and cosmetology and supervises the activities of the Student Scientific Society at the Department of Pharmacognosy. He is actively engaged in the scientific community as a reviewer and Editorial Board Member of several journals. For his achievements in research and teaching, he has received numerous awards and distinctions.

In parallel, he is actively involved in the professional community of pharmacists, serving as President of the Regional Pharmaceutical Chamber and as a member of the Polish Pharmaceutical Chamber.

Preface

Contemporary science faces the challenge of fully utilizing the potential of the secondary metabolites of plants, fungi, lichens, and microorganisms. Although these compounds do not directly participate in primary life processes, they play a key role in chemical communication, defense against pathogens and environmental stress, and, from the human perspective, they exhibit a wide spectrum of valuable biological activities. They serve as inspiration both for pharmacotherapy and for innovative solutions in food technology, agriculture, and industrial chemistry.

This monograph, published as a Reprint of the Special Issue of *Molecules* titled "Secondary Metabolites from Natural Products: Extraction, Isolation and Biological Activities", gathers current research results from this dynamically developing field. The collected articles cover diverse aspects: from the optimization of extraction and isolation methods, through the use of modern analytical and bioinformatics techniques, to the evaluation of bioactivity and potential practical applications. The volume includes both original studies presenting new chemical structures and their biological activities, as well as review papers that organize knowledge and indicate directions for further exploration.

The thematic scope includes, the following topics, among others: polyphenols and flavonoids, terpenoids and meroterpenoids, fungal and endophytic metabolites, lichen compounds, Maillard reaction products, and hemp extracts. The authors present the possibilities of applying these compounds in pharmacology, nutraceuticals, functional foods, crop biocontrol, and innovative environmentally friendly technologies.

The monograph is addressed to researchers, practitioners, and students engaged in phytochemistry, pharmacognosy, food technology, biotechnology, and related disciplines. The studies included here indicate that research on secondary metabolites can provide solutions that respond to global health, environmental challenges, and economic challenges, fitting into the concept of sustainable development.

We extend our sincere thanks to all authors for their valuable scientific contributions and to the reviewers for their insightful comments. Our gratitude also goes to the editorial team of *Molecules* for their support and professionalism at every stage of preparing this publication.

Radosław Kowalski and Tomasz Baj
Guest Editors

molecules

MDPI

Editorial

Secondary Metabolites from Natural Products: Extraction, Isolation and Biological Activities

Radosław Kowalski [1,*] and Tomasz Baj [2]

1 Department of Analysis and Evaluation of Food Quality, University of Life Sciences in Lublin, 8 Skromna Str., 20-704 Lublin, Poland
2 Department of Pharmacognosy with Medicinal Plants Garden, Medical University of Lublin, 1 Chodźki Str., 20-093 Lublin, Poland; tomasz.baj@umlub.pl
* Correspondence: radoslaw.kowalski@up.lublin.pl

1. Introduction

Secondary metabolites constitute an extremely diverse class of compounds produced by plants, fungi, lichens, and microorganisms. Although they do not directly participate in basic life processes, they perform key ecological and adaptive functions: they facilitate chemical communication, defense against pathogens and herbivores, tolerance to environmental stress, and allelopathy. From the human perspective, many of them exhibit valuable biological activity (antibacterial, antifungal, antiviral, anti-inflammatory, antioxidant, and cytotoxic), which for decades has made natural products one of the most important sources of inspiration for pharmacotherapy and food technology. Synthetic reviews confirm that nature consistently provides new leading structures (leads) and drugs, ranging from antibiotics and anticancer agents to modern antibody–drug conjugates and therapies for lifestyle diseases [1–3].

From a scientific point of view, research on "Secondary Metabolites from Natural Products: Extraction, Isolation and Biological Activities" has several complementary goals. First, it allows for the scientific substantiation of traditional and ethnopharmacological uses of plants—from verifying historical indications to identifying the compounds responsible for the observed effect. Second, it enables the "rediscovery" of species described in historical sources and the exploration of biodiversity in new ecosystems (equatorial forests, Arctic and desert extremes, the deep sea, etc.). Third, it supports the search for alternative drugs, nutraceuticals, and functional ingredients through the integration of phytochemistry, microbiology, and molecular biology with modern analytics (LC–HRMS/MS, metabolomics, etc.). Finally, it offers pathways to sustainable applications in agriculture (biopesticides, allelochemicals), the food industry (natural preservatives, antioxidants, colorants, and emulsifiers), and even in industrial chemistry (biocatalysts, "green" reaction media, etc.). These lines of research are consistent with the European "Farm to Fork" strategy within the European Green Deal, which assumes, among other things, a 50% reduction in the use of and subsequent risk posed by chemical pesticides by 2030 [4–9].

In this Special Issue, we present cross-sectional examples of the implementation of these assumptions: from the optimization of extraction and comparison of methods for polyphenol-rich plant materials (e.g., *Scutellaria baicalensis*), through the use of supercritical techniques to obtain fractions with biological activity (e.g., *Sorbus aucuparia*), to the isolation and characterization of new structures from plants and fungi (meroterpenoids from *Garcinia caudiculata*, secobutanolides from *Lindera obtusiloba*, and metabolites of *Dentipellis fragilis* and

 https://doi.org/10.3390/molecules30214233

endophytic *Xylaria*). Complementing these are review works (e.g., *Calendula officinalis* and *Foeniculum vulgare*, penicillides of *Penicillium/Talaromyces*, etc.), which organize knowledge and indicate research gaps, as well as studies in the area of ecophysiology and safety (allelopathy of lichens; Maillard reaction products as antioxidants). Together, they outline a map of current opportunities, from methodology and dereplication to applications in human health and agroecology.

1.1. Why Do We Study Secondary Metabolites?

Verification and consolidation of ethnopharmacology. Traditional knowledge is often a starting point for identifying bioactive molecules and mechanisms of action (e.g., alkaloids, terpenoids, and polyphenols). These studies transform empirical observations into evidence based on the standards of pharmacology and toxicology, serving the rationalization of the use of plant raw materials and the design of safe application forms [1–3].

Biodiversity as a resource for innovation. Extreme ecosystems (marine, Arctic, desert, etc.) and symbioses (endophytes, lichens, etc.) are "incubators" of unique structural scaffolds and biosyntheses. Maintaining, studying, and responsibly utilizing this biodiversity increases the chances of breakthrough discoveries [1–3].

Therapeutic and antimicrobial alternatives. In an era of antibiotic resistance and chronic noncommunicable diseases, natural products provide new points of engagement (targets) and pharmacophores, often with a multidirectional activity profile, which is advantageous in complex diseases [3].

Biocontrol and agroecology. Natural phytotoxins and biopesticides (including those of microbiological origin) can fit into integrated pest management (IPM), reducing environmental burden and the risk of pesticide residues in food. The development of this area supports the implementation of the Green Deal's goals [4–9].

Functional ingredients and "clean labels." The growing consumer demand for natural antioxidants, colorants, and preservatives directs attention to polyphenols, carotenoids, and phenolic metabolites with confirmed antioxidant and health-promoting activity, which are capable of replacing synthetic additives [10].

Sustainable media and catalysis. Research seeks "green" solvents and user-friendly technologies (e.g., pressurized water, supercritical CO_2, and natural eutectics) that reduce emissions, energy consumption, and waste, without compromising extract quality [11–14].

1.2. From the Biological Sample to the Molecule: Modern Approaches to Extraction and Isolation

A key stage is the extraction and fractionation of complex biological matrices. Conventional techniques (maceration, percolation, Soxhlet) remain useful but can be material- and solvent-intensive. Therefore, "green extraction" techniques are increasingly used, including extraction with supercritical CO_2 (SFE), ultrasound-assisted extraction (UAE) and microwave-assisted extraction (MAE), pressurized liquid/accelerated solvent extraction (PLE/ASE), and pressurized hot water extraction (PHWE). These techniques all display improved efficiency and selectivity with a smaller environmental footprint and better compatibility with downstream analysis [11–14]. For example, SFE–CO_2 is nontoxic and nonflammable and allows the "tuning" of solvent strength with pressure/temperature or a small addition of a polar modifier, which favors the isolation of both volatile and medium-polarity fractions. PHWE uses water above 100 °C under pressure that keeps it in a liquid state. Under such conditions, the dielectric constant and viscosity decrease, while diffusivity increases, meaning that water "behaves" more like an organic solvent, effectively extracting numerous metabolites (including phenolic ones) without the use of classical solvents [12,13].

Natural deep eutectic solvents (NADESs) are another promising solvent platform, characterized as mixtures of inexpensive, biodegradable components (e.g., choline, organic acids, sugars, etc.) with a wide solubility window, capable of selectively extracting polyphenols or alkaloids, often without the need for subsequent desolventization [11,15,16]. In this context, glycerol is sometimes considered an environmentally friendly component of extraction systems—mentioned in this study only as an example among "green" media.

The choice of extraction method and medium should be based on the research objective (chemical profile vs. bioactive fraction), the nature of the matrix (content of lipids, hydrolytic enzymes), the thermal sensitivity of the compounds, and downstream requirements (e.g., compatibility with LC–MS/MS). In practical terms, the principles of "green extraction" call for minimizing solvent toxicity, energy consumption, and waste while maintaining the quality and reproducibility of analytical data.

1.3. From Extract to Annotated Chemical Space: Analytics and Dereplication

Advances in analytical methods enable the rapid "mapping" of chemical composition and the prioritization of fractions. LC–HRMS/MS combined with bioinformatics has revolutionized metabolite annotation, driving a shift from classical dereplication (comparing spectra with databases, structure prediction) to molecular networking (GNPS) and its feature-based version (FBMN), which connect spectra into similarity networks, facilitating the identification of isomers and biosynthetic families and the integration with quantitative data. These tools shorten the path from the "dark chemical matter" to structural and biological hypotheses and also limit the repeated "discovery" of known compounds [17–20].

1.4. Biological Activity and Pathways to Applications

The evaluation of bioactivity includes standardized antibacterial/antifungal tests, antioxidant assays (e.g., radical systems, chemiluminescence, etc.), anti-inflammatory assays (mediators, cytokines, NO, etc.), cytotoxicity tests, and functional tests (target enzymes, receptors, etc.). In agriculture, biopesticides and allelochemicals are attracting increasing attention, which—with appropriate standardization and risk assessment—can support integrated plant protection programs and the implementation of public policy goals to reduce chemical pesticides [4–9]. In food technology, the priority is to replace synthetic additives with natural antioxidants and preservatives, which is facilitated by a growing body of data on the effectiveness and stability of polyphenols in various matrices [10].

2. Plant Secondary Metabolites

2.1. Polyphenols and Flavonoids

Polyphenols are widely distributed in the plant world. Tenório et al. [Contribution 1] isolated, among other substances, myricitrin, gallic acid, and ellagic acid from the leaves of *Eugenia uniflora* L. (Myrtaceae), which showed significant antifungal activity against *Candida albicans*, *C. glabrata*, and *C. auris* strains. The strongest effect was exhibited by fractions rich in myricitrin and ellagic acid (MIC 62.5–500 μg/mL). The isolated single constituents were characterized by weaker activity than the fractions of constituents, which indicates an important role of interactions among constituents in the complex mixture. *Eugenia uniflora* L. is a source of natural constituents characterized by anti-*Candida* potential, which is important in the context of the growing resistance to classical antifungal drugs.

Dzięcioł et al. [Contribution 2], in a study on the roots of *Scutellaria baicalensis* Georgi. (Lamiaceae), applied five extraction techniques, which made it possible to obtain ethanolic extracts differing in terms of the content of phenolic compounds and flavonoids. The highest antioxidant activity was exhibited by extracts after reflux extraction and Soxhlet extraction, and this activity was correlated with the highest content of the studied groups of biologically active compounds, while the extraction yield depended on the conditions used. The most important compounds identified in the extracts were flavonoids characteristic of this species, wogonin and oroxylin A, as well as other hydroxyflavone derivatives. Dzięcioł et al. demonstrated that in extraction using a Soxhlet apparatus, 5-hydroxymethylfurfural was formed, which constitutes a potentially undesirable constituent, and therefore they recommend reflux extraction with ethanol for 2 h as the most effective and safe method to obtain extracts.

Ryszczyńska et al. [Contribution 3] assessed the potential of rowanberry (*Sorbus aucuparia* L., Rosaceae) fruits as a natural agent for the biological protection of plants against pathogens of the genus *Fusarium* (*Fusarium proliferatum* and *F. culmorum*), including the effect on mycotoxin biosynthesis by these phytopathogens. Extracts were prepared by means of supercritical fluid extraction (SFE, using CO_2 with methanol as a co-solvent) under a variety of different temperature and pressure conditions. The rowanberry extracts showed diverse activity; for example, extracts obtained under conditions of 70 °C/300 bar exhibited the strongest inhibitory effect on the growth of *F. proliferatum* and on the reduction in ergosterol content, whereas in the case of *F. culmorum*, no clear reduction in colony growth was recorded, but the decrease in ergosterol content indicated a limitation of mycelial biomass. The *Sorbus* extracts exhibited different inhibitory–stimulating effects on the biosynthesis of various mycotoxins depending on the *Fusarium* species, indicating the potential of this plant in the biological protection of crops. The tested extracts from *Sorbus aucuparia* may be a promising tool for the biological control of *Fusarium* in cereal crops.

2.2. *Terpenoids and Meroterpenoids*

Valmiki et al. [Contribution 4] isolated two meroterpenoids from the leaves of *Garcinia caudiculata* Ridl. (Clusiaceae) from the island of Borneo, including, for the first time, caudiquinol containing a full geranylgeranyl chain, and a previously described benzofuranone lactone. The structures of the two isolated constituents were established (NMR and MS). The crude dichloromethane extract showed moderate antibacterial activity against a *Staphylococcus aureus* MSSA strain; however, the pure constituents did not exhibit significant antibacterial activity, and no cytotoxic effect was observed against a lung cancer cell line.

Yang et al. [Contribution 5], from the stems of *Lindera obtusiloba* Blume (Lauraceae), used traditionally in Oriental medicine as an anti-inflammatory and hepatoprotective agent, isolated new secobutanolides and determined their structures based on the analysis of spectroscopic data. The biological activity of the isolated secobutanolides was evaluated against bone-marrow-derived dendritic cells stimulated with lipopolysaccharide (LPS). The effect on the production of pro-inflammatory cytokines IL-6 and IL-12 p40 with anti-inflammatory action was assessed. Three of the compounds studied showed a strong inhibitory effect on cytokine synthesis and at the same time did not show cytotoxicity at the tested concentrations. The isolated constituents constitute a starting point for the development of new anti-inflammatory compounds.

Vella et al. [Contribution 6] prepared a review devoted to *Calendula officinalis* L. (Asteraceae) and *Foeniculum vulgare* Mill. (Apiaceae), which, from a synthetic perspective, indicates a rich set of biologically active constituents in these species: phenols (including phenolic acids and flavonoids), terpenes (monoterpenes, sesquiterpenes, triterpenes, and carotenoids), and alkaloids. Both species, present in phytotherapy, cosmetics, and dietetics for centuries, are currently included in many pharmacopeias. *C. officinalis* is characterized by antioxidant, anti-inflammatory, antimicrobial, cytotoxic, hypoglycemic, nootropic, and anticancer properties. Of particular importance in this species are flavonoids (quercetin, rutin, luteolin, etc.) and triterpenoids (including faradiol and calendula saponins)—these compounds are responsible for anti-inflammatory effects, wound healing, and liver and kidney protection, as well as potential antiviral properties (HSV, HIV). In turn, *F. vulgare* is valued for its aromatic fruits and essential oil. It shows antioxidant, antibacterial, antifungal, antiviral (HSV-1, PI-3), anti-inflammatory, anticancer, hepatoprotective, cardioprotective, gastroprotective, antidiabetic, and estrogen-like activity. Traditionally, it has been used in the treatment of digestive disorders, respiratory diseases, and ocular ailments and as a spice. Both *C. officinalis* and *F. vulgare* are plants with great potential for further research and fuller use in health prevention and in the development of nutraceuticals, dietary supplements, and herbal medicines.

3. Fungal and Endophytic Metabolites

3.1. Wood-Inhabiting and Endophytic Fungi

Sum et al. [Contribution 7] studied metabolites occurring in cultures of the rare wood-inhabiting fungus *Dentipellis fragilis* (Pers.) Donk (Hericiaceae), among which the structures of five compounds were described, including two new substances, dentifragilone A and dentifragilone B, along with three known derivatives of benzoic acid and drimane. In microbiological tests, only the methyl ester of 4-chloro-3,5-dimethoxybenzoic acid showed moderate activity against *Staphylococcus aureus*, whereas studies on specific cell lines (mouse L929 fibroblasts and the KB3.1 cervical cancer line) did not observe cytotoxicity for these constituents. The results indicate that even rare temperate-zone Basidiomycota fungi can provide new, hitherto undescribed metabolites, although their biological activity is sometimes limited and requires further verification.

Zhang et al. [Contribution 8] conducted chemical studies on the endophytic fungus *Xylaria* sp. Z184 isolated from the leaves of *Fallopia convolvulus* (L.) Á. Löve of the Polygonaceae family. From a methanolic extract of the culture of this strain, three new pyranone derivatives—fallopiaxylaresters A, B, and C—and a new bisabolane-type sesquiterpenoid named fallopiaxylarol A were isolated and characterized. In addition, known compounds were identified in the extracts studied, including other pyronones, sesquiterpenoids, isocoumarin derivatives, and an allylaryl ether. Studies of biological activity showed that the crude extract of the fungus strongly inhibited nitric oxide (NO) production in RAW264 macrophage cells. Selected compounds (e.g., pestalotiopyrone M, xylariaopyrone A and H) exhibited only weak antibacterial activity against *Staphylococcus aureus*. *Xylaria* sp. Z184 constitutes a rich source of compounds with diverse chemical structures. The results obtained confirm the potential of endophytic fungi in the search for new metabolites with anti-inflammatory and antimicrobial properties.

3.2. Penicillium and Talaromyces

The review by Salvatore et al. [Contribution 9] discusses penicillides, a unique group of secondary metabolites of these fungi, whose prototypical representative is penicillide, first described in the 1970s in species of *Penicillium*. These compounds occur mainly in fungi of the genus *Talaromyces*, and sporadically in other *Ascomycota* and even in plant tissues. Penicillides are characterized by a depsidone structure with an eight-membered heterocyclic ring and various side substituents. In terms of their biological activity, penicillides exhibit a broad spectrum of effects, including antibacterial, antifungal, antimalarial, and antiplasmodial activity, as well as having cytotoxic effects against various cancer cell lines. They also exhibit other biochemical actions such as calpain inhibition (with potential relevance in the treatment of muscular dystrophies and neurodegenerative diseases), elastase inhibition (COPD therapy), the modulation of oxytocin binding, while also exhibiting anti-inflammatory activity. The authors emphasize the potential of penicillides as starting structures for creating semi-synthetic derivatives with greater potency, especially in the context of increasing antibiotic resistance and the growing importance of respiratory diseases.

3.3. Marine and Extremophilic Microorganisms

The study by Wu et al. [Contribution 10] concerned the Arctic strain *Streptomyces* sp. MNP-1, isolated from a sample of marine origin. In order to increase the diversity of metabolites, the OSMAC ("One Strain, Many Compounds") strategy was used, consisting of culturing the microorganism under different conditions and on different media. In this study, 20 secondary metabolites were isolated and identified, some of which were reported for the first time in microorganisms. Biological tests confirmed that several of these compounds possess significant antibacterial (including against *Staphylococcus aureus* and *Escherichia coli*) and antifungal activity (e.g., against *Candida albicans*), whereas two compounds (phenazine and staurosporine) additionally showed moderate anticancer effects against lung (A549), breast (MCF-7) and liver (HepG2) cancer cell lines. The *Streptomyces* sp. MNP-1 strain is a promising producer of bioactive secondary metabolites, which may be of interest in the development of new antibiotics and potential anticancer drugs.

4. Lichen Metabolites and Allelopathy

Lichens are interesting symbiotic systems in which secondary metabolites play allelopathic roles. Bačkor et al. [Contribution 11] examined the effect of secondary metabolites of Australian lichens *Ramalina celastri* (usnic acid) and *Stereocaulon ramulosum* (a mixture of atranorin and perlatolic acid) on the growth and metabolism of the lichen photobiont *Asterochloris erici* grown under aposymbiotic conditions. Their study showed that the extracts acted phytotoxically, reducing the growth of photobiont cultures. The mixture of atranorin and perlatolic acid showed a stronger effect than usnic acid alone. Extracts from *S. ramulosum* caused greater damage to the photosynthetic system, a decrease in pigment content, an increase in lipid peroxidation, and a decrease in the levels of endogenous antioxidants. The changes also concerned the metabolism of organic acids, which was considered a sensitive marker of phytotoxicity. This study confirms that lichen secondary metabolites act as allelochemicals, regulating the balance between the partners of the lichen symbiosis and they can significantly affect ecosystem functioning.

5. Maillard Reaction Products as a Source of Bioactive Compounds

Products of the Maillard reaction (MRPs), formed during the process of reducing sugars with amino acids, are known for their role in forming the color and flavor of food. Bolchini et al. [Contribution 12] developed and optimized a method for the determination of antioxidant Maillard reaction products (KAMs—"known antioxidant MRPs") using liquid chromatography coupled with high-resolution mass spectrometry (HPLC-HRMS), applying different combinations of 20 amino acids and 6 reducing sugars depending on pH and reaction time. The authors showed that neutral and alkaline pH favored the formation of compounds with antioxidant activity, with optimal conditions obtained at pH 7. The highest production of antioxidant compounds was recorded in reactions involving threonine and disaccharides (maltose, lactose, etc.). The results of this study indicate the potential possibility of using the Maillard reaction to obtain natural antioxidants for use in the production of functional foods.

6. Hemp Extracts and Functional Applications

Industrial hemp varieties (*Cannabis sativa* L., Cannabaceae Endl.)—traditionally used for the production of fibers and seeds—can be a valuable source of bioactive compounds with potential applications in the prevention of neurodegenerative diseases and in the production of functional foods. Sip et al. [Contribution 13] studied extracts from *C. sativa* obtained by means of supercritical CO_2 extraction (SFE) and then combined them with prebiotic carriers, namely dextran, inulin and trehalose, creating three delivery systems for active substances. HPLC analyses showed the presence of key cannabinoids (CBD) and trace amounts of tetrahydrocannabinol (THC). All extracts exhibited the ability to neutralize free radicals, whereas in neuroprotective tests, the extracts inhibited acetyl- and butyrylcholinesterase (AChE and BChE). The obtained systems of hemp extracts with prebiotic carriers supported the growth of beneficial gut bacteria (*Bifidobacterium*, *Lactobacillus*, *Faecalibacterium*), while maintaining antioxidant and neuroprotective properties. The combination of hemp extracts with prebiotic substances opens up new possibilities for developing innovative nutraceutical delivery systems.

7. Synthesis of Knowledge and Perspectives

The analysis of the entire set of studies allows several key trends to be distinguished:

- Diversity of sources—plants, fungi, lichens, endophytes, extremophilic microorganisms, etc.;
- New structures—meroterpenoids, secobutanolides, pyranones, penicillides, etc.;
- Methodological progress—supercritical CO_2 extraction, optimization of conventional solvent extraction, HPLC-HRMS profiling, etc.;
- A wealth of biological activities—antibacterial, antifungal, antioxidant, anti-inflammatory, anticancer, etc.;
- Application possibilities—functional foods, nutraceuticals, crop biocontrol, new drugs, etc.

It is worth emphasizing the role of the review papers presented in this Special Issue, which organize knowledge about selected species and classes of compounds, supplemented by original research, presenting new experimental data.

8. Conclusions

Secondary metabolites remain an inexhaustible source of inspiration for the natural and applied sciences. Advances in extraction techniques, bioanalysis and structural characterization accelerate the discovery of compounds with potential pharmaceutical, agricultural, and industrial applications. This Special Issue of *Molecules* emphasizes the importance of integrating phytochemistry, microbiology, and biotechnology in order to fully utilize natural resources in innovations consistent with the principles of sustainable development. The studies collected in this Special Issue implement the objectives indicated in the Introduction: they verify ethnopharmacological premises (the richness of polyphenols and terpenoids), explore biodiversity (tropical plants, wood-inhabiting fungi and endophytes, marine microorganisms, etc.), develop 'green' extraction and dereplication methods (SFE, LC–HRMS/MS, network approaches, etc.), and also propose directions for practical implementations, from biocontrol and agroecology to functional foods and potential lead compounds in drug discovery.

Author Contributions: Conceptualization, R.K. and T.B.; methodology, R.K.; investigation, R.K. and T.B.; resources, R.K. and T.B.; writing—original draft preparation, R.K.; writing—review and editing, R.K. and T.B.; supervision, R.K.; project administration, R.K. and T.B. All authors have read and agreed to the published version of the manuscript.

Funding: The author(s) declare that no financial support was received for the research and/or publication of this article.

Institutional Review Board Statement: Not applicable.

Informed Consent Statement: Not applicable.

Data Availability Statement: No new data were created or analyzed in this study. Data sharing is not applicable to this article.

Acknowledgments: We acknowledge the academic editor for handling, editing, and promoting this Special Issue.

Conflicts of Interest: The authors declare no conflicts of interest.

List of Contributions:

1. Tenório, C.J.L.; Dantas, T.d.S.; Abreu, L.S.; Ferreira, M.R.A.; Soares, L.A.L. Influence of Major Polyphenols on the Anti-*Candida* Activity of *Eugenia uniflora* Leaves: Isolation, LC-ESI-HRMS/MS Characterization and In Vitro Evaluation. *Molecules* **2024**, *29*, 2761. https://doi.org/10.3390/molecules29122761.
2. Dzięcioł, M.; Wala, K.; Wróblewska, A.; Janda-Milczarek, K. The Effect of the Extraction Conditions on the Antioxidant Activity and Bioactive Compounds Content in Ethanolic Extracts of *Scutellaria baicalensis* Root. *Molecules* **2024**, *29*, 4153. https://doi.org/10.3390/molecules29174153.
3. Ryszczyńska, S.; Gumulak-Wołoszyn, N.; Urbaniak, M.; Stępień, Ł.; Bryła, M.; Twarużek, M.; Waśkiewicz, A. Inhibitory Effect of *Sorbus aucuparia* Extracts on the *Fusarium proliferatum* and *F. culmorum* Growth and Mycotoxin Biosynthesis. *Molecules* **2024**, *29*, 4257. https://doi.org/10.3390/molecules29174257.
4. Valmiki, M.; Teo, S.P.; de Resende, P.E.; Gibbons, S.; Ganesan, A. Caudiquinol: A Meroterpenoid with an Intact C20 Geranylgeranyl Chain Isolated from *Garcinia caudiculata*. *Molecules* **2024**, *29*, 3613. https://doi.org/10.3390/molecules29153613.
5. Yang, H.J.; Koh, Y.-S.; Na, M.; Li, W. Secobutanolides Isolated from *Lindera obtusiloba* Stem and Their Anti-Inflammatory Activity. *Molecules* **2024**, *29*, 4292. https://doi.org/10.3390/molecules29184292.
6. Vella, F.M.; Pignone, D.; Laratta, B. The Mediterranean Species *Calendula officinalis* and *Foeniculum vulgare* as Valuable Source of Bioactive Compounds. *Molecules* **2024**, *29*, 3594. https://doi.org/10.3390/molecules29153594.
7. Sum, W.C.; Ebada, S.S.; Ibrahim, M.A.A.; Kellner, H.; Stadler, M. Dentifragilones A–B and Other Benzoic Acid Derivatives from the European Basidiomycete *Dentipellis fragilis*. *Molecules* **2024**, *29*, 2859. https://doi.org/10.3390/molecules29122859.
8. Zhang, Y.; Jin, Y.; Yan, W.; Gu, P.; Zeng, Z.; Li, Z.; Zhang, G.; Wei, M.; Xue, Y. New Pyranone Derivatives and Sesquiterpenoid Isolated from the Endophytic Fungus *Xylaria* sp. Z184. *Molecules* **2024**, *29*, 1728. https://doi.org/10.3390/molecules29081728.
9. Salvatore, M.M.; Nicoletti, R.; Fiorito, F.; Andolfi, A. Penicillides from *Penicillium* and *Talaromyces*: Chemical Structures, Occurrence and Bioactivities. *Molecules* **2024**, *29*, 3888. https://doi.org/10.3390/molecules29163888.
10. Wu, M.; Liu, Z.; Wang, J.; Hu, W.; Zhang, H. Bioactive Secondary Metabolites from an Arctic Marine-Derived Strain, *Streptomyces* sp. MNP-1, Using the OSMAC Strategy. *Molecules* **2025**, *30*, 1657. https://doi.org/10.3390/molecules30081657.
11. Bačkor, M.; Kecsey, D.; Drábová, B.; Urminská, D.; Šemeláková, M.; Goga, M. Secondary Metabolites from Australian Lichens *Ramalina celastri* and *Stereocaulon ramulosum* Affect Growth and Metabolism of Photobiont *Asterochloris erici* through Allelopathy. *Molecules* **2024**, *29*, 4620. https://doi.org/10.3390/molecules29194620.
12. Bolchini, S.; Larcher, R.; Morozova, K.; Scampicchio, M.; Nardin, T. Screening of Antioxidant Maillard Reaction Products Using HPLC-HRMS and Study of Reaction Conditions for Their Pro-

duction as Food Preservatives. *Molecules* **2024**, *29*, 4820. https://doi.org/10.3390/molecules29204820.

13. Sip, S.; Stasiłowicz-Krzemień, A.; Sip, A.; Szulc, P.; Neumann, M.; Kryszak, A.; Cielecka-Piontek, J. Development of Delivery Systems with Prebiotic and Neuroprotective Potential of Industrial-Grade *Cannabis sativa* L. *Molecules* **2024**, *29*, 3574. https://doi.org/10.3390/molecules29153574.

References

1. Newman, D.J.; Cragg, G.M. Natural Products as Sources of New Drugs over the Nearly Four Decades from 01/1981 to 09/2019. *J. Nat. Prod.* **2020**, *83*, 770–803. [CrossRef] [PubMed]
2. Newman, D.J. Natural products and drug discovery. *Natl. Sci. Rev.* **2022**, *9*, nwac206. [CrossRef] [PubMed]
3. Atanasov, A.G.; Zotchev, S.B.; Dirsch, V.M.; International Natural Product Sciences Taskforce; Supuran, C.T. Natural products in drug discovery: Advances and opportunities. *Nat. Rev. Drug Discov.* **2021**, *20*, 200–216. [CrossRef] [PubMed]
4. European Commission. Farm to Fork Strategy (EU Green Deal). Available online: https://food.ec.europa.eu/horizontal-topics/farm-fork-strategy_en (accessed on 11 September 2025).
5. Schneider, K.; Barreiro-Hurle, J.; Rodriguez-Cerezo, E. Pesticide reduction amidst food and feed security concerns. *Nat. Food* **2023**, *4*, 746–750. [CrossRef] [PubMed]
6. European Environment Agency. How Pesticides Impact Human Health and Ecosystems in Europe. 2023. Available online: https://www.eea.europa.eu/en/analysis/publications/how-pesticides-impact-human-health (accessed on 11 September 2025).
7. Ayilara, M.S.; Adeleke, B.S.; Akinola, S.A.; Fayose, C.A.; Adeyemi, U.T.; Gbadegesin, L.A.; Omole, R.K.; Johnson, R.M.; Uthman, Q.O.; Babalola, O.O. Biopesticides as a promising alternative to synthetic pesticides. *Front. Microbiol.* **2023**, *14*, 1040901. [CrossRef] [PubMed]
8. Khursheed, A.; Rather, M.A.; Jain, V.; Wani, A.R.; Rasool, S.; Nazir, R.; Malik, N.A.; Majid, S.A. Plant based natural products as potential ecofriendly and safer biopesticides: A comprehensive overview of their advantages over conventional pesticides, limitations and regulatory aspects. *Microb. Pathog.* **2022**, *173 Pt A*, 105854. [CrossRef] [PubMed]
9. Daraban, G.M.; Hlihor, R.-M.; Suteu, D. Pesticides vs. Biopesticides: From Pest Management to Toxicity and Impacts on the Environment and Human Health. *Toxics* **2023**, *11*, 983. [CrossRef]
10. Shahidi, F.; Ambigaipalan, P. Phenolics and polyphenolics in foods, beverages and spices: Antioxidant activity and health effects—A review. *J. Funct. Foods* **2015**, *18*, 820–897. [CrossRef]
11. Chemat, F.; Vian, M.A.; Cravotto, G. Green Extraction of Natural Products: Concept and Principles. *Int. J. Mol. Sci.* **2012**, *13*, 8615–8627. [CrossRef] [PubMed]
12. Plaza, M.; Turner, C. Pressurized hot water extraction of bioactives. *TrAC Trends Anal. Chem.* **2015**, *71*, 39–54. [CrossRef]
13. Chemat, F.; Abert-Vian, M.; Fabiano-Tixier, A.-S.; Strube, J.; Uhlenbrock, L.; Gunjevic, V.; Cravotto, G. Green extraction of natural products. Origins, current status, and future challenges. *TrAC Trends Anal. Chem.* **2019**, *118*, 248–263. [CrossRef]
14. Herrero, M.; Cifuentes, A.; Ibáñez, E. Sub- and supercritical fluid extraction of functional ingredients from different natural sources: Plants, food-by-products, algae and microalgae: A review. *Food Chem.* **2006**, *98*, 136–148. [CrossRef]
15. Dai, Y.; van Spronsen, J.; Witkamp, G.-J.; Verpoorte, R.; Choi, Y.H. Natural deep eutectic solvents as new potential media for green technology. *Anal. Chim. Acta* **2013**, *766*, 61–68. [CrossRef] [PubMed]
16. Jiménez-Amezcua, I.; López Martínez, M.I.; Ruiz Matute, A.I.; Sanz, M.L. Natural Deep Eutectic Solvents for Solubility and Selective Fractionation of Bioactive Low Molecular Weight Carbohydrates. *Foods* **2023**, *12*, 4355. [CrossRef] [PubMed]
17. Gaudêncio, S.P.; Pereira, F. Dereplication: Racing to speed up the natural products discovery process. *Nat. Prod. Rep.* **2015**, *32*, 779–810. [CrossRef] [PubMed]
18. Nothias, L.-F.; Petras, D.; Schmid, R.; Dührkop, K.; Rainer, J.; Sarvepalli, A.; Protsyuk, I.; Ernst, M.; Tsugawa, H.; Fleischauer, M.; et al. Feature-based molecular networking in the GNPS analysis environment. *Nat. Methods* **2020**, *17*, 905–908. [CrossRef] [PubMed]
19. Wang, M.; Carver, J.; Phelan, V.; Sanchez, L.M.; Garg, N.; Peng, Y.; Nguyen, D.D.; Watrous, J.; Kapono, C.A.; Luzzatto-Knaan, T.; et al. Sharing and community curation of mass spectrometry data with Global Natural Products Social Molecular Networking. *Nat. Biotechnol.* **2016**, *34*, 828–837. [CrossRef] [PubMed]
20. Dührkop, K.; Fleischauer, M.; Ludwig, M.; Aksenov, A.A.; Melnik, A.V.; Meusel, M.; Dorrestein, P.C.; Rousu, J.; Böcker, S. SIRIUS 4: A rapid tool for turning tandem mass spectra into metabolite structure information. *Nat. Methods* **2019**, *16*, 299–302. [CrossRef] [PubMed]

molecules

MDPI

Article

Influence of Major Polyphenols on the Anti-*Candida* Activity of *Eugenia uniflora* Leaves: Isolation, LC-ESI-HRMS/MS Characterization and In Vitro Evaluation

Camylla Janiele Lucas Tenório [1,2], Thainá dos Santos Dantas [1,3], Lucas Silva Abreu [4], Magda Rhayanny Assunção Ferreira [1,5] and Luiz Alberto Lira Soares [1,2,3,*]

1 Laboratory of Pharmacognosy, Department of Pharmaceutical Sciences, Federal University of Pernambuco, Recife 50670-901, PE, Brazil; camylla.tenorio@ufpe.br (C.J.L.T.); thaina.dantas@ufpe.br (T.d.S.D.); magda.raferreira@ufpe.br (M.R.A.F.)
2 Post-Graduate Program in Pharmaceutical Sciences, Federal University of Pernambuco, Recife 50670-901, PE, Brazil
3 Post-Graduate Program in Therapeutic Innovation, Federal University of Pernambuco, Recife 50670-901, PE, Brazil
4 Chemistry Institute, Fluminense Federal University, Niterói 24020-150, RJ, Brazil; abreu_lucas@id.uff.br
5 Pharmaceutical Abilities Laboratory, Pharmacy, School of Health and Life Sciences, Catholic University of Pernambuco, Recife 50050-900, PE, Brazil
* Correspondence: luiz.albertosoares@ufpe.br

Abstract: The content of chemical constituents in *Eugenia uniflora* leaf extracts correlates positively with biological activities. The experimental objective was to carry out the phytochemical screening and purification of the major polyphenols from the leaves of *E. uniflora*. In addition, the anti-*Candida* activity of the hydroalcoholic extract, fraction, subfractions and polyphenols purified were evaluated. After partitioning of the extract with ethyl acetate, the fractions were chromatographed on Sephadex® LH-20 gel followed by RP-flash chromatography and monitored by TLC and RP-HPLC. The samples were characterized by mass spectrometry (LC-ESI-QTOF-MS^2) and subjected to the microdilution method in 96-well plates against strains of *C. albicans*, *C. auris*, and *C. glabrata*. Myricitrin (93.89%; w/w; m/z 463.0876), gallic acid (99.9%; w/w; m/z 169.0142), and ellagic acid (94.2%; w/w; m/z 300.9988) were recovered. The polyphenolic fraction (62.67% (w/w) myricitrin) and the ellagic fraction (67.86% (w/w) ellagic acid) showed the best antifungal performance (MIC between 62.50 and 500 μg/mL), suggesting an association between the majority constituents and the antifungal response of *E. uniflora* derivatives. However, there is a clear dependence on the presence of the complex chemical mixture. In conclusion, chromatographic strategies were effectively employed to recover the major polyphenols from the leaves of the species.

Keywords: flavonoids; tannins; purification; *Candida* spp.

1. Introduction

Chromatography is a physical, chemical, and/or mechanical separation method based on the interactions and distribution of the components of a sample between two immiscible phases. It has become a very effective and ubiquitous technique in various areas for analytical and preparatory purposes, the latter being the main general separation strategy for the purification and recovery of chemical species, among them polyphenols, one of the largest groups of substances distributed in the plant kingdom [1,2].

However, the recovery of polyphenols from herbal species remains a challenge due to the complex multi-component constitution of these biological matrices. Most of the tests consider the development and application of combined methodologies within the world of chromatography [3,4]. The recovery of these constituents is desirable for the structural and pharmacological elucidation of bioactives, contributing to the collection of information available for this class of metabolites [5,6].

Polyphenols, in general, are widely recognized for their therapeutic properties in plant species [7,8]. Gallic and ellagic acids, together with myricitrin, are examples of polyphenolic structures that stand out as the main constituents of *Eugenia uniflora* Linn leaves. Similar to other species, the concentration of these components in the extracts and fractions exhibited a positive correlation with the biological activities described for the plant, particularly regarding antifungal activity against strains of *Candida* spp. through various mechanisms [9,10].

Eugenia uniflora belongs to one of the largest genera in the Myrtaceae family and holds significant socio-economic potential due to the commercial exploitation of its plant parts and its therapeutic use in traditional medicine [11,12]. Although native to Brazil, Eugenia uniflora is also commonly found in other South American countries and is remarkably adaptable [13], positioning the species as a potential plant source to address the increasing demand for new therapeutic agents derived from plants.

The growing interest in this species is evidenced by the increase in the number of publications in recent years. Publications using the leaves focus on in vitro and/or in vivo research into the pharmacological properties of its extracts and fractions [14–16]; evaluation of their toxicological safety [17,18]; development of efficient methodologies and strategies for obtaining, identifying, characterizing, and quantifying their constituents [14,19–21]. In addition, there are reports on the development of intermediate products and pharmaceutical formulations containing leaf extracts of the species [22,23].

In this context, it is crucial to identify, develop, evaluate, and optimize separation strategies for the main secondary metabolism compounds present in *E. uniflora* leaves. This will ensure the availability of these substances, enabling both individual and combined pharmacological investigations as well as monitoring their presence in different matrices of the extract and its fractions. This approach not only meets the need for a wide range of research possibilities on the constituents of the species but also presents itself as a viable alternative to commercialized reference standards.

2. Results

2.1. Spray-Dried Crude Extract and Enriched Fraction

To produce a total of 810 g of Spray-dried Crude Extract (SDCE), approximately 9.36 kg of herbal material is required, assuming the same extraction and drying processes described in this study are used. Given the significantly large amount needed, the process was evaluated in triplicate, and the obtained yields were calculated (Table 1). The Ethyl Acetate Fraction (EAF) constituted about 2.27% (*w*/*w*) of the SDCE.

Table 1. Yield of fractions obtained from SDCE by LLE.

	Sample (g)			**Yields %**		
	SDCE	**AqF**	**EAF**	**AqF**	**EAF/AqF**	**EAF/SDCE**
1	360	177	7.87	49.16	4.44	2.19
2	270	110	6.14	40.74	5.58	2.27
3	180	78	4.27	43.33	5.47	2.37
Mean ± SD (RSD%)				43.33 ± 4.31 (9.95)	5.47 ± 0.62 (11.49)	2.27 ± 0.09 (3.97)

SDCE—spray-dried crude extract; AqF—aqueous fraction; EAF—ethyl acetate fraction; SD—standard deviation; RSD%—relative standard deviation.

Analysis of the samples using High-Performance Liquid Chromatography with a diode array detector (HPLC-DAD) enabled the identification of the predominant peaks in the SDCE, corresponding to Gallic Acid (GA) at 7.6 min, myricitrin (MyR) at 22.3 min, and Ellagic Acid (EA) at 23.9 min (Figure S1A) [19]. GA exhibited absorption maxima at 214 nm and 271 nm, which corroborates literature reports that describe maxima ranging between 214 and 215 nm and between 271 to 273 nm. GA, along with its various polymeric derivatives (gallotannins), whether glycosylated or substituted, is extensively documented in species of the *Eugenia* spp. genus [14,21]. For MyR, absorption maxima were observed at 207 nm, 259 nm, and 353 nm, which are characteristic of flavonols. This is consistent with literature descriptions of absorption maxima ranging between 207 and 212 nm, 259 and 260 nm, and 352 and 358 nm [19,24]. For EA, absorbance maxima were noted at 253 nm and 367 nm, aligning with literature findings that describe absorption maxima between 253 and 255 nm and 360 and 368 nm. Additionally, this constituent is also frequently reported in the *Eugenia* spp. genus [19,25,26].

Comparing the results of SDCE and EAF, both evaluated at 1 mg/mL, the major polyphenols in EAF are up to 20 times more concentrated than in SDCE. While GA, MyR, and EA show peak areas of 3.44, 9.58, and 2.38 mAU in SDCE, respectively, peak areas of 34.31, 78.17, and 45.51 mAU were found in EAF (Figure S1B). Despite the yield of 2.27 ± 0.09% (3.97% *w*/*w*), the Liquid–Liquid Extraction (LLE) strategy efficiently concentrated the three major constituents from the leaves of the species. Enriched fractions are preferred samples for application in polyphenol recovery methodologies in herbal matrices [2,27]. When applied as an initial step, LLE reduces the excess of apolar interferents present in the complex matrix, and, consequently, with the removal of undesirable components, the compounds of interest are concentrated at the end of the process. In general, the isolation of polyphenols from enriched fractions, compared to the direct use of crude extracts, shows better separation and recovery efficiency [2,27,28].

2.2. Screening Tests for Subfractionation

2.2.1. EAF Processing by Reversed-Phase Flash Chromatography (RP-FC)

The optimization of methodologies and conditions was a critical step in devising a successful strategy. According to the literature, the Reversed-Phase Flash Chromatography (RP-FC) system is effectively used for the analysis and separation of polyphenolic structures, facilitating the separation of a broad spectrum of solutes with medium to high polarity that exhibit diverse interactions with the octadecylsilane of the stationary phase [5,29]. When the technique was applied with the transposition of the analytical method, no effective chromatographic separation of any individual polyphenol in the EAF was observed (Figure S2A). Prolonging the run time enhanced the chromatographic profile (Figure 1A), and GA was semi-purified in subfraction 1, showing the same retention factor (RF) as the standard (0.65) when monitored by Thin Layer Chromatography (TLC) (Figure 1B), which was confirmed by HPLC-DAD, evaluated at 1 mg/mL, at 7.6 min (Figure 1C). A total of 10.20 mg of semi-purified GA was recovered, amounting to 5.09% (*w*/*w*) of the utilized EAF. The RP-FC method proved inefficient for the recovery of semi-purified or concentrated MyR and EA in a single subfraction. In complex samples, such as extracts and concentrated fractions, other matrix components, besides the targeted major polyphenols, interact with the stationary phase and/or possess characteristics incompatible with the UV-Vis detection system, potentially disrupting the separation process [30]. This screening indicated that the technique is more suitably applied to less complex subfractions, thereby reducing the number of fractions processed and affirming its efficient use in the final stages [18].

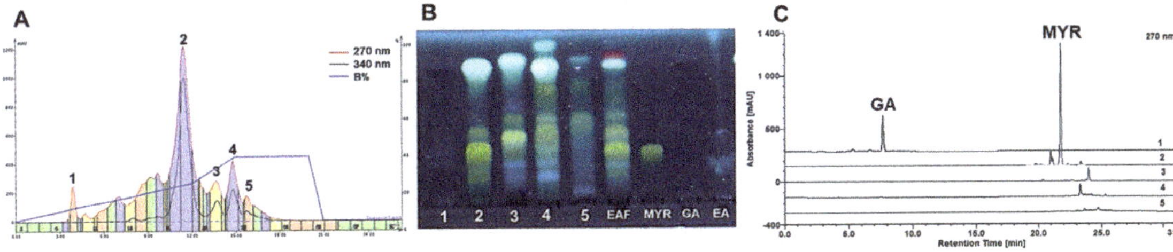

Figure 1. Chromatogram of test 2 of the RP-FC screening (**A**) and TLC of the five collected sub-fractions derivatized with 5% (w/v) $AlCl_3$ (**B**). Chromatograms of the sub-fractions from test 2 by HPLC at 270 nm (**C**).

2.2.2. EAF Processing by Size Exclusion Chromatography

In the initial Size Exclusion Chromatography (SEC) test (Test 1), 41 subfractions were obtained, with blue bands observed in subfractions 1–17, predominantly yellow bands in subfractions 18–35, and in subfraction 39, an apparently semi-purified fluorescent band. (Figure 2A). The results indicate an effective separation by metabolite class. Beyond size exclusion separation, the Sephadex® LH-20 gel, a dextran cross-linked with hydroxypropylated spheres, is also notable for adsorption separation, creating a chromatographic medium with amphiphilic and adsorptive properties modulated by the mobile phase [31,32]. Dual separation is advantageous for the fractionation and isolation of polyphenolic constituents, accommodating a broad range of low- to medium-molecular-weight structures with varying polarities [33,34].

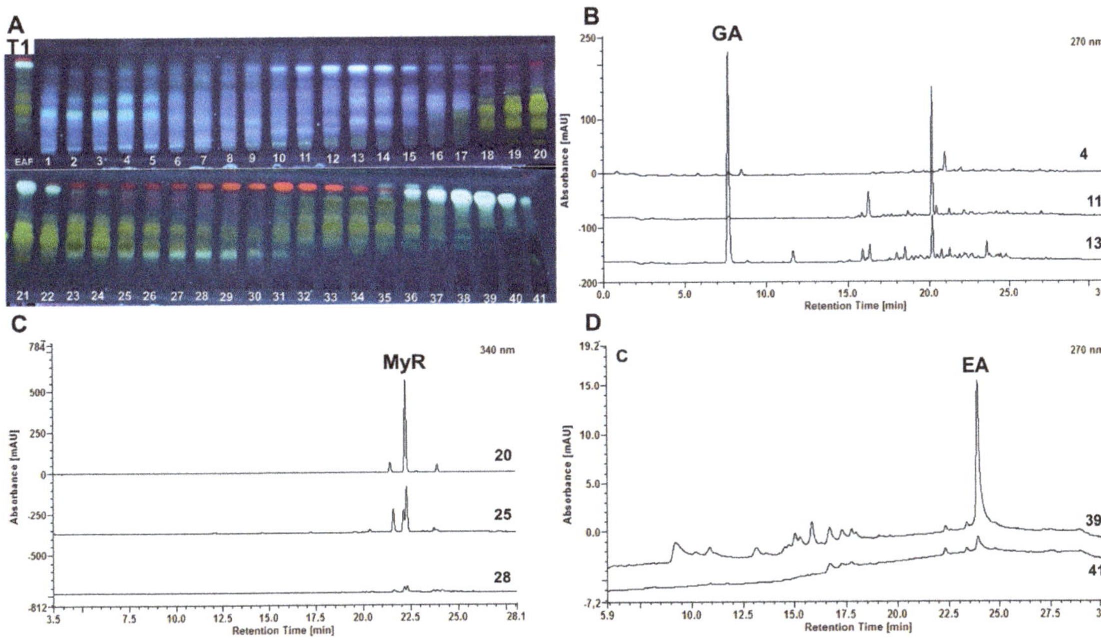

Figure 2. TLC of the subfractions collected in test 1 of the SEC screening derivatized with 5% (w/v) $AlCl_3$ (**A**), Chromatograms of sub-fractions 4, 11, 13 at 270 nm (**B**), 20, 25, and 28 at 340 nm (**C**) and 39, 41 at 270 nm (**D**) from test 1 of the SEC screening.

The two mechanisms acting concurrently in the system, retention by adsorption, are attributed to the formation of hydrogen bonds with the hydroxyl groups of the polyphenols.

The affinity strength is directly proportional to the number of hydrogen atoms, which also correlates with the molecular weight of the compound. Consequently, the general rule of SEC, where elution occurs in reverse order to the size of the analyte, is modified in tests with Sephadex® LH-20. In this experiment, simple phenolic constituents, despite their low molecular weight, eluted first, likely due to their interaction with the initially more aqueous medium [1,35].

During the tests, adjusting the mobile phase from higher to lower polarity using a gradient between water and methanol resulted in the sequential elution of phenolic acids, glycosylated flavonoids, flavonoid aglycones, and tannins, displaying blue, yellow, and fluorescent bands (Figure 2A), traits typical of polyphenolic components [36]. Fractions 4, 11, 13, 20, 25, 28, 39, and 41 were chosen for injection into the HPLC at 0.25 mg/mL. The selection criteria took into account the TLC profile of each fraction and the amount recovered that was viable for weighing and preparing the sample for analysis.

Hydroalcoholic mixtures with methanol (MeOH) were initially chosen based on methodologies found in the literature for the isolation of polyphenols by Sephadex® LH-20 [31]. Ethanol (EtOH) is less widely reported but is also used [35]. In this sense, EtOH was tested as a comparative criterion (Test 2), and the separation process observed showed advantages over MeOH. The total elution time was 24 h, spread over 3 days with a flow rate of approximately 0.17 mL/min, while that of MeOH was 40 h, spread over 5 days with a flow rate of approximately 0.10 mL/min.

Reducing the procedure time is desirable since the aim of the study is to develop a methodology for optimized major polyphenol recovery. In this sense, the shorter the time, the more processes can be carried out in less time. Additionally, in terms of process costs, EtOH has a lower market price. When monitoring, it was possible to observe that the initial and final compounds were better separated (Figure S4). The volume of the bed and the exclusion limit vary depending on the solvent used in the process. For example, the gradual increase in EtOH not only has a greater capacity to reduce pore size when compared to MeOH but also subtly reduces the polarity of the bed [35,37].

In the SEC test with EtOH, GA was identified at 7.62 min in subfraction 16. The majority peak of the subfraction, despite the difference in retention time (20.19 min), resembles the GA spectrum and may be a derivative of GA (Figure S5A). In subfraction 20, MyR was identified at 22.23 min. Peaks 1 (21.42 min) and 3 (23.95 min) displayed spectra similar to MyR, suggesting the presence of other flavonoids (Figure S5B). In subfractions 40, 44, and 48, the spectra of the majority peaks were identified as ellagic acid at 23.84 min. Peaks 1 and 2, with retention times of 18.35 and 19.00 min, respectively, showed spectra similar to EA, a characteristic absorption of flavanols and flavanones [38].

Scaling the method from a smaller column to a larger one (Test 3) demonstrated reproducibility in the chromatographic profile, as monitored by TLC (Figure S6). A total of 64 sub-fractions were collected during the 38-h procedure, which was spread over 5 days with a flow rate of approximately 0.5 mL/min. The main objective of the study was to isolate the primary constituents of the species (GA, MyR, and EA) in sufficient quantities for other planned tests. To optimize the experimental time, the hydroethanolic ratio was adjusted, reducing the procedure time to 28 h, spread over 3.5 days, with a flow rate of approximately 0.54 mL/min. Sixty sub-fractions were collected (Figure S7), and the eluates containing MyR and GA started to be eluted from sub-fraction 18. The yields of the subfractions are described in Table S2.

2.3. Chromatographic Strategies for Major Polyphenol Recovery

2.3.1. Fractionation by Sephadex® LH-20

Based on the selection tests, SEC strategies were established for the initial processing of the EAF. Collections were made in volumes according to the profile of components observed during monitoring, reducing the time to 22 h, spread over 2 days, with 22 collections.

As in the tests, the eluate profiles were maintained, with the presence of blue and yellow bands [36] (Figure S8). Using the same elution profile, Tian, Liimatainen [39] reported similar results: simple phenolic acids were eluted in the first subfractions, followed by glycosylated flavonoids, flavones, and flavanones, and, lastly, tannins and ellagitannins, including EA. The authors also point out that the use of acetone as the final eluent is essential to reduce the adsorptive interaction of polymeric forms with the gel matrix [39], explaining why ellagitannins were eluted last in the system used.

Visual monitoring from SP1 to SP4 demonstrates the reproducibility of the chromatographic runs (Figure S8). Regarding the yields, in SP1, approximately 23.57% (*w*/*w*) of the yield corresponds to the MyR and GA-rich subfractions, and 2.7% (*w*/*w*) to the subsequent EA-rich subfractions (Table 2). The similar subfractions from all tests were combined according to their chromatograms, and the main subfractions were designated as the simple phenolics fraction (SPF), the main polyphenol fraction (MPF), and the last fractions (LF).

Table 2. Yields of subfractions collected in SP1 from SEC.

		SP1		
nFr	**Eluate (mL)**	**Weight (mg)**	**Y%**	**RY%**
1	50	32.90	3.21	3.26
2	30	144.10	14.05	14.30
3	20	63.70	6.21	6.32
4	20	30.00	2.92	2.98
5	20	23.50	2.29	2.33
6	20	51.70	5.04 #	5.13
7	20	42.90	4.18 #	4.26
8	20	79.20	7.72 #	7.86
9	20	47.70	4.65 #	4.73
10	25	20.30	1.98 #	2.01
11	20	19.80	1.93	1.96
12	20	14.10	1.37	1.40
13	20	11.60	1.13	1.15
14	20	12.00	1.17	1.19
15	25	14.60	1.42	1.45
16	25	15.40	1.50	1.53
17	25	17.10	1.67	1.70
18	50	25.90	2.53	2.57
19	55	53.40	5.21	5.30
20	65	19.00	1.85 *	1.89
21	50	8.70	0.85 *	0.86
22	100	36.80	3.59	3.65

yield% of the subfractions rich in gallic acid and myricitrin (MPF); * yield of the subfractions with ellagic acid (LF). nFr—subfraction number; EAF—ethyl acetate fraction; Y%—yield in percentage considering the amount of EAF; RY%—yield in percentage considering the sum weight of the subfractions.

2.3.2. Fractionation and Isolation of Major Polyphenols by Flash Chromatography

- Ellagic acid (EA)

For the EA component, utilizing LF and the most optimized methodology, the peaks were partially separated (Figure 3A). Subfraction 4 (EA I), which corresponds to the semi-purified EA, was subsequently resubmitted to RP-FC (Figure 3B). From subfraction 3 (EA II), the purified EA was obtained (Figure 3C). The recovered substance was a fine, yellowish, odorless powder that was insoluble in water. EA can manifest as cream-colored needles or as a yellow powder with a water solubility of less than 1 mg/mL at 21 °C. It is odorless, weakly acidic, and incompatible with strong reducing agents [40].

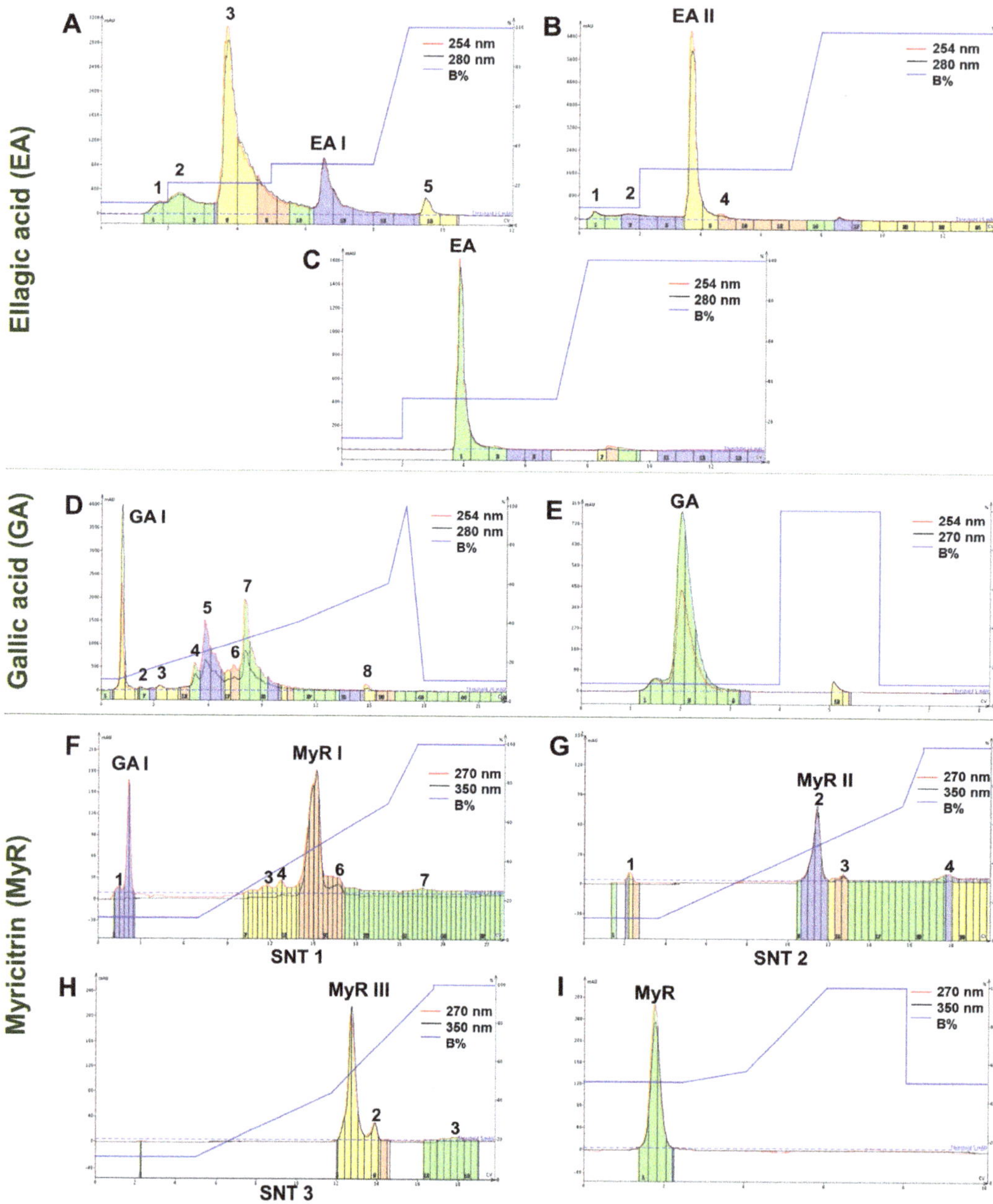

Figure 3. RP-FC chromatograms of LF subfractionation to obtain EA I (**A**), EA I to obtain EA II (**B**) and EA II to recover EA (**C**); MPF subfractionation (**D**) to obtain GA I (**D**) and GA I to recover GA (**E**); Subfrac-tionation of SNT 1 of MPF to obtain MyR I (**F**), SNT 2 to obtain MyR II (**G**), SNT 3 to obtain MyR III (**H**) and recovery of MyR from MyR III (**I**).

In terms of yield, the yields of the EA-containing subfractions were low compared to the other major constituents, necessitating multiple processing steps to obtain optimal quantities of the purified product. The processing of LF yielded between 1.23 and 3.52% (*w*/*w*) of EA I, and 11.85% (*w*/*w*) of this was recovered as EA II (Table S2). Ultimately, the recovery of EA from EA II had a yield of 67.86% (*w*/*w*), with 1.90 mg collected from the processing of 2.8 mg of EA I.

- Gallic acid (GA)

The RP-FC methodology applied proved efficient in the semi-purification of GA from MPF. Peak 1 showed a band corresponding to GA on TLC (RF = 0.74) and was thus named GA I (Figure 3D). GA I was subjected to RP-FC, resulting in the recovery of GA (Figure 3E). The content recovered in GA I was brown, and after reprocessing the sample, an odorless, hygroscopic fine powder was obtained. This powder exhibited crystalline characteristics and was shaped like needles with a light brownish color. Purified GA is described as white or light brown needle-shaped crystals or powders [41]. The yields of GA I from MPF were up to 35.75% (*w*/*w*) (Table S3). From GA I, approximately 58.12% (*w*/*w*) of GA was recovered, totaling 29.12 mg.

- Myricitrin (MyR)

After processing the MPF using RP-FC in several chromatographic runs, it was observed that a portion of the sample, which was previously completely solubilized, precipitated at the start of elution. This resulted in the chromatographic profile and MyR recovery not being reproducible, leading to subsequent sample losses. Less concentrated samples were tested, and precipitation was observed again. As a result, an additional strategy of centrifugation was adopted. The chromatograms showed different chromatographic profiles for each processing of the supernatant (SNT), and the majority of peaks corresponded to MyR (Figure 3F–H).

During the drying process of the MyR I and II subfractions, an abundant light yellow powder and a darker colored powder were observed in the middle and at the edges of the flask, respectively. Monitoring by TLC revealed that separation still occurred during the evaporation of the solvent present in the eluate (Figure S9). A possible explanation is the saturation precipitation of MyR in MyR I and II, as well as its low solubility in aqueous media, given that the organic solvent evaporates more quickly. The precipitate is easily collected during drying. The powders collected in this process were also named MyR III.

The recovered powder was light yellow in color, had an unidentified characteristic smell, and was insoluble in water. MyR is characterized as a yellow powder that is practically insoluble in water, slightly soluble in ethanol, and has a characteristic bayberry aroma [42,43]. The recovery of MyR-rich fractions from MPF, including I and II, ranged from 50.52%, 61.60%, and 62.67% (*w*/*w*). The edges separated from the precipitate accounted for approximately 14.94% (*w*/*w*) of the total MyR I and II collected. Around 17.8 mg of MyR I yielded 10.23 mg (57.47% *w*/*w*) of MyR II (Table S4), and when this was subjected to RP-FC again, it showed yields of 89.32% (*w*/*w*) of MyR III and 93.89% (*w*/*w*) of purified MyR.

When the MPF was processed using this methodology, GA was also recovered, reserved, and named GA I. However, the yields of this fraction were subtly reduced when the detection range was changed: previously, in the recovery tests for GA, with detection set at 254 and 280 nm, the yields were 29.20% (*w*/*w*), while in these tests, with detection set at 270 and 350 nm, the yield was 20.47% (*w*/*w*).

2.4. Monitoring the RP-Flash Subfractionation and Isolation Process by HPLC-DAD

In the HPLC-DAD analysis, the purified EA showed a retention time of 23.52 min (Figure 4A) with a peak purity of 94.2% at 4.26 mAU. The absorption spectrum exhibits absorbance maxima at 253 and 367 nm [19,25,26]. The recovered GA displayed a retention time of 7.51 min (Figure 4B) and a peak purity of 99.90% at 4.93 mAU. Its absorption spectrum revealed absorption maxima at 214 and 271 nm [14,21]. As for the MyR recovered, it had a retention time of 21.99 min (Figure 4C), slightly reduced compared to that observed in MPF (23 min), and a peak purity of 99.9% with 4.49 mAU. The absorption spectrum recorded absorption maxima at 207, 259, and 353 nm [19,24]. Chromatograms obtained from the injection of LF at 250 µg/mL, EA I, GA I and MyR I at 25 µg/mL, and AE, AG, and MyR at 2 µg/mL.

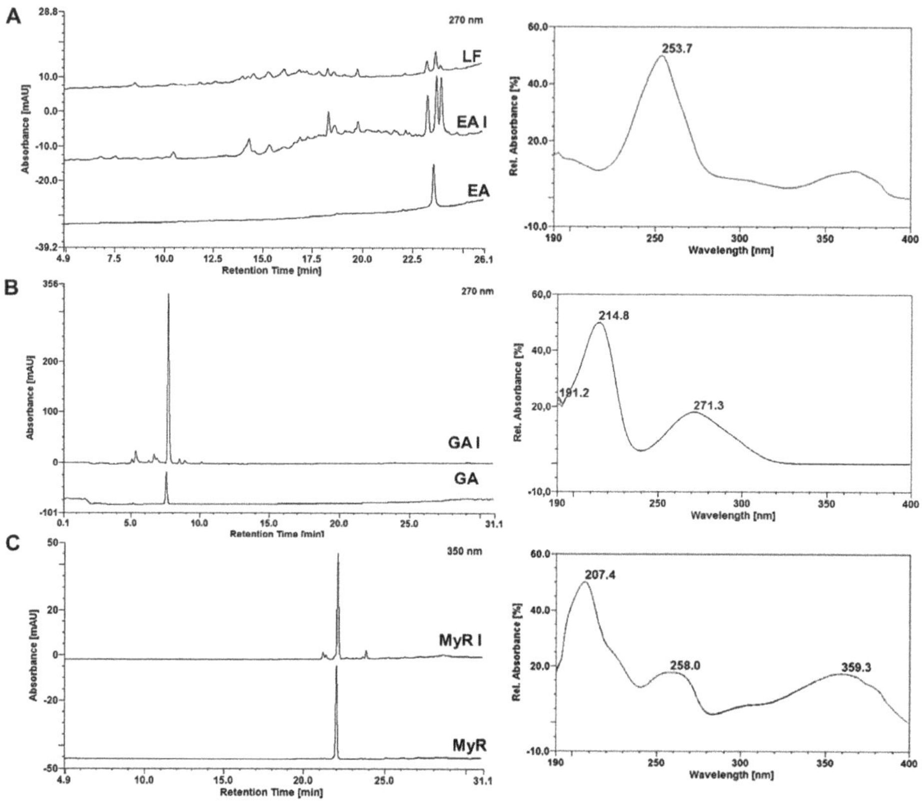

Figure 4. HPLC-DAD chromatograms of the Last Fraction (LF), Ellagic Acid after RP-LF processing of the last fraction (EA I) and purified Ellagic Acid at 270 (**A**); Gallic Acid I after RP-FC processing of the main polyphenols fraction (MPF) and purified Gallic Acid (GA) at 270 (**B**); Myricitrin after RP-FC pro-cessing of supernatant I of the MPF (MyR I) and purified Myricitrin (MyR) at 350 nm (**C**).

2.5. Phytochemical Characterization of the Extract, Fractions, Subfractions and Isolates

2.5.1. Phytochemical Profile

The 22 compounds were identified in *E. uniflora* leaf extract and fractions by the interpretation of their fragmentation patterns obtained from mass spectra (HRMS/MS experiments). Data provided by reference standards and the literature information was also employed for the comprehensive evaluation of samples. The retention times and mass spectrum data, along with peak assignments for compounds identified using negative ionization, are described in Table 3.

Table 3. Characterization of the compounds identified by HPLC-DAD-ESI-HRMS/MS in *Eugenia uniflora*.

Peak No.	t_R (min.)	m/z $[M-H]^-$	Molecular Formula	Error (ppm)	MS/MS (Relative Abundance%)	Tentative Assignment	Fraction	References
1	2.2	191.0562	$C_7H_{12}O_6$	−0.6	173.0954 (23.81); 165.0385 (19.05); 127.0367 (45.24)	Quinic acid	SPF/EAF	[15,44–46]
2	4.6	191.0197	$C_6H_8O_7$	0.2	173.0078 (73.33); 111.0086 (60.00)	Citric acid	SPF/MPF/EAF	[47]
3	6.3	169.0149	$C_7H_6O_5$	−3.8	125.0244 (100.00)	Galic acid*	SPF/MPF/EAF	[15,44,46]
4	15.2	483.0788	$C_{20}H_{20}O_{14}$	−1.7	313.0595 (17.37); 271.0433 (0.47); 211.0230 (2.11); 169.0149 (77.43); 125.0264 (0.43)	Digalloyl-hexoside	SPF	[15,44–46]
5	15.2	353.0889	$C_{16}H_{18}O_9$	−3.2	191.0543 (100.00)	5-*O*-Caffeoylquinic acid	SPF/EAF	[44,46]
6	19.2	337.0928	$C_{16}H_{18}O_8$	0.1	191.0529 (100.00); 163.0380 (14.34); 119.0491(8.93)	5-*O*-Coumaroylquinic acid	SPF/EAF	[15,46]
7	24.2	653.2130	$C_{30}H_{38}O_{16}$	2.4	501.1996 (44.29); 483.1829 (19.36); 313.0577(14.16); 271.0507 (3.87); 211.0194 (7.37); 193.0133 (4.53); 169.0153 (54.58); 125.0216 (6.99)	Digalloyl-hexoside derivative	SPF/EAF	[48]
8	25.1	479.0848	$C_{21}H_{20}O_{13}$	−3.5	317.0245 (24.83); 316.0230 (100.00); 287.0196 (7.65); 271.0269 (13.79); 178.9984 (2.67); 151.0062 (3.39)	Myricetin-*O*-hexoside	MPF	[15,44–46]
9	25.2	653.2084	$C_{30}H_{38}O_{16}$	0.5	501.2041 (43.42); 483.1850 (30.92); 313.0568 (23.68); 241.0370 (11.84); 169.0177 (100.00)	Digalloyl-hexoside derivative	MPF	[48]
10	25.4	501.1983	$C_{23}H_{34}O_{12}$	−1.1	451.3264 (25.87); 313.0593 (15.85); 271.0444 (12.63); 211.0243 (9.84); 169.0138 (59.41); 125.0236 (10.19)	Digalloyl-hexoside derivative	SPF	[48]
11	25.4	539.2142	$C_{23}H_{36}O_{12}$	−1.4	501.1896 (9.52); 313.0560 (6.54); 271.0440 (4.93); 169.0147 (46.44); 125.0235 (15.25)	Digalloyl-hexoside derivative	SPF	[48]
12	25.5	449.0731	$C_{20}H_{18}O_{12}$	−1.2	317.0267 (25.26); 316.0229 (100.00); 287.0193 (13.96); 271.0245 (21.05); 178.9980 (4.98); 151.0067 (2.28)	Myricetin-*O*-pentoside	MPF	[15,44–46]
13	26.6	449.0728	$C_{20}H_{18}O_{12}$	−0.6	317.0251 (21.11); 316.0225 (100.00); 287.0193 (14.18); 271.0242 (27.36); 178.9992 (2.01); 151.0019 (3.45)	Myricetin-*O*-pentoside	MPF	[15,44–46]
14	27.1	463.0900	$C_{21}H_{20}O_{12}$	−3.9	317.0265 (20.75); 316.0223 (100.00); 287.00201 (8.96); 271.0243 (19.39); 178.9995 (3.86); 151.0046 (3.38)	Myricetin-*O*-rhamnoside * (Myricitrin)	MPF	[15,44,46]
15	28.0	463.0875	$C_{21}H_{20}O_{12}$	1.6	301.0322 (33.21); 300.0263 (100.00); 271.0232 (32.62); 255.0293 (14.11); 179.0010 (4.51); 151.0071 (4.72)	Quercetin-*O*-hexoside	MPF	[15,45]
16	28.3	463.0876	$C_{21}H_{20}O_{12}$	1.3	301.0319 (35.81); 300.0264 (100.00); 271.0228 (37.49); 255.0310 (25.03); 178.9933 (10.66); 151.0052 (8.38)	Quercetin-*O*-hexoside	MPF	[15,45]

Table 3. *Cont.*

Peak No.	t_R (min.)	*m/z* $[M-H]^-$	Molecular Formula	Error (ppm)	MS/MS (Relative Abundance%)	Tentative Assignment	Fraction	References
17	28.9	433.0773	$C_{20}H_{18}O_{11}$	0.8	301.0308 (28.23); 300.0270 (100.00); 271.0214 (40.46); 255.0279 (27.53); 178.9937 (3.17); 151.0023 (4.13)	Quercetin-*O*-pentoside	MPF	[15,45]
18	30.1	433.0771	$C_{20}H_{18}O_{11}$	1.2	301.0337 (33.29); 300.0274 (100.00); 271.0261 (33.82); 255.0314 (23.84); 179.0014 (3.36); 151.0080 (8.93)	Quercetin-*O*-pentoside	MPF	[15,45]
20	30.5	447.0941	$C_{21}H_{20}O_{11}$	−1.8	301.0324 (60.54); 300.0266 (100.00); 271.0244 (42.29); 255.0294 (18.59); 178.9983 (6.63); 151.0042 (11.36)	Quercetin-*O*-deoxyhexoside	MPF	[15,45]
21	33.2	431.0992	$C_{21}H_{20}O_{10}$	−1.8	285.0404 (75.39; 284.0299 (100.00); 255.0314 (86.57); 227.0335 (81.93)	Kaempferol-*O*-deoxyhexoside	MPF	[15,45]
22	33.3	471.1305	$C_{24}H_{24}O_{10}$	−1.7	285.0368 (10.07); 284.0314 (14.76); 255.0298 (1.50); 227.0376 (2.45)	Kaempferol- derivative	MPF	-

* Compared with standard.

2.5.2. Identification of Flavonols

The identification of these compounds was facilitated by the analysis of fragmentation pathways of ions in the negative modes and the observation that glycosidic residues (pentosyl (132 Da), rhamnosyl (146 Da), glucosyl (162 Da), and rutinosyl (308 Da) were cleaved sequentially and generated characteristic aglycone fragments compared to the available literature. Among these compounds, five compounds were identified as quercetin glycoside (16–20), eight were identified as myricetin glycosides (8, 12, 13), and among them the major constituent myricitrin (14) and two kaempferol glycosides (21 and 22) were identified. In addition, kaempferol, quercetin, and myricetin derivatives were observed on the basis of the main ion fragments produced on the MS/MS experiments, appearing at m/z 284 and 285 for kaempferol derivatives, m/z 300 and 301 for quercetin derivatives, and m/z 316 and 317 for myricetin derivatives; these pairs of ion fragments corresponded to the respective homolytic and heterolytic cleavage of the glycosidic bonds in these compounds [49]. Moreover, the ion fragments produced with m/z 179 and 151 were compared with literature data and attributed to the confirmation of the flavonol core (kaempferol, quercetin, or myricetin) (Table 3). All these compounds were identified previously in *E. uniflora* [15,44–46,48].

2.5.3. Identification of Other Phenolic Compounds

Gallotannins were detected and were distinguished by their characteristic fragment ion spectra, yielding sequential losses of galloyl (m/z 152) and gallate (m/z 170) residues. The digalloyl-hexoside derivatives (4, 7, 9–11) were assinalated based on the $[M-H]^-$ ion at m/z 483 or your derivatives and MS/MS produced as typical product ions at m/z 313, 271, 211, 193, 169, and 125 [15,44–46]. Additionally, the compounds citric acid (2), gallic acid (3), 5-*O*-Caffeoylquinic acid (5), 5-*O*-Coumaroylquinic acid (6) were identified based on $[M-H]^-$ ions at m/z 191, 169, 353, 337, respectively. MS/MS data of these ions were compared with the literature data [15,44–47].

2.5.4. Isolated Compounds from *E. uniflora*

Myricitrin, myricetin-*O*-rhamnoside, was identified with m/z of 463.0871 (1.07 ppm) and m/z 316.02 of its myricetin aglycone after glycosidic loss (Figure S6A). Its dimer was also found at m/z 927.1825 (−2.48 ppm). Peaks at m/z 151.0053, 178.9986, and 271.0234 are characteristic of flavonoid fragments, with the m/z 151 fragment being characteristic of the cleavage of the C ring [45]. Gallic acid was identified at m/z 169.0142 (6.5 ppm) (Figure S7B). In the same sample, its digallic acid dimer was identified at m/z 339.0358 (3.53 ppm) and m/z 483.1851 and 313.0577, corresponding to fragments of glycosylated diglycosyl derivatives [48]. Additionally, m/z 300.9986 was suggestive of ellagic acid (2.65 ppm). Ellagic acid itself was identified at m/z 300.9982 (1.01 ppm) (Figure S7C). The dimer at m/z 603.0044 (0.63 ppm) and the trimer at m/z 905.0109 (0.52 ppm) were also found in the same sample. Among the fragments described in the literature for ellagic acid, m/z of 183.0110, 249.0387, and 275.0167 were observed [50].

2.6. *Antifungal Activity*

The selection of *Candida* spp. strains for this study was based on the findings of an-tifungal tests previously published for the extract and fractions of *Eugenia uniflora*. The minimum inhibitory and fungicidal concentrations against *C. albicans* and *C. glabrata* showed promise [9,10]. It has also been observed that fractions enriched with the main constituents of the species exhibit better antifungal performance against strains of *Candida albicans* [9]. In this study, SDCE and EAF showed similar MIC (Minimum Inhibitory Concentration) and MFC (Minimum Fungicidal Concentration) values, which align with observations in the literature. Most of the other samples tested also exhibited antifungal activity against the evaluated strains.

The Hexane Fraction (HF) and the SPF demonstrated low activity, with MIC and MFC values greater than 1000 μg/mL. However, the MPF, rich in myricitrin and gallic acid, as well as the LF, rich in ellagic acid, yielded the best results. Their MIC values ranged between 62.50 and 500 μg/mL (Table 4). Interestingly, when evaluated individually, the three polyphenols exhibited higher MIC and MFC values compared to their respective subfractions. Surprisingly, the combinations of these polyphenols did not demonstrate additive or synergistic effects.

Table 4. MIC and MFC results of the extract, fractions, subfractions, and isolates of *E. uniflora* leaves.

	Minimum Inhibitory Concentration (MIC) and Minimum Fungicidal Concentration (MFC) (μg/mL)					
	Candida albicans		*Candida glabrata*		*Candida auris*	
Samples	**MIC**	**MFC**	**MIC**	**MFC**	**MIC**	**MFC**
SDCE	250 *	500	125 *	≥1000	31.2 *	1000
Fractions						
HF	1000	1000	1000	1000	62.5 *	1000
EAF	250*	500	125 *	1000	31.2 *	1000
rFaq	1000	1000	250	1000	125	1000
SPF	1000	≥1000	1000	≥1000	500	1000
MPF	125 *	125	62.5 *	250	500	1000
MyR I	250 *	500	250 *	500	500	≥1000
GA I	500	500	250 *	500	500	≥1000
LF	125 *	125	125 *	250	125 *	1000
EA I	250	250	62.5 *	250	125 *	1000
Phytochemicals						
GA	500	≥1000	250 *	500	500	500
MyR	500	1000	250 *	250 *	500	1000
EA	500	1000	125 *	250 *	1000	1000
Synergic samples						
GA + MyR	500	500	125 *	500	1000	1000
MyR + EA	500	1000	125 *	500	500	≥1000
EA + GA	1000	≥1000	1000	≥1000	1000	≥1000
GA + EA + MyR	1000	≥1000	500	≥1000	1000	≥1000

MIC—Minimum Inhibitory Concentration; MFC—Minimum Fungicidal Concentration; CE—crude extract; HF—hexane fraction; EAF—ethyl acetate fraction; rFaq—residual aqueous fraction; SPF—fraction of cinnamic derivatives; MPF—flavonoid fraction; MyR I—myricitrin-rich fraction; LF—last fractions collected on Sephadex® LH-20; EA I—ellagic acid-rich fraction; GA—gallic acid isolate; MyR—myricitrin isolate; EA—ellagic acid isolate. * MIC with considerably significant results.

The findings suggest that the presence of flavonoids significantly contributes to the antifungal properties of *E. uniflora* leaf extracts. In general, various antifungal mechanisms have been attributed to this important class of secondary metabolites. When tested against *Candida* spp. strains, these flavonoids inhibit processes such as cell wall formation, cell division, and the synthesis of RNA and proteins responsible for virulence factors.

Among the flavonoids individually evaluated in the literature, derivatives of myricitrin and kaempferol have demonstrated antifungal activity against strains of *C. albicans* [51] and *C. glabrata* [52,53]. Interestingly, the MPF contains derivatives of myricitrin and kaempferol, as well as quercetin and its derivatives (Table 3).

The other phenolic components present in LF, similar to EA, also yield considerable results. In general, phenols are associated with the antifungal action of herbal species due to their ability to induce membrane damage, resulting in an increase in cell permeability [54].

EA, which is present in LF, along with its glycosidic derivatives evaluated individually, has demonstrated antifungal activity against *C. albicans* and *C. auris* [55,56].

Species of the genus *Candida* spp. have significant clinical relevance due to their role as opportunistic pathogens in humans. Among the species evaluated, *C. albicans* is the most common, associated with frequent and recurrent vulvovaginal and oral infections, often exhibiting resistance to conventional antifungal drugs [9]. Although less prevalent, *C. glabrata* has emerged in healthcare settings, also demonstrating resistance to certain antifungal agents. At last, *C. auris*, a relatively recent addition to the list of pathogenic *Candida* species, has raised concern due to its multidrug-resistant nature, posing a serious threat to public health. In this context, the results of this study reinforce the promising potential of *E. uniflora* leaves, as already evidenced in the literature.

Therefore, the findings suggest that the concentration of the major polyphenols, myricitrin and ellagic acid, correlates with better antifungal responses. However, both compounds may act in conjunction with other components present in the phytocomplex, especially those with similar structures. Furthermore, the conducted in vitro tests offer crucial insights and suggest the potential for additional investigations into the specific mechanisms of antifungal action, including in vivo studies. These endeavors would yield more comprehensive data and validate the effectiveness of extracts, fractions, and isolates from *E. uniflora* leaves.

3. Material and Methods

3.1. Herbal Material, Extract, and Enriched Fraction

The leaves of *Eugenia uniflora* Linn were collected in the city of Paulista-PE and identified at the Agronomic Institute of Pernambuco (IPA) under registration number 93732 and registered with the National System for the Management of Genetic Heritage and Traditional Knowledge (Sisgen, Federal Republic of Brazil, Brazil) (number A449575). The material was dried in an air circulation oven (Luca-82-480®, Lucadema, São José do Rio Preto, São Paulo, Brazil) for 48 h at 40 °C. After drying, the leaves were ground in a Willye-type knife mill (TE-680®, Tecnal, Piracicaba, São Paulo, Brazil). The extractive solution was obtained by turboextraction at 10% (*w*/*v*), according to the published methodology [19] and spray-dried in a mini spray dryer (MSDi 1.0®, LabMaq, Ribeirão Preto, São Paulo, Brazil) under the following drying conditions: an inlet temperature of 150 °C, an airflow rate of 40 L/h, and a feed flow of 0.9 L/h to obtain the SDCE (18). The enriched fraction was obtained according to the LLE methodology described by Ramos, Bezerra [20]. The resulting fractions were concentrated under reduced pressure (RV10 Basic, IKA®), reconstituted with 50% (*w*/*v*) ethanol, and spray-dried in a mini spray dryer (MSDi 1.0, LabMaq®).

3.2. Screening Tests for Processing the Enriched Fraction

3.2.1. Reversed-Phase Flash Chromatography

Around 0.2 g of the EAF was solubilized in 4 mL of distilled water and subjected to separation in an Isolera™ system (P/N 411829, Biotage®, Uppsala, Sweden) coupled to a variable-wave UV-Vis detector, set at 270 and 340 nm, and an automatic collector. The chromatographic separation was carried out using a Biotage® SNAP-C_{18} column (cartridge size 25 g, average mass 30 g, and volume 33 mL) equipped with a pre-column of the same material. Two conditions were tested following the transposition of a methodology developed and validated on an analytical scale [17] and another with an extension of the analytical time. The sub-fractions were collected every 20 mL under a starting signal at a minimum detection of 5 mUA.

3.2.2. Exclusion Chromatography

Initially, tests were carried out on an open glass column (h = 31 cm, Ø = 2 cm) packed with Sephadex® LH-20 gel (MERCK®, Darmstadt, Germany) dispersed in an aqueous solution acidified with 2% (*v*/*v*) acetic acid. The final volume of the stationary phase was 50 mL, initially conditioned with 50% (*v*/*v*) methanol. Around 0.5 g of EAF was solubilized in 5 mL of the initial mobile phase and added to the top of the column. The mobile phase conditions evaluated in steps were: Test 1—40, 60, 80 and 100% (*v*/*v*) hydromethanolic solution; Test 2—40, 60, 80, and 100% (v/v) hydroethanolic solution; fractions of 5 mL were collected in Tests 1 and 2; Test 3—consisted of scaling up the technique to a glass column (h = 47 cm, Ø = 4 cm) with an stationary phase volume of 230 mL, previously conditioned with 40% (*v*/*v*) ethanol. Approximately 1 g of EAF solubilized in 5 mL of initial mobile phase was used and the elutions followed a gradient in steps: hydroethanolic solution 40, 60, 80 and 100% (*v*/*v*); Test 4—hydroethanolic solution 50, 80 and 100% (*v*/*v*). The sub-fractions from Tests 3 and 4 were collected at intervals of 10 mL of eluate. At the end of each test, the system was washed with acetone: water (7:3, *v*/*v*).

3.3. Chromatographic Strategies for Major Polyphenol Recovery

3.3.1. EAF Processing by Subfractionation by Sephadex® LH-20

Sephadex® LH-20 (Sigma-Aldrich, San Luis, MO, USA) was suspended in 2% (*v*/*v*) acetic acid, and the resulting gel was packed into a glass column (h = 47 cm, Ø = 4 cm) to a volume of 230 mL, then the system was conditioned with 50% (*v*/*v*) ethanol. Chromatographic runs, called SPn, were carried out using approximately 1 g of EAF solubilized in 5 mL of 50% (*v*/*v*) ethanol for each run. The elutions followed the gradient of Test 5. Eluates of between 10 and 50 mL were collected. Similar subfractions, according to their evaluation during monitoring, were merged for processing in the following steps.

3.3.2. EAF Processing by Subfractionation and Purification by Isolera™

The sub-fractions obtained in item 3.3.1 were subjected to separation by RP-FC using an Isolera™ system (Biotage—P/N 411829) coupled to a variable-wave UV-Vis detector and automatic collector. The chromatographic separations were carried out using a Biotage® SNAP-C_{18} column (cartridge size of 10 or 25 g, average mass of 12 or 30 g, and volume of 15 or 33 mL, respectively), with a pre-column suitable for each cartridge size and made of the same material. The eluent consisted of distilled water and methanol, acidified with 0.1% (*v*/*v*) acetic acid, as mobile phases A and B, respectively. Chromatographic conditions such as column size, gradient, and flow were modified and optimized for each constituent during the processing of the subfractions, as were the set wavelength, starting detection limits, and collection volume (Table 5).

Table 5. Chromatographic conditions in RP-FC for the recovery of majority polyphenols.

		Recovery of EA				Recovery of GA				Recovery of MyR					
	Processed Samples	LF		EA I		MPF		GA I		MPF (SNT1)		MyR I (SNT2)		MyR II (SNT3)	
Chromatographic conditions in RP-FC		B%	min	B%	min	B%	min	B%	min	B%	min	B%	min	B%	min
	Gradient	10	2	10	3	10	4	10	4	10	7	30	1	50	3
		20	3	30	5	10–40	10	100	2	10–100	20	30–70	6	50–56	2
		30	3	30–100	1	40–100	6	10	2	100–10	1	100	3	56–100	2
		30–100	1	100	5	100–10	1	-	-	10	3	30	1	100	2
		100	3	-	-	10	4	-	-					50	2
	Flow (mL/min)	12		12		15		20		20		20		20 e 15	
	Column (g)	12		12		12		30		30		30		30 e 12	
	λ (nm)	254 and 280				254 and 270				270 and 350					
	ST (mAU)	10		10		20		5		10		5		5	
	max. vol (mL)	10		10		15		10		15		10		10	

B%—proportion of mobile phase B; LF—last fractions; EA I—ellagic acid subfraction; EA—purified ellagic acid after processing EA I in isolera; GA I—gallic acid subfraction after processing MPF in isolera; GA—purified gallic acid after processing GA I in isolera; MPF—main polyphenols subfraction; MyR I—myricitrin-rich subfraction; MyR II—subfraction obtained after processing MyR I; SNT—supernatant; λ—wavelength; ST—starting detection; max vol—maximum collection volume.

Exceptionally, an additional sample preparation procedure was adopted for the recovery of the flavonoid myricitrin. Approximately 50 mg of MPF was solubilized in 5 mL of the initial mobile phase B 10% (v/v) using an ultrasonic bath (Cristófoli®, Campo Mourão, Paraná, Brazil). The solution was then subjected to 2000 rpm for 5 min in a centrifuge (EEQ9004A-2, 9TEC®, Curitiba, Paraná, Brazil), and the resulting supernatant was subsequently subjected to RP-FC. The precipitate was solubilized under the described conditions and centrifuged again; this procedure was repeated until no more precipitate was observed.

3.3.3. Yields

All the subfractions obtained were placed at 40 ± 2 °C to evaporate the solvent. After drying, each subfraction was resuspended in methanol, and the solution was transferred to a pre-weighed vial. The vials were then stored uncapped at room temperature (30 °C) until the solvent had completely evaporated. The vials containing the dried subfractions were weighed, and the yields were calculated as a percentage. This calculation took into account the weight of each subfraction, the weight of the sample that was used, named yield (Y%), and the total weight of all the subfractions, named relative yield (RY%), for the tests in SEC. For the RP-FC tests, the weights of the samples used were compared with the weight of the sub-fraction collected.

3.4. Monitoring the Sub-Fractionation and Isolation Process

3.4.1. Thin Layer Chromatography (TLC)

The samples obtained during the process were resuspended in methanol, and small volumes were applied to 60-F_{254} silica gel plates with 10—12 μm particles (Merck KGaA®, Darmstadt, Germany) using a glass capillary. The chromatograms were developed in a vertical glass chamber with a double trough (20 cm × 10 cm, Camag®, Muttenz, Switzerland) after being saturated for 30 min with the 90:5:5 mobile phase (ethyl acetate: formic acid: water, $v/v/v$). Following this process, the TLC plates were derivatized with the reagent aluminum chloride ($AlCl_3$) at 5% (w/v) and evaluated under UV light at 365 nm for flavonoids. They were also derivatized with ferric chloride ($FeCl_3$) at 5% (w/v) for hydrolysable tannins and evaluated under visible light. Image acquisition and UV observations were carried out using the MultiDoc—It Imaging System® (Model 125) with UVP® software and a Canon®-coupled camera (Rebel T3, EOS 1100 D, Canon, Tokyo, Japan).

3.4.2. High Performance Liquid Chromatography (HPLC-DAD)

The samples were analyzed by HPLC using an Ultimate 3000 system (Thermo Scientific®, Waltham, MA, USA), equipped with a diode array detector (DAD; Thermo Fisher Scientific®, Waltham, MA, USA). Chromeleon software (Dionex®, version 6.0) was utilized for data acquisition and processing. Chromatographic separation was performed using a C_{18} column (250 mm × 4.6 mm i.d., 5 μm; Supelco®, Sigma-Aldrich, San Luis, MO, USA), which was equipped with a pre-column of the same material (4 mm × 3.9 μm, Phenomenex®, Torrance, CA, USA). The analysis conditions that were described and validated previously were employed [19].

3.5. Characterization LC-ESI-HRMS/MS

EAF, SPF, MPF, and LF were analyzed by Ultra-Performance Liquid Chromatography (UPLC) using a Prominence UFLC system (Shimadzu®, Quioto, Japan), equipped with a diode array detector (SPD-M20A, Shimadzu®), and coupled to a MicrO-TOF-QII™ mass detector (Bruker®, Billerica, MA, USA). Chromatographic separation was performed using a C_{18} column (100 mm × 4.6 mm i.d., 3.5 μm, WaterS Symmetry®, Milford, MA, USA) at an oven temperature of 25 °C. The mobile phase comprised solvent A (purified water, Milli-Q®, Merck Group, Darmstadt, Germany) and solvent B (methanol, LC-MS grade), both acidified with 0.1% formic acid (HPLC grade), and flowed at a rate of 0.5 mL/min^{-1}, following the gradient: 0–2 min (10% B), 2–40 min (40–75% B), 40–48 min (75% B), 48–48.5 min (75–10% B), and 48.5–50 min (10% B). The MS parameters used were: electrospray ionization (ESI) at negative ion polarity; capillary voltage 2.3 kV and end plate offset −500 V; nebulizer pressure 2.0 Bar; dry gas flow 10.0 L.min^{-1} and dry heater 200 °C. The radio frequency in the collision cell was maintained at 200 Vpp, and the MS/MS collision energy was 35 eV. The isolates were evaluated by direct injection into a MicrO-TOF-QII™ mass spectrometer (Bruker®, Billerica, MA, USA). The equipment utilized electrospray ionization (ESI) under the following operating conditions: negative ion polarity; capillary voltage of 2.3 kV and end plate offset of −500 V; nebulizer pressure of 1.0 Bar; dry gas flow of 4.0 L.min^{-1} and dry heater at 180 °C. The radio frequency in the collision cell was maintained at 250 Vpp. The samples were introduced with the help of an automatic syringe pump, using approximately 10 μg/L samples at a constant flow of 180 μL/h of MS-grade methanol. The spectra were obtained in the 100–1000 m/z scanning range, processed in the Bruker DataAnalysis® 4.0 software, and expressed as m/z.

3.6. Minimum Inhibitory Concentration (MIC) and Minimum Fungicidal Concentration (MFC)

The antifungal evaluation of SDCE, fractions, subfractions, and isolates was conducted using standard strains of *Candida albicans* (90028), *Candida glabrata* (9001), and *Candida auris* (CDC B11903) from the American Type Culture Collection (ATCC). The yeasts were manipulated under sterile experimental conditions and grown on Sabouraud agar for 48 h at 37 °C. The fungal suspension was obtained on the 0.5 McFarland scale by adjusting the turbidity to 530 nm on a spectrophotometer (Micronal®, AJX1900, Tecnal, Piracicaba, São Paulo, Brazil). After the inoculums were calibrated, two dilutions were made: one of 1:50 in sterile saline solution and a further dilution of 1:20 in culture medium, yielding a final concentration of 10^3 CFU/mL. The tests were carried out in a 96-well microplate, using 100 μL of the culture medium in all the wells and 100 μL of the sample, each sample in the well of its corresponding row. Serial dilutions were made in the following wells, discarding 100 μL from the well of column 10, obtaining concentrations of 1000 to 1.95 μg/mL. A total of 100 μL of the inoculum was added to the wells of columns 1 to 11. In column 11, the presence of inoculum and absence of a sample served as a positive control, facilitating the observation of inoculum viability. In column 12, 200 μL of culture medium without sample or inoculum was added, constituting the negative control. The plates were incubated for 48 h at a temperature of 30–37 °C, after which they were examined for the presence or absence of growth. The MIC was determined by the lowest concentration at which each sample could inhibit fungal growth, in comparison to the positive control. To determine the MFC, an aliquot of 5 μL was taken from each well of the MIC test and transferred to a Petri dish containing Sabouraud agar, then incubated for 48 h at 37 °C. The MFC was evaluated at the lowest concentration at which each sample could prevent visible fungal growth (CLSI).

4. Conclusions

The fractions and crude extract of *E. uniflora* leaves demonstrated anti-*Candida* activity against *C. albicans*, *C. glabrata*, and *C. auris*, consistent with findings in the literature. The partitioning and purification strategy facilitated the extraction of primary phytochemicals, revealing that certain compounds, such as hydrolysable tannin ellagic acid and the flavonoid myricitrin, were associated with the biological response of these ex-tracts. This discovery carries significant implications, offering valuable insights and guiding future research endeavors aimed at developing and standardizing new antifungal products derived from plant species rich in these classes of metabolites. The fractions enriched with myricitrin and ellagic acid exhibited yields of 10.36% and 1.19%, respectively, indicating the feasibility of obtaining sub-fractions and purified compounds from *E. uniflora* leaves for research purposes, although not immediately scalable for incorporation as active ingredients in new products. The MICs ranged from 62.5 to 125.0 for the fractions enriched with myricitrin and ellagic acid, respectively, underscoring the potential and influence of these constituents on the antifungal activity of the extracts. Therefore, further studies are warranted to elucidate the mechanisms of action involved and explore the potential contributions of other phytochemicals described in the matrix.

Supplementary Materials: The following supporting information can be downloaded at: https://www.mdpi.com/article/10.3390/molecules29122761/s1, Figure S1. Chromatographic profile of spray-dried crude extract (SDCE) from E. uniflora leaves at 270 nm (A) and chomatograms of the Ethyl Acetate Fraction (EAF) and the crude extract (SDCE) compared at 270 nm. Figure S2. Chromatogram of test 1 of the RP-FC screening (A) and TLCs of the collected sub-fractions derivatized with 5% (*w*/*v*) AlCl3 (B) and FeCl3 (C). Figure S3. Absorption spectrum of the peak with retention time of 23,91 min from Test 1 in SEC screening corresponding to EA. Figure S4. TLC of the subfractions collected in test 2 of the SEC screening derivatized with 5% (*w*/*v*) AlCl3. Figure S5. Chromatograms of subfraction 16 at 270 nm (A), along with the respective UV spectra of GA and peak 2 (A); chromatograms of subfraction 20 at 340 nm, and the respective UV spectra of MyR and peaks 1 and 3 (B); chromatograms of subfractions 40, 44, and 48 at 254 nm, and the respective UV spectra of EA and peaks 1 and 2. Test 2 samples in SEC. Figure S6. Monitoring TLCs of the subfractions collected from test 3 by SEC derivatized with 5% (*w*/*v*) AlCl3. Figure S7. Monitoring TLCs of the subfractions collected from test 4 by SEC derivatized with 5% (*w*/*v*) AlCl3. Figure S8. Monitoring TLCs of the subfractions collected in SEC fractionations derivatized with 5% (*w*/*v*) AlCl3. Table S1. Yields of sub-fractions collected in Test 4 from SEC. Table S2. Yields of the sub-fractions collected in the process of obtaining EA II. Table S3. Yields of the subfractions collected in the process of obtaining GAF from FLAVF. Figure S9. TLC of precipitate (P) and edge (B) samples from MyR I and II suffractions derivatized with 5% (*w*/*v*) AlCl3. Table S4. Yields of subfractions collected when processing MyR I and II subfractions. Figure S10. Spectrometric profile of myricitrin (A), gallic acid (GA) and ellagic acid (C) isolates from E. uniflora leaves evaluated by direct injection in ESI-HRMS/MS.

Author Contributions: C.J.L.T.: Conceptualization, Data curation, Formal analysis, Investigation, Methodology, Writing—original draft; T.d.S.D.: Formal analysis, Investigation, Methodology; L.S.A.: Formal analysis; M.R.A.F.: Conceptualization, Formal analysis, Project administration, Supervision, Writing—review and editing; L.A.L.S.: Acquisition of funding, Project administration, Supervision, Validation, Writing—review and editing. All authors have read and agreed to the published version of the manuscript.

Funding: This study was funded in part by Conselho Nacional de Desenvolvimento Científico e Tecnológico (CNPq–Financial Code 405297/2018-1; 304234/2021-4 and 408863/2022-6) and Fundação de Amparo à Ciência e Tecnologia de Pernambuco (FACEPE: IBPG-0489-4.03/21).

Institutional Review Board Statement: Not applicable.

Informed Consent Statement: Not applicable.

Data Availability Statement: Data are contained within the article.

Acknowledgments: The authors would also like to thank the Laboratory of Structural Molecular Biology (LABIME) of Federal University of Santa Catarina (UFSC) for providing the UPLC-ESI-HRMS/MS system.

Conflicts of Interest: The authors declare that they have no conflict of interest.

References

1. Abu Khalaf, R.; Alhusban, A.A.; Al-Shalabi, E.; Al-Sheikh, I.; Sabbah, D.A. Chapter 10—Isolation and structure elucidation of bioactive polyphenols. *Stud. Nat. Prod. Chem.* **2019**, *63*, 267–337.
2. Oldoni, T.L.C.; Merlin, N.; Bicas, T.C.; Prasniewski, A.; Carpes, S.T.; Ascari, J.; de Alencar, S.M.; Massarioli, A.P.; Bagatini, M.D.; Morales, R.; et al. Antihyperglycemic activity of crude extract and isolation of phenolic compounds with antioxidant activity from *Moringa oleifera* Lam. leaves grown in Southern Brazil. *Food Res. Int.* **2021**, *141*, 110082. [CrossRef] [PubMed]
3. Guo, X.; Long, P.; Meng, Q.; Ho, C.-T.; Zhang, L. An emerging strategy for evaluating the grades of *Keemun black* tea by combinatory liquid chromatography-Orbitrap mass spectrometry-based untargeted metabolomics and inhibition effects on α-glucosidase and α-amylase. *Food Chem.* **2018**, *246*, 74–81. [CrossRef] [PubMed]
4. Luo, L.; Cui, Y.; Zhang, S.; Li, L.; Li, Y.; Zhou, P.; Sun, B. Preparative separation of grape skin polyphenols by high-speed counter-current chromatography. *Food Chem.* **2016**, *212*, 712–721. [CrossRef] [PubMed]
5. Mukherjee, P. Chapter 10—High-Performance Liquid Chromatography for Analysis of Herbal Drugs. In *Quality Control and Evaluation of Herbal Drugs*; Elsevier: Amsterdam, The Netherlands, 2019; pp. 421–458. [CrossRef]
6. Sarker, S.D.; Nahar, L. Chapter 19—Applications of High Performance Liquid Chromatography in the Analysis of Herbal Products. In *Evidence-Based Validation of Herbal Medicine*; Mukherjee, P.K., Ed.; Elsevier: Boston, MA, USA, 2015; pp. 405–425. [CrossRef]

7. Di Lorenzo, C.; Colombo, F.; Biella, S.; Stockley, C.; Restani, P. Polyphenols and Human Health: The Role of Bioavailability. *Nutrients* **2021**, *13*, 273. [CrossRef] [PubMed]
8. Rasouli, H.; Farzaei, M.H.; Khodarahmi, R. Polyphenols and their benefits: A review. *Int. J. Food Prop.* **2017**, *20*, 1700–1741. [CrossRef]
9. Silva-Rocha, W.P.; de Azevedo, M.F.; Ferreira, M.R.A.; da Silva, J.d.F.; Svidzinski, T.I.E.; Milan, E.P.; Soares, L.A.L.; Rocha, K.B.F.; UchÔa, A.F.; Mendes-Giannini, M.J.S.; et al. Effect of the Ethyl Acetate Fraction of *Eugenia uniflora* on Proteins Global Expression during Morphogenesis in *Candida albicans*. *Front. Microbiol.* **2017**, *8*, 1788. [CrossRef] [PubMed]
10. Silva-Rocha, W.P.; de Brito Lemos, V.L.; Ferreira, M.R.A.; Soares, L.A.L.; Svidzisnki, T.I.E.; Milan, E.P.; Chaves, G.M. Effect of the crude extract of Eugenia uniflora in morphogenesis and secretion of hydrolytic enzymes in Candida albicans from the oral cavity of kidney transplant recipients. *BMC Complement. Altern. Med.* **2015**, *15*, 6. [CrossRef] [PubMed]
11. Sardi, J.d.C.O.; Freires, I.A.; Lazarini, J.G.; Infante, J.; de Alencar, S.M.; Rosalen, P.L. Unexplored endemic fruit species from Brazil: Antibiofilm properties, insights into mode of action, and systemic toxicity of four *Eugenia* spp. *Microb. Pathog.* **2017**, *105*, 280–287. [CrossRef]
12. Franzon, R.C.; Carpenedo, S.; Viñoly, M.D.; Raseira, M.d.C.B. Pitanga—*Eugenia uniflora* L. In *Exotic Fruits*; Rodrigues, S., Silva, E.d.O., de Brito, E.S., Eds.; Academic Press: Cambridge, MA, USA, 2018; pp. 333–338. [CrossRef]
13. Denardin, C.C.; Hirsch, G.E.; da Rocha, R.F.; Vizzotto, M.; Henriques, A.T.; Moreira, J.C.F.; Guma, F.T.C.R.; Emanuelli, T. Antioxidant capacity and bioactive compounds of four Brazilian native fruits. *J. Food Drug Anal.* **2015**, *23*, 387–398. [CrossRef]
14. Sobeh, M.; El-Raey, M.; Rezq, S.; Abdelfattah, M.A.O.; Petruk, G.; Osman, S.; El-Shazly, A.M.; El-Beshbishy, H.A.; Mahmoud, M.F.; Wink, M. Chemical profiling of secondary metabolites of *Eugenia uniflora* and their antioxidant, anti-inflammatory, pain killing and anti-diabetic activities: A comprehensive approach. *J. Ethnopharmacol.* **2019**, *240*, 111939. [CrossRef] [PubMed]
15. Sobeh, M.A.-O.; Hamza, M.S.; Ashour, M.A.-O.; Elkhatieb, M.A.-O.X.; El Raey, M.A.-O.; Abdel-Naim, A.A.-O.; Wink, M.A.-O. A Polyphenol-Rich Fraction from *Eugenia uniflora* Exhibits Antioxidant and Hepatoprotective Activities *in vivo*. *Pharm.* **2020**, *13*, 84. [CrossRef]
16. Sobral-Souza, C.E.; Silva, A.R.P.; Leite, N.F.; Rocha, J.E.; Costa, J.G.M.; Menezes, I.R.A.; Cunha, F.A.B.; Rolim, L.A.; Sousa, A.K.; Coutinho, H.D.M. The role of extracts from *Eugenia uniflora* L. against metal stress in eukaryotic and prokaryotic models. *S. Afr. J. Bot.* **2020**, *131*, 360–368. [CrossRef]
17. da Cunha, F.A.B.; Waczuk, E.P.; Duarte, A.E.; Barros, L.M.; Elekofehinti, O.O.; Matias, E.F.F.; da Costa, J.G.M.; Sanmi, A.A.; Boligon, A.A.; da Rocha, J.B.T.; et al. Cytotoxic and antioxidative potentials of ethanolic extract of *Eugenia uniflora* L. (Myrtaceae) leaves on human blood cells. *Biomed. Pharmacother.* **2016**, *84*, 614–621. [CrossRef]
18. de Souza, C.E.S.; da Silva, A.R.P.; Rocha, J.E.; Vega Gomez, M.C.; Rolóm, M.; Coronel, C.; Martins da Costa, J.G.; Netto, M.L.C.; Rolim, L.A.; Coutinho, H.D.M. LC–MS characterization, anti-kinetoplastide and cytotoxic activities of natural products from *Eugenia jambolana* Lam. and *Eugenia uniflora*. *Asian Pac. J. Trop. Biomed.* **2017**, *7*, 836–841. [CrossRef]
19. Bezerra, I.C.F.; Ramos, R.T.d.M.; Ferreira, M.R.A.; Soares, L.A.L. Chromatographic profiles of extractives from leaves of *Eugenia uniflora*. *Rev. Bras. Farmacogn.* **2018**, *28*, 92–101. [CrossRef]
20. Ramos, R.T.M.; Bezerra, I.C.F.; Ferreira, M.R.A.; Soares, L.A.L. Spectrophotometric Quantification of Flavonoids in Herbal Material, Crude Extract, and Fractions from Leaves of *Eugenia uniflora* Linn. *Pharmacogn. Res.* **2017**, *9*, 253. [CrossRef]
21. Souza, O.A.; Furlani, R.P.; Ramalhão, V.G.d.S.; Borges, M.S.; Funari, C.S.; Bolzani, V.d.S.; Rinaldo, D. Eco-friendly and inexpensive food grade bioethanol for *Eugenia uniflora* L. chromatographic fingerprinting: A trade-off between separation and sustainability. *Phytochem. Lett.* **2021**, *43*, 200–207. [CrossRef]
22. Chakravartula, S.S.N.; Lourenço, R.V.; Balestra, F.; Quinta Barbosa Bittante, A.M.; Sobral, P.J.d.A.; Dalla Rosa, M. Influence of pitanga (*Eugenia uniflora* L.) leaf extract and/or natamycin on properties of cassava starch/chitosan active films. *Food Packag. Shelf Life* **2020**, *24*, 100498. [CrossRef]
23. Donadel, G.; Dalmagro, M.; de Oliveira, J.A.; Zardeto, G.; Pinc, M.M.; Hoscheid, J.; Alberton, O.; Belettini, S.T.; Jacomassi, E.; Gasparotto Junior, A.; et al. Safety Investigations of Two Formulations for Vaginal Use Obtained from *Eugenia uniflora* L. Leaves in Female Rats. *Pharmaceuticals* **2022**, *15*, 1567. [CrossRef]
24. Kurkin, V.A.; Zimenkina, N.y.I. HPLC Determination of Myricitrin in *Juglans nigra* L. Bark. *Pharm. Chem. J.* **2021**, *55*, 881–885. [CrossRef]
25. Agrawal, O.D.; Kulkarni, Y.A. Mini-Review of Analytical Methods used in Quantification of Ellagic Acid. *Rev. Anal. Chem.* **2020**, *39*, 31–44. [CrossRef]
26. Silva, L.; de Oliveira, M.; Martins, C.; Borges, L.; Fiuza, T.; da Conceição, E.; de Paula, J. Validation HPLC-DAD Method for Quantification of Gallic and Ellagic Acid from *Eugenia punicifolia* Leaves, Extracts and Fractions. *J. Braz. Chem. Soc.* **2023**, *34*, 401–413. [CrossRef]
27. Nantarat, N.; Mueller, M.; Lin, W.-C.; Lue, S.-C.; Viernstein, H.; Chansakaow, S.; Sirithunyalug, J.; Leelapornpisid, P. Sesaminol diglucoside isolated from black sesame seed cake and its antioxidant, anti-collagenase and anti-hyaluronidase activities. *Food Biosci.* **2020**, *36*, 100628. [CrossRef]
28. Pawłowska, K.A.; Hałasa, R.; Dudek, M.K.; Majdan, M.; Jankowska, K.; Granica, S. Antibacterial and anti-inflammatory activity of bistort (*Bistorta officinalis*) aqueous extract and its major components. Justification of the usage of the medicinal plant material as a traditional topical agent. *J. Ethnopharmacol.* **2020**, *260*, 113077. [CrossRef] [PubMed]

29. Žuvela, P.; Skoczylas, M.; Jay Liu, J.; Bączek, T.; Kaliszan, R.; Wong, M.W.; Buszewski, B. Column Characterization and Selection Systems in Reversed-Phase High-Performance Liquid Chromatography. *Chem. Rev.* **2019**, *119*, 3674–3729. [CrossRef] [PubMed]
30. LaCourse, M.E.; LaCourse, W.R. Chapter 17—General instrumentation in HPLC. In *Liquid Chromatography*, 2nd ed.; Fanali, S., Haddad, P.R., Poole, C.F., Riekkola, M.-L., Eds.; Elsevier: Amsterdam, The Netherlands, 2017; pp. 417–429. [CrossRef]
31. Mottaghipisheh, J.; Iriti, M. Sephadex® LH-20, Isolation, and Purification of Flavonoids from Plant Species: A Comprehensive Review. *Molecules* **2020**, *25*, 4146. [CrossRef] [PubMed]
32. Yanagida, A.; Shoji, T.; Shibusawa, Y. Separation of proanthocyanidins by degree of polymerization by means of size-exclusion chromatography and related techniques. *J. Biochem. Biophys. Methods* **2003**, *56*, 311–322. [CrossRef] [PubMed]
33. Iwashina, T.; Tanaka, N.; Aung, M.M.; Devkota, H.P.; Mizuno, T. Phenolic compounds from parasitic *Sapria himalayana* f. *albovinosa* and *Sapria myanmarensis* (Rafflesiaceae) in Myanmar. *Biochem. Syst. Ecol.* **2020**, *93*, 104179. [CrossRef]
34. Wen, C.; Song, D.; Zhuang, L.; Liu, G.; Liang, L.; Zhang, J.; Liu, X.; Li, Y.; Xu, X. Isolation and identification of polyphenol monomers from celery leaves and their structure-antioxidant activity relationship. *Process Biochem.* **2022**, *121*, 69–77. [CrossRef]
35. Ovchinnikov, D.V.; Bogolitsyn, K.G.; Druzhinina, A.S.; Kaplitsin, P.A.; Parshina, A.E.; Pikovskoi, I.I.; Khoroshev, O.Y.; Turova, P.N.; Stavrianidi, A.N.; Shpigun, O.A. Study of Polyphenol Components in Extracts of Arctic Brown Algae of *Fucus vesiculosus* Type by Liquid Chromatography and Mass-Spectrometry. *J. Anal. Chem.* **2020**, *75*, 633–639. [CrossRef]
36. Stanek, N.; Jasicka-Misiak, I. HPTLC Phenolic Profiles as Useful Tools for the Authentication of Honey. *Food Anal. Methods* **2018**, *11*, 2979–2989. [CrossRef]
37. Striegel, A.M. Chapter 10—Size-exclusion chromatography. In *Liquid Chromatography*, 2nd ed.; Fanali, S., Haddad, P.R., Poole, C.F., Riekkola, M.-L., Eds.; Elsevier: Amsterdam, The Netherlands, 2017; pp. 245–273. [CrossRef]
38. Vidal-Casanella, O.; Núñez, O.; Granados, M.; Saurina, J.; Sentellas, S. Analytical Methods for Exploring Nutraceuticals Based on Phenolic Acids and Polyphenols. *Appl. Sci.* **2021**, *11*, 8276. [CrossRef]
39. Tian, Y.; Liimatainen, J.; Puganen, A.; Alakomi, H.-L.; Sinkkonen, J.; Yang, B. Sephadex LH-20 fractionation and bioactivities of phenolic compounds from extracts of *Finnish berry* plants. *Food Res. Int.* **2018**, *113*, 115–130. [CrossRef] [PubMed]
40. Rios, J.L.; Giner, R.M.; Marin, M.; Recio, M.C. A Pharmacological Update of Ellagic Acid. *Planta Med.* **2018**, *84*, 1068–1093. [CrossRef] [PubMed]
41. Bai, J.; Zhang, Y.; Tang, C.; Hou, Y.; Ai, X.; Chen, X.; Zhang, Y.; Wang, X.; Meng, X. Gallic acid: Pharmacological activities and molecular mechanisms involved in inflammation-related diseases. *Biomed. Pharmacother.* **2021**, *133*, 110985. [CrossRef] [PubMed]
42. Zhang, X.; Zhang, K.; Wang, Y.; Ma, R. Effects of Myricitrin and Relevant Molecular Mechanisms. *Curr. Stem Cell Res. Ther.* **2020**, *15*, 11–17. [CrossRef]
43. PUBCHEM. PubChem Compound Summary for CID 5281673, Myricitrin: National Center for Biotechnology Information. 2023. Available online: https://pubchem.ncbi.nlm.nih.gov/compound/Myricitrin (accessed on 30 January 2023).
44. de Oliveira, F.M.G.; Romão, W.; Kuster, R.M. Identification of phenolic compounds in *Eugenia uniflora* leaves by FTICR MS in association with different ionization sources. *Anal. Methods* **2018**, *10*, 1647–1655. [CrossRef]
45. Souza, P.; Santos, M.; Monteiro, R.; Espindola, M.; Souza, H.; Monteiro, A.; Camara, C.; Silva, T. Taninos e Flavonóides das flores de *Eugenia uniflora* (Myrtaceae). *Quim. Nova* **2022**, *45*, 1083–1091. [CrossRef]
46. Bagatini, L.; Zandoná, G.P.; Hoffmann, J.F.; de Souza Cardoso, J.; Teixeira, F.C.; Moroni, L.S.; Junges, A.; Kempka, A.P.; Stefanello, F.M.; Rombaldi, C.V. Evaluation of *Eugenia uniflora* L. leaf extracts obtained by pressurized liquid extraction: Identification of chemical composition, antioxidant, antibacterial, and allelopathic activity. *Sustain. Chem. Pharm.* **2023**, *35*, 101214. [CrossRef]
47. García, Y.M.; Ramos, A.A.-O.; de Oliveira Júnior, A.A.-O.; de Paula, A.; de Melo, A.A.-O.; Andrino, M.A.; Silva, M.R.; Augusti, R.A.-O.; de Araújo, R.L.B.; de Lemos, E.A.-O.; et al. Physicochemical Characterization and Paper Spray Mass Spectrometry Analysis of *Myrciaria floribunda* (H. *West ex Willd.) O. Berg Molecules* **2021**, *26*, 7206. [CrossRef] [PubMed]
48. Rattmann, Y.D.; de Souza Lm Fau-Malquevicz-Paiva, S.M.; Malquevicz-Paiva Sm Fau-Dartora, N.; Dartora N Fau-Sassaki, G.L.; Sassaki Gl Fau-Gorin, P.A.J.; Gorin Pa Fau-Iacomini, M.; Iacomini, M. Analysis of Flavonoids from *Eugenia uniflora* Leaves and Its Protective Effect against Murine Sepsis. In *Evidence-Based Complementary and Alternative Medicine*; Wiley: Hoboken, NJ, USA, 2012. [CrossRef]
49. Jaiswal, R.; Jayasinghe, L.; Kuhnert, N. Identification and characterization of proanthocyanidins of 16 members of the Rhododendron genus (Ericaceae) by tandem LC-MS. *J. Mass Spectrom.* **2012**, *47*, 502–515. [CrossRef] [PubMed]
50. Bowers, J.J.; Gunawardena, H.P.; Cornu, A.; Narvekar, A.S.; Richieu, A.; Deffieux, D.; Quideau, S.; Tharayil, N. Rapid Screening of Ellagitannins in Natural Sources via Targeted Reporter Ion Triggered Tandem Mass Spectrometry. *Sci. Rep.* **2018**, *8*, 10399. [CrossRef] [PubMed]
51. Ogundele, A.V.; Yadav, A.; Das, A.M. Antimicrobial and α-Amylase Inhibitory Activities of Constituents from *Elaeocarpus floribundus*. *Rev. Bras. Farmacogn.* **2021**, *31*, 330–334. [CrossRef]
52. Salazar-Aranda, R.; Granados-Guzmán, G.; Pérez-Meseguer, J.; González, G.M.; de Torres, N.W. Activity of Polyphenolic Compounds against *Candida glabrata*. *Molecules* **2015**, *20*, 17903. [CrossRef]
53. Gadetskaya, A.V.; Tarawneh, A.H.; Zhusupova, G.E.; Gemejiyeva, N.G.; Cantrell, C.L.; Cutler, S.J.; Ross, S.A. Sulfated phenolic compounds from *Limonium caspium*: Isolation, structural elucidation, and biological evaluation. *Fitoterapia* **2015**, *104*, 80–85. [CrossRef]

54. Gatto, L.J.; Veiga, A.; Gribner, C.; Moura, P.F.; Rech, K.S.; Murakami, F.S.; Dias, J.d.F.G.; Miguel, O.G.; Miguel, M.D. *Myrcia hatschbachii: Antifungal* activity and structural elucidation of ellagic and 3-*O*-methyl ellagic acids. *Nat. Prod. Res.* **2021**, *35*, 5540–5543. [CrossRef]
55. Rossatto, F.C.P.; Tharmalingam, N.; Escobar, I.E.; d'Azevedo, P.A.; Zimmer, K.R.; Mylonakis, E. Antifungal Activity of the Phenolic Compounds Ellagic Acid (EA) and Caffeic Acid Phenethyl Ester (CAPE) against Drug-Resistant *Candida auris*. *J. Fungi* **2021**, *7*, 763. [CrossRef]
56. Vigbedor, B.Y.; Akoto, C.O.; Neglo, D. Isolation and characterization of 3,3′-di-*O*-methyl ellagic acid from the root bark of *Afzelia africana* and its antimicrobial and antioxidant activities. *Sci. Afr.* **2022**, *17*, e01332. [CrossRef]

molecules

MDPI

Article

The Effect of the Extraction Conditions on the Antioxidant Activity and Bioactive Compounds Content in Ethanolic Extracts of *Scutellaria baicalensis* Root

Małgorzata Dzięcioł [1,*], Klaudia Wala [1], Agnieszka Wróblewska [2] and Katarzyna Janda-Milczarek [3,*]

[1] Department of Chemical Organic Technology and Polymeric Materials, Faculty of Chemical Technology and Engineering, West Pomeranian University of Technology in Szczecin, Piastów Ave. 42, 71-065 Szczecin, Poland; klaudiawala94@gmail.com

[2] Department of Catalytic and Sorbent Materials Engineering, Faculty of Chemical Technology and Engineering, West Pomeranian University of Technology in Szczecin, Piastów Ave. 42, 71-065 Szczecin, Poland; agnieszka.wroblewska@zut.edu.pl

[3] Department of Human Nutrition and Metabolomics, Pomeranian Medical University in Szczecin, 24 Broniewskiego Street, 71-460 Szczecin, Poland

* Correspondence: malgorzata.dzieciol@zut.edu.pl (M.D.); katarzyna.janda.milczarek@pum.edu.pl (K.J.-M.)

Abstract: Ethanolic extracts of Baikal skullcap (*Scutellaria baicalensis*) root were obtained using various techniques, such as maceration, maceration with shaking, ultrasound-assisted extraction, reflux extraction, and Soxhlet extraction. The influence of the type and time of isolation technique on the extraction process was studied, and the quality of the obtained extracts was determined by spectrophotometric and chromatographic methods to find the optimal extraction conditions. Radical scavenging activity of the extracts was analyzed using DPPH assay, while total phenolic content (TPC) was analyzed by the method with the Folin–Ciocalteu reagent. Application of gas chromatography with mass selective detector (GC-MS) enabled the identification of some bioactive substances and a comparison of the composition of the particular extracts. The Baikal skullcap root extracts characterized by both the highest antioxidant activity and content of phenolic compounds were obtained in 2 h of reflux and Soxhlet extraction. The main biologically active compounds identified in extracts by the GC-MS method were wogonin and oroxylin A, known for their broad spectrum of biological effects, including antioxidant, anti-inflammatory, antiviral, anticancer, and others.

Keywords: antioxidant activity; bioactive compounds; Baikal skullcap; ethanolic extracts; DPPH scavenging activity; extraction techniques; GC-MS; *Scutellaria baicalensis*; total phenolic content

1. Introduction

Both scientists and increasingly aware consumers are looking for natural raw materials that would have a health-promoting effect on the human body and could be used to diversify the daily diet. Very often, researchers' attention focuses on plants that have been used for centuries, e.g., in local folk medicine in various parts of the world. Such plants are characterized by a specific chemical composition, which determines interesting and unique health-promoting properties. The biologically active compounds contained in them are mostly not used in modern medicine, which mainly uses the biologically active compounds obtained by chemical synthesis. Very often, individual biologically active substances are not as active as extracts obtained from plants with medicinal and nutritional properties. This is due to the fact that the presence of various compounds in the extract contributes to their synergistic effect, which enhances the health-promoting effects.

An example of such a plant is the Baikal skullcap (synonyms: Chinese skullcap, Hooded skullcap) (*Scutellaria baicalensis* Georgi). It is a perennial herbaceous plant from the family *Lamiaceae* whose root (*Radix scutellariae*) has been used in Chinese medicine for over 2000 years. This plant comes from Korea, China, the Russian regions of the Far East,

Mongolia, and Siberia, but it has also been introduced into experimental crops in Europe, including Poland and Ukraine [1–3]. In its natural environment, at altitudes from 60 m to 2000 m above sea level, it grows on dry, sunny, and grassy slopes. In traditional Chinese medicine, this raw material was used in colds, lung and liver diseases, in the treatment of diarrhea, dysentery, hypertension, hemorrhages, insomnia, inflammation, and respiratory infections [4]. Typically, the roots are harvested in spring or autumn, then dried and used in their natural form or processed into powders, tinctures, or pills [4]. Roots contain many compounds of nutritional and pharmacological importance, with a total of about 130 compounds having been isolated so far [5]. The dominant compounds in the root are flavonoids and glycosides [6]. So far, over 40 different polyphenols have been identified, including flavonoids and their flavonols, dihydroflavones and their dihydroflavonols, chalcones, and bioflavonoids. Among them, the most characteristic of this plant's raw material are baicalin, baicalein, wogonoside, and wogonin [6–15]. Biologically active compounds are found in the roots, with antioxidant and anti-inflammatory effects, among others, contributing to multidirectional health-promoting effects, as demonstrated by studies conducted on animal models (rats, mice), cell lines, and clinical trials. Baikal skullcap root extracts are characterized by, among others, the following effects: (1) anti-inflammatory [10,16–20], including within the pancreas [21], uterine mucosa [22], stomach [23], and lungs [24]; (2) antioxidant [2,7,25]; (3) hepatoprotective [26,27]; (4) protective for bones and joints [28–30] and the circulatory system [31]; (5) autoimmune [32]; (6) antiviral [33]; (7) antibacterial [34], including against bacteria causing tooth decay [35] and a resistance to antibiotics [36]; and (8) anti-cancer [37], e.g., in the case of nasopharyngeal tumors [38], cervical cancer [39], the liver [40], the large intestine [15], and ovaries [41].

Moreover, Yao et al. indicated the potential use of Baikal skullcap root in the prevention and treatment of Covid-19 [42]. Research by Ma et al. indicates that *Scutellaria baicalensis* may also alleviate depressive behaviors [43]. The potential of this raw material in the treatment of neurological diseases and improvement of cognitive functions [44–46], including Parkinson's [47] and Alzheimer's disease [48], has also been demonstrated.

The extraction process of bioactive compounds from plants depends on many factors, including the type of solvent used, the ratio of solvent to plant material, the applied extraction technique conditions, and all pre- and post-treatment operations. In the literature, there are some reports related to the extraction of Baikal skullcap root, conducted in various conditions. Ni et al. [49] described the extraction process of baicalin from *S. baicalensis* using water, optimized by orthogonal test and controlled by an HPLC analysis of the baicalin yield. An interesting, non-conventional approach has been described by Choi et al. [50], who applied alkaline reduced water as a solvent for Baikal skullcap extraction at 100 °C for 24 h, which resulted in a slight improvement of extraction efficiency and higher DPPH radical scavenging activity than using distilled water or 70% ethanol in similar conditions. Generally, among organic solvents, methanol and ethanol in various concentrations were described in the literature as the most effective solvents in the isolation of different flavonoids and other bioactive compounds from *S. baicalensis* [51–53]. As an alternative to the conventional organic solvents, natural deep eutectic solvents (NADES) were also proposed, and their effectiveness in the extraction of various flavonoids from this plant was compared to 70% ethanol and 80% methanol aqueous solutions [54]. An application of a supercritical fluid extraction (SFE) for the extraction of the flavonoids from the root of *S. baicalensis* was described as being preferable in comparison to the conventional extraction using ultrasounds [55]. However, the availability of this method is limited and the costs are high, as well as microwave-assisted extraction (MAE) [52].

The aim of this study was to investigate the influence of the isolation technique conditions on the antioxidant activity and content of bioactive compounds in *Scutellaria baicalensis* root ethanolic extracts and to indicate the best conditions for extraction and obtaining dry extracts of this plant. It was possible to conduct the comprehensive research by applying a uniform methodology for different extraction techniques in variable time:

maceration, maceration with shaking, ultrasound-assisted extraction, reflux extraction, and Soxhlet extraction, which has not been described in the literature so far.

2. Results and Discussion

2.1. Antioxidant Potential of Baikal Skullcap Root Extracts

Five popular extraction techniques were used in studies, starting from the simple maceration (M) and maceration with shaking (MSH) to the more advanced and effective processes, including ultrasound-assisted extraction (UAE), reflux extraction (RE), and Soxhlet extraction (SE). All experiments were performed using 96% ethanol as a solvent in variable extraction time (0.5, 1, and 2 h). The results of antioxidant activity were measured spectrophotometrically using the DPPH method and expressed as radical scavenging activity (RSA) for extracts obtained in various conditions. They are shown in Table 1.

Table 1. Radical scavenging activity (RSA) of Baikal skullcap (*S. baicalensis*) root extracts obtained using ethanol as a solvent in various conditions.

Extraction Technique	Extraction Time [h]	RSA [%] (Mean ± SD *)
Maceration (M)	0.5	15.46 ± 0.50
	1	18.40 ± 0.43
	2	24.04 ± 0.52
Maceration with shaking (MSH)	0.5	43.42 ± 0.47
	1	45.65 ± 0.27
	2	48.85 ± 0.46
Ultrasound-assisted extraction (UAE)	0.5	52.41 ± 0.53
	1	70.19 ± 0.80
	2	87.46 ± 0.22
Reflux extraction (RE)	0.5	91.88 ± 0.50
	1	94.16 ± 0.25
	2	94.30 ± 0.13
Soxhlet extraction (SE)	0.5	92.82 ± 0.43
	1	95.59 ± 0.18
	2	95.22 ± 0.17

* SD–standard deviation (n = 3).

As can be seen in Table 1, the radical scavenging activity of Baikal skullcap root extracts obtained by the maceration process was relatively low and varied from 15.46% to 24.04%. Application of shaking resulted in an approximate doubling, while using ultrasounds almost quadrupled these values. However, the highest antioxidant activity was observed in the extracts obtained using reflux- and Soxhlet-extraction techniques, where, even for 0.5 h processes, the obtained RSA values exceeded 90%.

In order to illustrate the tendencies observed and facilitate the proper interpretation of the results, the graphical visualization of the extraction conditions' effect on the antioxidant activity and the total phenolic content were conducted. The antioxidant potential, expressed as the Trolox Equivalents Antioxidant Capacity (TEAC) for extracts obtained in various conditions, is presented in Figure 1, while the total phenolic content (TPC) was expressed in Gallic Acid Equivalents in Figure 2. Both parameters were calculated in relation to 1 g of plant material subjected to extraction in various conditions.

Analyzing the antioxidant activity of the obtained extracts (Figure 1), it was found that the TEAC values varied from 1.68 μM TE/g for 0.5-h maceration (M 0.5 h) to the similar highest values above 11 μM TE/g for all reflux- (RE) and Soxhlet- (SE) extraction processes. Generally, prolongation of the extraction time from 0.5 h to 1 h resulted in an increase of antioxidant potential in the case of all extraction techniques, but for maceration (M), maceration with shaking (MSH), reflux (RE), and Soxhlet (SE) extraction, this effect was not very strong. For the 2 h processes, observable enhancement was found in the case of maceration (2.76 μM TE/g) and maceration with shaking (5.90 μM TE/g), but especially for ultrasound-assisted extraction (10.78 μM TE/g). The influence of extraction time was the most noticeable for the UAE technique, which can be caused by the observable rise of the temperature from room temperature to 52 °C after 2 h. In the case of reflux (RE) and Soxhlet (SE) techniques, extending the extraction time from 1 h to 2 h did not result in a further increase in antioxidant activity.

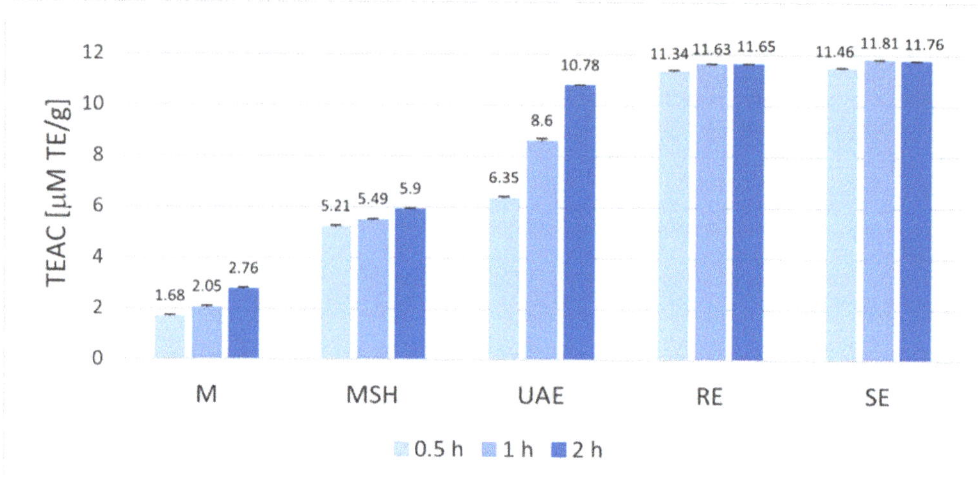

Figure 1. Influence of extraction technique and time on Trolox Equivalents Antioxidant Capacity (TEAC) of Baikal skullcap root ethanolic extracts (M—maceration; MSH—maceration with shaking; UAE—ultrasound-assisted extraction; RE—reflux extraction, SE—Soxhlet extraction).

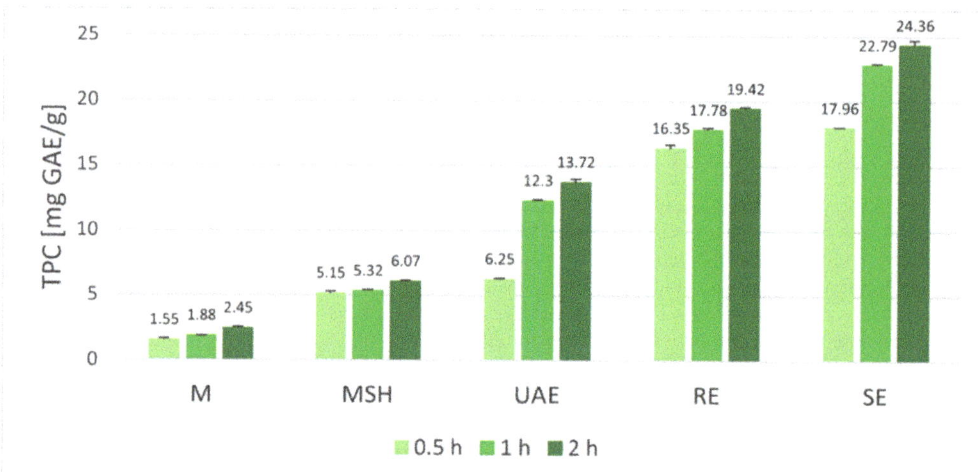

Figure 2. Influence of extraction technique and time on total phenolic content (TPC) of Baikal skullcap root ethanolic extracts (M—maceration; MSH—maceration with shaking; UAE—ultrasound-assisted extraction; RE—reflux extraction, SE—Soxhlet extraction).

Analyzing the total phenolic content in the extracts (Figure 2), some similarities can be seen between the observed trends in comparison to antioxidant activity. TPC values obtained for ethanolic extracts varied from 1.55 mg GAE/g (0.5 h M) to 24.36 mg GAE/g

(RE 2 h). Regardless of the technique used, the phenolic content increased with prolongation of the extraction time from 0.5 to 2 h. The effectiveness of maceration was low, resulting after 2 h in TPC = 2.45 mg GAE/g for simple maceration (M 2 h) and 6.07 mg GAE/g when shaking was applied (MSH 2 h). Extracts obtained in a 0.5 h ultrasonic-assisted extraction (UAE 0.5 h) are also characterized by a relatively low content of phenolic compounds (6.25 mg GAE/g), but extending the time to 1 and 2 h allowed for the obtainment of a significantly higher effectiveness of phenolic compound extraction (12.30 and 13.72 mg GAE/g, respectively). Reflux extraction was an even more effective technique, resulting in TPC values ranging from 16.35 mg GAE/g (RE 0.5 h) to 19.42 mg GAE/g (RE 2 h). Nevertheless, the best technique for the isolation of phenolic compounds was Soxhlet extraction (SE), allowing for the obtainment of the TPC values from 17.96 mg GAE/g (0.5 h SE) to 24.36 mg GAE/g (2 h SE).

On the basis of both the antioxidant activity and total phenolic content measurements, it can be indicated that the most effective techniques were Soxhlet (SE 2 h) and reflux extraction (RE 2 h) conducted in the longest extraction time. The antioxidant potentials of extracts obtained in these conditions were similar (11.76 and 11.65 and µM TE/g, respectively). The application of 2 h of Soxhlet extraction allowed for the isolation of the highest amount of phenolic compounds (24.36 mg GAE/g), but the result obtained by using 2 h of reflux extraction was not much lower (19.42 mg GAE/g). Although there are some articles available in the literature related to Baikal skullcap extraction [49–54], no analogous studies were found, comparing the effectiveness of the various extraction techniques and conditions using 95% ethanol, which were the subject of this work. For this reason, as well as because other authors studied various parts of the plant, including roots and leaves [52] or hairy root culture [8], using different research methodologies and different units, it is difficult to compare the presented results with the others that can be found in the scientific publications.

2.2. Statistical Analysis of Antioxidant Potential and Total Phenolic Content

Descriptive statistics of the antioxidant potential of all analyzed Baikal skullcap root extracts, measured by the DPPH method (as RSA and TEAC) and total phenolic content (TPC), are summarized in Table 2.

Table 2. Descriptive statistics of the tested variables of antioxidant potential and total phenolic content for Baikal skullcap root extracts.

Variables (n = 45)	Mean	Median	Min.	Max.	Lower Quartile	Upper Quartile	Standard Deviation
RSA [%]	64.656	70.500	14.936	95.763	43.921	93.978	29.942
TEAC [µM TE/g]	7.898	8.637	1.610	11.832	5.275	11.606	3.786
TPC [mg GAE/g]	11.535	12.260	1.414	24.630	5.191	17.937	7.702

Statistical analysis of the data showed that, for the particular time variants (0.5 h, 1 h, or 2 h), the differences in the antioxidant potential expressed as RSA (Table 1) and TEAC (Figure 1) between the various extraction techniques used were significant ($p \leq 0.05$). The only exception was a statistically insignificant difference between the antioxidant properties of the extracts obtained in 0.5 h processes using the RE and SE techniques ($p = 0.202$). If we consider the individual extraction techniques, the antioxidant potentials of extracts obtained after 0.5 h, 1 h, and 2 h differed significantly in the case of M, MSH, and UAE techniques ($p < 0.05$). No significant differences were found in the antioxidant potential of extracts obtained after 1 h and 2 h using the RE and SE methods, ($p = 0.874$ and $p = 0.325$, respectively).

In the case of the total phenolic contents (Figure 2), statistical analysis of the data showed that, for all particular time variants (0.5 h, 1 h, and 2 h), the differences between TPC for the extracts obtained by the various extraction techniques were significant ($p < 0.05$).

Importantly, considering the individual techniques of extraction, data for the extracts obtained after 0.5 h, 1 h, and 2 h differed significantly ($p < 0.05$) for each applied method (Figure 2).

The correlations between the parameters studied, including antioxidant potential (TEAC), total phenolic content (TPC), and extraction time for each extraction technique, were also statistically analyzed (Table 3). In most cases (M, MSH, UAE, and RE), a highly significant correlation was found between extraction time and antioxidant potential (correlation coefficient (r) between 0.76 and 0.99). This indicates that the prolongation of the extraction process led to higher values of the antioxidant potential. The only exception was the Soxhlet-extraction (SE) technique, where no significant effect of extraction time on the antioxidant potential was shown. Importantly, regardless of the extraction technique used, positive significant correlations were found between extraction time and total phenolic content (correlation coefficient (r) between 0.85 and 0.97). Furthermore, the statistical correlation between antioxidant potential and total phenolic content (TEAC vs. TPC) was also investigated. For each extraction technique used, positive significant correlations were found between these parameters (correlation coefficient (r) between 0.85 and 0.98). It means that the antioxidant potential of the extracts, measured as the DPPH scavenging activity, may result from the content of various hydroxyflavones (including identified wogonin and oroxylin A) and other compounds belonging to the group of phenolic compounds. High correlation between DPPH scavenging activity and TPC was described in the literature by other authors who studied various plants [56,57].

Table 3. Correlation coefficients (r) between tested parameters.

Extraction Technique	TEAC vs. Time	TPC vs. Time	TPC vs. TEAC
Maceration (M)	0.9937 *	0. 9715 *	0.9869 *
Maceration with shaking (MSH)	0.9849 *	0.9579 *	0.9641 *
Ultrasound-assisted extraction (UAE)	0.9798 *	0.8597 *	0.9427 *
Reflux extraction (RE)	0.7651 *	0.9663 *	0.8515 *
Soxhlet extraction (SE)	0.6570	0.8871 *	0.9166 *

* Significant differences at $p \leq 0.05$.

The performed statistical analysis confirmed that due to significant differences between the particular extraction conditions, the reflux and Soxhlet extraction performed in the longest extraction time (2 h) allowed for the obtainment of the extracts that were characterized by both the highest antioxidant potential and total phenolic content. Therefore, these conditions can be indicated as the best among all that were applied in the studies.

2.3. *Dry Extracts of Baikal Skullcap Root*

For the best extraction conditions selected on the basis of analyses, with the results described above, the research was extended by obtaining the dry extracts from *S. baicalensis* root, determining their yields and antioxidant properties. After removal of the solvent, gentle drying at 40 °C, and stabilization in the desiccator, the yields of dry extracts were calculated considering the obtained masses. The next step was determining the half-maximal inhibitory concentration, the IC_{50} parameter, which required preparing the series of solutions of the known concentrations and analyzing their RSA [%] values. The values of IC_{50}, which inversely correlated with the radical scavenging activity, were found from the RSA = f(C) equation in the linear concentration range (0.1–0.5 mg/mL). The yields and antioxidant properties of dry extracts obtained by 2 h of reflux extraction and Soxhlet extraction are compared in Table 4.

Table 4. Characteristic of dry extracts obtained from Baikal skullcap (*S. baicalensis*) root using ethanol as a solvent in the optimized conditions.

Extraction Conditions	Yield [%]	C [mg/mL]	RSA [%]	RSA(y) vs. C(x) Equation	R^2	IC_{50} [mg/mL]
RE 2 h	17.0	0.1 0.2 0.3 0.4 0.5	12.21 25.35 38.58 52.10 63.28	y = 128.89x − 0.363	0.9990	0.39
SE 2 h	21.7	0.1 0.2 0.3 0.4 0.5	14.19 25.35 38.58 52.10 63.28	y = 127.95x + 2.393	0.9982	0.37

Analyzing the results obtained, it can be noticed that 2 h of Soxhlet extraction of Baikal skullcap root using ethanol allowed for the obtainment of a higher yield of dry extract (21.7%) than reflux extraction (17.0%). The extract obtained in the Soxhlet apparatus was characterized by a slightly higher antioxidant potential (lower value of IC_{50}) than obtained by boiling under reflux, but, in fact, their values of IC_{50} are very similar (0.37 and 0.39 mg/mL, respectively). Comparing the obtained yields of extraction to the results of other authors who obtained 20.3% yield during extraction with 70% ethanol at 80 °C for 24 h [50], it can be noticed that an application of 95% ethanol in our research enabled the achievement of similar results in a much shorter time (2 h). Generally, it can be stated that the extracts obtained using these two different techniques were obtained with good yields, and that they were characterized by similar antioxidant potential. Nevertheless, considering practical aspects related to the possible scale-up of the process of obtaining of dry extracts, the use of extraction at the boiling temperature under reflux may be more convenient than the use of the Soxhlet apparatus.

2.4. GC-MS Analysis

The composition of Baikal skullcap root ethanolic extracts was analyzed by gas chromatography with the mass selective detector (GC-MS) method. The Total Ion Chromatograms (TICs) of the extracts obtained by 2 h of extraction using all applied techniques are shown in Figure 3, while the obtained chromatographic data are summarized in Table 5. GC-MS analysis allowed for the identification of nine compounds, which were present in the extracts and for the comparison of their contents, although some compounds remained unidentified (Peaks No. 2 and 6), and some others are not possible to analyze using this method, including the non-volatile and very polar compounds. The mass spectra of the main bioactive compounds characteristic of the Baikal skullcap, which were identified in the extracts (wogonin and oroxylin A), are shown in Figure 4. The composition of dry extracts was also analyzed by the GC-MS method, and it was found to be in good agreement with the composition of the primary ethanolic extracts obtained in the same conditions.

The main components found in the extracts were wogonin and oroxylin A (Figure 4), belonging to the flavonoid compounds characteristic of the Baikal skullcap [12,13]. The identification of wogonin was performed by comparison of the mass spectrum with the data of wogonin from the NIST 04 library, while its isomer, oroxylin A, was identified based on the comparison with the mass spectrum presented in the literature [58]. In addition to these main compounds, two other compounds, also belonging to the flavonoids, were identified based on the mass spectra analysis: dihydroxydimethoxyflavone and dihydroxytetramethoxyflavone. Their mass spectra show similarity to the mass spectra of wogonin and oroxylin A (M = 284), and the differences in the molecular masses probably correspond to additional 1 and 3 methoxy groups (CH_3O-) in the structure of dihydroxydimethoxyflavone (M = 314) and dihydroxytetramethoxyflavone (M = 374). The presence of these compounds in the Baikal skullcap was described in the literature [59]. Unfortunately, the exact determination of the positions of the particular substituents was not possible and requires further in-depth analysis.

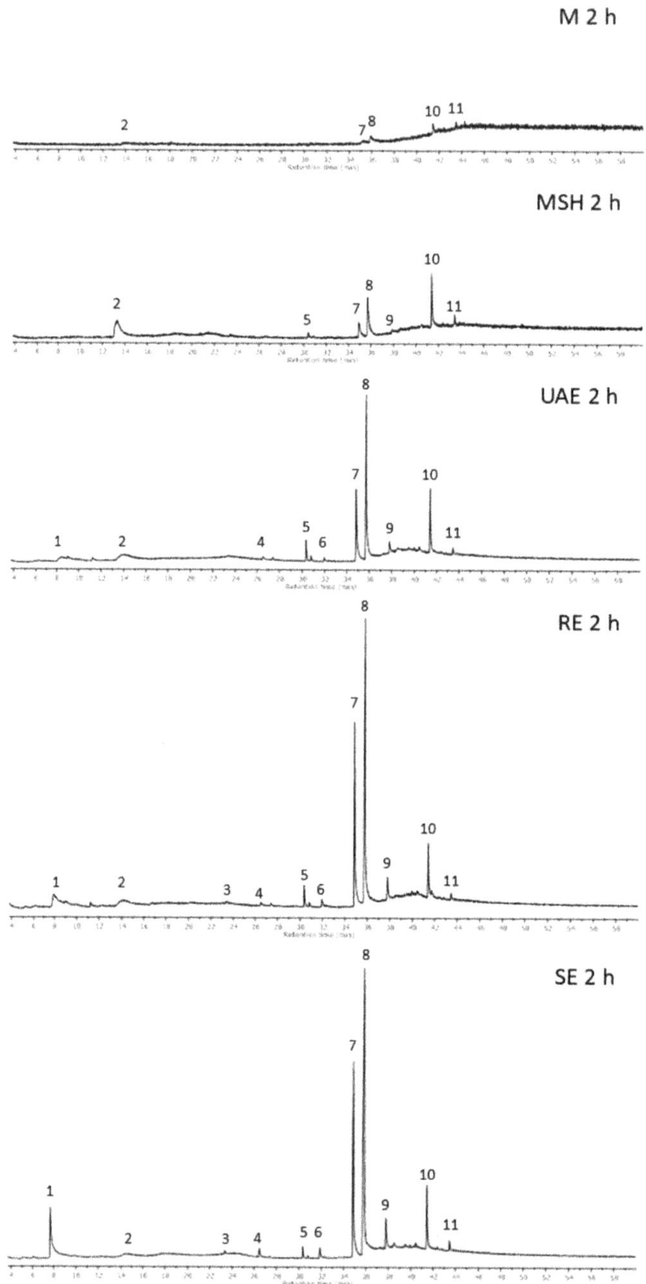

Figure 3. Total Ion Chromatograms of ethanolic extracts of Baikal skullcap (*S. baicalensis*) root obtained by various extraction techniques (peaks numbered according to Table 5).

Table 5. Compounds identified in ethanolic extracts of Baikal skullcap (*S. baicalensis*) root by GC-MS analysis.

No.	Compound	RT [min]	RI	Identification Methods	MS Signals *, *m/z*		Presence in Extract **				
					[M•]$^+$	Characteristic Fragment Ions	M	MSH	UAE	RE	SE
1	5-Hydroxy-methylfurfural	7.72	1251	MS-NIST	126	<u>97</u>, 41, 126, 69, 29	nd	nd	tr	+	++
2	unidentified	14.05	1524	-	164(?)	<u>31</u>,29,57, 43, 73	tr	+	+	+	tr
3	Palmitic acid	23.36	1964	RI, MS-NIST	256	<u>43</u>,73, 60, 41, 57	nd	nd	nd	tr	tr
4	Linoleic acid	26.49	2133	RI, MS-NIST	280	<u>67</u>,81, 82, 95, 55	nd	nd	nd	tr	+
5	(Z)-9-Octadecen-amide	30.35	2360	MS-NIST	281	<u>59</u>,72,55, 41, 43	nd	tr	+	+	+
6	unidentified	31.90	2456	-	286(?)	<u>167</u>, 69, 182, 78, 103	nd	nd	tr	tr	+
7	Oroxylin A	34.83	2649	MS-LIT	284	<u>241</u>, 69, 269, 266, 139	tr	+	++	+++	+++
8	Wogonin	35.80	2716	MS-NIST	284	<u>269</u>, 139, 241, 69, 167	tr	++	+++	+++	+++
9	Dihydroxydimethoxyflavone	37.78	2858	MS-analysis	314	<u>299</u>, 271, 169, 69, 102	nd	tr	+	+	++
10	Dihydroxytetramethoxyflavone	41.44	3139	MS-analysis	374	<u>359</u>,211,183,69, 127	tr	++	++	++	++
11	β-Sitosterol	43.47	3306	RI, MS-NIST	414	<u>43</u>,57,55,41,145	tr	tr	+	+	+

* Base peak ion underlined; (?)—supposed molecular ion; ** presence in extract, based on peak area in total ion chromatogram of extracts obtained in 2 h processes of maceration (M), maceration with shaking (MSH), ultrasound-assisted extraction (UAE), reflux extraction (RE) and Soxhlet extraction (SE): nd—not detected, tr—trace signal, +—minor component, ++—intermediate component, +++—major component; RT—retention time; RI—linear retention index determined experimentally on HP5-MS column; MS-NIST—match of mass spectrum (>95%) with the NIST 04 Library; MS-LIT—match of mass spectrum with presented in the literature [58]; MS-analysis—self-reliant analysis and interpretation of mass spectrum.

The compounds from the flavonoid group belong to popular antioxidants, and their presence and amounts are often correlated with the antioxidant properties of the plant extracts. In our studies, it was also observable that the increased amounts of flavonoids were correlated with the rise of antioxidant potential, and that their highest contents were found in the optimized extraction conditions: RE 2 h and SE 2 h. Comparing the composition of these extracts (Figure 3, Table 5), a big similarity can be seen. In addition to the described flavonoids, these extracts contained small amounts of biologically active compounds, such as β-sitosterol and (Z)-9-octadecenamide, as well as trace amounts of linoleic and palmitic acid.

As a result, extracts rich in compounds with high and interesting biological activity were obtained, mainly wogonin and oroxylin A, characterized by a broad spectrum of action. Wogonin is known for its anti-inflammatory, antiviral, anticancer, antioxidant, and hepatoprotective effects [6–15], and oroxylin A is attributed with the anti-inflammatory, antioxidant, and strongly neuroprotective properties [60]. β-Sitosterol lowers cholesterol levels and has high antioxidant and anti-inflammatory activity [61]. Linoleic acid is known for its hypolipidemic antioxidant and antiglycemic effects [62], while (Z)-9-octadecenamide has anxiolytic and hypnotic properties, and, therefore, it can be used in preparations that facilitate falling asleep [63].

Unexpectedly, in the most severe extraction conditions, the presence of 5-hydroxymethylfurfural (HMF) was also found, with the largest amount in the extract being obtained during 2 h of extraction in the Soxhlet apparatus. This compound is probably not a metabolite of the tested plant because it is known from the literature that it is formed during the processing and storage of various foods and plants rich in reducing sugars and aminoacids as the product of the Maillard reaction [64]. There are contradictory reports in the literature regarding the activity and toxicity of HMF. According to some reports, it may be responsible for cytotoxic and carcinogenic effects [65], while other researchers believe that it does not pose a serious health risk and may even possess valuable properties, such as antioxidant, antiproliferative, anti-inflammatory, or anti-allergic [66,67]. In our studies, its formation may be due to the effect of the high temperature associated with the use of an electric bath during extraction in the Soxhlet apparatus, in contrast to extraction under reflux, where a water bath was used and a significantly smaller amount of this compound was found. This leads to an important practical premise to avoid any possible overheating during the extraction process, which can lead to the formation of this by-product.

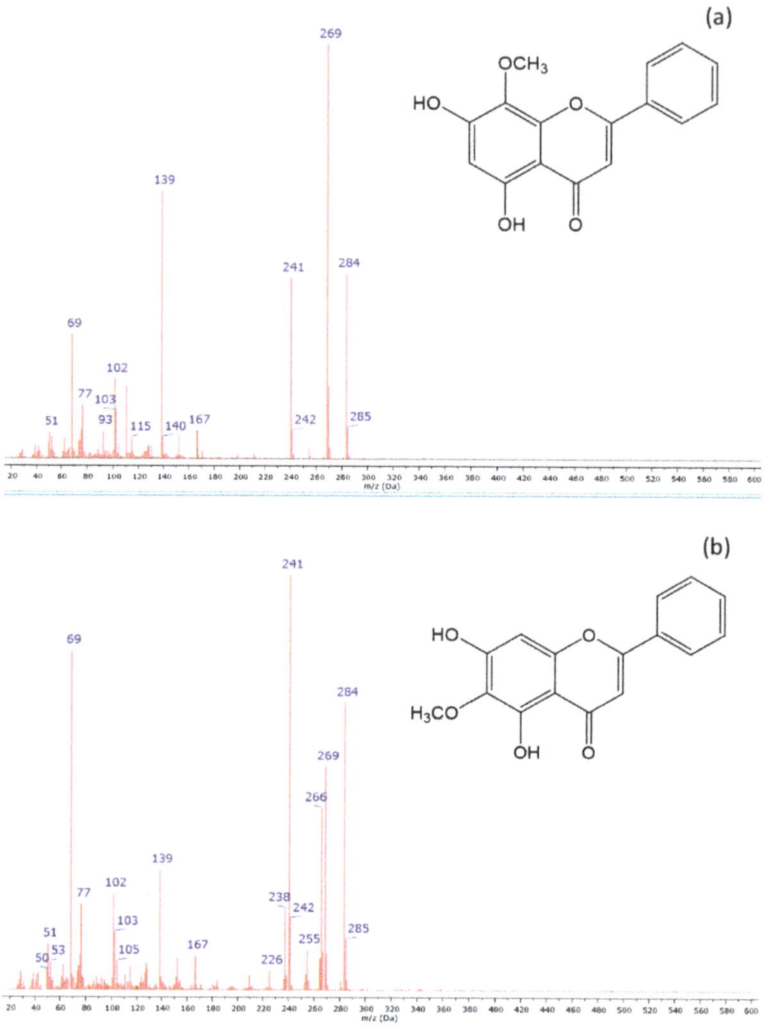

Figure 4. Mass spectra of wogonin (**a**) and oroxylin A (**b**) identified in ethanolic extracts of Baikal skullcap (*S. baicalensis*) root.

Considering all of the results obtained, for the larger-scale production of dry extracts from Baikal skullcap, 2 h of reflux extraction with ethanol, using a water bath or other effectively controlled sources of heat, can be recommended. The extracts obtained in these conditions were characterized by similar high antioxidant potential and chemical composition in comparison to 2 h of Soxhlet extraction. The yield of the dry extract was a little lower than using the Soxhlet apparatus, but the formation of by-product 5-hydroxymethylfurfural was significantly reduced.

3. Materials and Methods

3.1. Plant Material

Commercially available finely ground root of Baikal skullcap (*Radix Scutellariae baicalensis*) was used for the studies. The plant material originated from Poland (producer: FARMVIT, Szczecin, Poland). A form of intake suggested by producer is infusion prepared by boiling 2–3 g in 200 mL of water for 5 min, leaving for 30 min, and straining the obtained herbal tea. For better extraction of the bioactive compounds, previous generous sprinkling of plant material using 96% ethanol is recommended by the producer.

3.2. Extraction Techniques and Sample Preparation

Baikal skullcap root was extracted using 96% ethanol as a solvent (Stanlab, Poznań, Poland). Ethanol, being a safer alternative than toxic methanol, was selected as a good solvent for the extraction of bioactive compounds from Baikal skullcap, which are mainly polar compounds [51–55]. Using ethanol in a concentration of 96% as the extractant can also play an additional role in plant material sterilization. Five various extraction techniques were applied in studies: maceration (M), maceration with shaking (MSH), ultrasound-assisted extraction (UAE), reflux extraction (RE), and Soxhlet extraction (SE). For each technique used, the influence of the extraction time was also studied by performing processes by 0.5 h, 1 h, and 2 h. In the majority of the experiments, 5.00 g of Baikal skullcap root and 50 mL of solvent (96% ethanol) were used. The only exception was Soxhlet Extraction (SE), where the same proportions of material to solvent were applied, but their amounts were doubled (10.00 g and 100 mL, respectively) in order to adjust them to the size of the Soxhlet extractor. Maceration (M) and maceration with shaking (MSH) processes were conducted at room temperature (23 °C) in the 50 mL round bottom flasks equipped with ground glass stoppers. Reflux extraction (RE) and ultrasound-assisted extraction (UAE) were performed in the 50 mL round bottom flasks equipped with the reflux condensers, using a water bath (temperature setting: 95 °C) and a Sonis 4 (Iskra PIO, Šentjerna, Slovenia) ultrasonic bath (power: 75 W, frequency: 40 kHz). During UAE extraction, the temperature in ultrasonic bath increased from room temperature to 35 °C (0.5 h), 44 °C (1 h), and 52 °C (2 h), respectively. The electric bath (temperature setting: 120 °C) was used for heating during extraction in Soxhlet apparatus. After completing the extraction process, the samples were cooled using running water to room temperature and left for 5 min for sedimentation of a plant material followed by decantation. The decanted extracts were subjected to centrifugation for 10 min at 2320 RCF using MPW-223e Centrifuge (MPW Med. Instruments, Warszawa, Poland). Finally, 0.5 mL of the obtained clear extracts were filled up with ethanol to 10 mL in volumetric flask to prepare the samples of extracts for the analyses.

In the case of the extracts that were characterized by the best properties, dry extracts were additionally obtained and subjected to further research. For this purpose, the extracts were placed in Petri dishes and left under the fume hood for solvent evaporation at room temperature, protecting them from the light. Next, they were dried for 16 h at 40 °C using a laboratory dryer and then stabilized in the desiccator to a constant mass. The yields of the dry extracts were calculated from the relation of their masses to the masses of raw material used in the experiments.

3.3. Antioxidant Activity

The antioxidant properties of the extracts obtained from Baikal skullcap in various conditions were studied spectrophotometrically by means of DPPH radical scavenging assay [68]. The analyses were performed using a 1600PC UV–VIS spectrophotometer (VWR International, Leuven, Belgium) in 1 cm glass cuvettes. A radical scavenging activity of extracts was determined using the DPPH method with a 2,2-diphenyl-1-picrylhydrazyl radical (DPPH). Before the analysis, 0.002 mM/mL of DPPH stock solution in methanol was prepared, and then 3 mL of this solution were diluted with methanol to 50 mL in a volumetric flask in order to obtain a DPPH working solution. All DPPH solutions were freshly prepared and protected from light by using aluminum foil. Next, 3 mL of the DPPH working solution were added to 0.5 mL of the prepared sample of extract, mixed, and left for incubation in darkness for exactly 30 min. All analyses were conducted in three repetitions, and the absorbance was measured at 517 nm. A reference sample containing 0.5 mL of a solvent (96% ethanol) was prepared and analyzed analogously. The radical scavenging activity (RSA) of Baikal skullcap extracts was calculated from the absorbance of the sample (A_{30}) and the absorbance of reference sample A_0, as follows:

$$RSA\ [\%] = 100\ (A_0 - A_{30})/A_0, \quad (1)$$

The DPPH scavenging activity of the samples extracted in various conditions was also expressed as Trolox Equivalents Antioxidant Capacity (TEAC). For this purpose, the various concentrations of Trolox (Acros Organics, Geel, Belgium) standard solutions were prepared and analyzed analogously. On the basis of the obtained results, the calibration curve of RSA [%] versus C_{Trolox} [μM/L] was prepared. TEAC values for the studied extracts were calculated from the obtained linear regression equation in the concentration range of linearity (10–300 μM/L) and, finally, expressed in relation to 1 g of extracted plant material [μM TE/g].

Moreover, for the dry extracts obtained in the optimized conditions, the IC_{50} parameter was determined, referred to as the half-maximal inhibitory concentration, which results in 50% inhibition of the free radical activity. For the antioxidant activity tests of dry extracts, stock solutions in ethanol (C = 10 mg/mL) were first prepared and diluted with ethanol in volumetric flasks to obtain the series of working solutions in the concentration range of 0.1–0.9 mg/mL. On the basis of the analysis of their RSA [%] values, the linear range was estimated (from 0.1 to 0.5 mg/mL), which allowed for the determination of the IC_{50} parameter from the equation of the obtained trend line.

3.4. Total Phenolic Content (TPC)

Folin–Ciocalteu (F–C) method [69] was applied for spectrophotometric determination of the total amount of phenolic compounds extractable from the plant material by using ethanol in various conditions. For this purpose, 0.5 mL of the prepared samples of extracts, 0.5 mL of a Folin–Ciocalteu reagent (Chempur, Piekary Śląskie, Poland), and 1.5 mL of freshly prepared sodium carbonate (Na_2CO_3) solution in water (200 mg/mL) were placed in a 25 mL volumetric flask and filled up with a demineralized water. The solutions were mixed and left for 30 min, while the development of the blue color could be observed. Next, the absorbance at 760 nm was measured for the extract samples and the reference samples, which contained only solvents and reagents (all analyses in three repetitions). The calibration curve was prepared using gallic acid (GA) as a standard (Sigma–Aldrich, St. Louis, MO, USA) on the basis of analyses of standard solutions in a concentration range of 20–400 mg/L performed using the same procedure. The total phenolic content (TPC) was expressed as mg of Gallic Acid Equivalent extracted from 1 g of the plant material [mg GAE/g].

3.5. GC-MS Analysis

In order to identify the biologically active substances in the ethanolic extracts of Baikal skullcap, the gas chromatography with mass selective detector (GC-MS) method was applied. The analyses were performed using a 6890N gas chromatograph with a 5973 Network Mass Selective Detector (Agilent Technologies, Palo Alto, CA, USA) equipped with an HP-5MS capillary column (5%-phenyl 95%-methylpolysiloxane, 30 m × 0.25 mm × 0.25 μm). The column temperature was programmed and increased from 80 °C at 5 °C/min to 300 °C (kept by 16 min). The MSD temperatures were as follows: quadrupole 150 °C and ion source 230 °C. The carrier gas was helium (1.2 mL/min). The samples of extracts (3.0 μL) were injected into a column in a split mode (10:1) using a 7683 Series Injector Autosampler. Electron impact ionization (70 eV) mass spectra were recorded via SCAN mode in the range of 20–600 m/z. Identification of the particular compounds was conducted by the comparison of their mass spectra with the data of available standards, standards from the NIST 04 library, and for oroxylin A with the literature data [58]. The identification was confirmed by the comparison of the calculated linear retention indices (RI) with the values found in the literature [70] when the data were available. To determine the retention indices, the standard mixture of the C_7-C_{40} n-alkanes (1000 μg/mL) in heksane (Supelco,

Bellefonte, PA, USA) was analyzed under the same chromatographic conditions. In the case of dihydroxydimethoxyflavone and dihydroxytetramethoxyflavone, the identification was performed on the basis of the analysis of characteristic ions in the mass spectrum. The estimated contents of particular compounds were found based on their peak area percentage in the Total Ion Chromatogram (TIC) of the extracts obtained using the MestReNova 10.0.2 software.

3.6. Statistical Analysis

Statistical analysis was performed using StatSoft Statistica 13.0 (STATISTICA 13.0; StatSoft Inc., Palo Alto, CA, USA) and Microsoft Excel 2021. In all the experiments, three samples were analyzed, and all the assays were conducted at least in triplicate. The results are expressed as the mean values and standard deviation (mean $\pm$ SD). For antioxidant potential and total phenolic content, one-way analysis of variance (ANOVA) and Tukey post-hoc test were used. Correlation analysis was performed by *Pearson* coefficient. Differences were considered significant at $p \leq 0.05$.

4. Conclusions

Our studies of Baikal skullcap root ethanolic extracts obtained in various conditions enabled the determination of the impact of extraction conditions on their DPPH radical scavenging activity and total phenolic content. The effectiveness of the simple maceration process was low, and the enhancement possibilities of this process by shaking and application of the ultrasounds were not as effective as the application of Soxhlet and reflux extraction. The extracts obtained using these two techniques were characterized by high-antioxidant properties and content of valuable compounds, mainly wogonin and oroxylin A, possessing a broad spectrum of bioactive properties. The results of these studies can be applied in the development and optimization of extraction processes of active ingredients from Baikal skullcap roots in the production of dietary supplements characterized by the high-antioxidant potential and content of health-promoting compounds. On the basis of the results obtained, 2 h of reflux extraction using ethanol can be recommended as the convenient and effective technique for the larger-scale production of dry extracts from this precious plant.

Author Contributions: Conceptualization, M.D. and K.W.; methodology, M.D. and K.W.; software, K.J.-M.; validation, K.J.-M.; formal analysis, M.D., A.W. and K.J.-M.; investigation, K.W. and M.D.; resources, M.D. and K.W.; data curation, M.D. and K.W.; writing—original draft preparation, M.D., K.J.-M. and A.W.; writing—review and editing, M.D., K.J.-M. and A.W.; visualization, M.D. and K.J.-M.; supervision, M.D. and A.W.; project administration, M.D.; funding acquisition, M.D. All authors have read and agreed to the published version of the manuscript.

Funding: This research received no external funding.

Institutional Review Board Statement: Not applicable.

Informed Consent Statement: Not applicable.

Data Availability Statement: The raw data supporting the conclusions of this article will be made available by the authors upon request.

Conflicts of Interest: The authors declare no conflicts of interest.

References

1. Kosakowska, O. Intrapopulation variability of flavonoid content in roots of baikal skullcap (*Scutellaria baicalensis* Georgi). *Herba. Pol.* **2017**, *63*, 23–31. [CrossRef]
2. Vergun, O.; Svydenko, L.; Grygorieva, O.; Shymanska, O.; Rakhmetov, D.; Brindza, J.; Ivanišová, E. Antioxidant capacity of plant raw material of *Scutellaria baicalensis* Georgi. *Slovak J. Food Sci./Potravinarstvo* **2019**, *13*, 614–621. [CrossRef]
3. Kosakowska, O.; Baczek, K.; Przybył, J.; Pióro-Jabrucka, E.; Węglarz, Z. Chemical variability of common skullcap (*Scutellaria galericulata* l.) wild growing in the area of eastern poland. *Herba. Pol.* **2016**, *62*, 7–19. [CrossRef]

4. Zhao, T.; Tang, H.; Xie, L.; Zheng, Y.; Ma, Z.; Sun, Q.; Li, X. *Scutellaria baicalensis* Georgi. (Lamiaceae): A review of its traditional uses, botany, phytochemistry, pharmacology and toxicology. *J. Pharm. Pharmacol.* **2019**, *71*, 1353–1369. [CrossRef] [PubMed]
5. Wang, Z.-L.; Wang, S.; Kuang, Y.; Hu, Z.-M.; Qiao, X.; Ye, M. A comprehensive review on phytochemistry, pharmacology, and flavonoid biosynthesis of *Scutellaria baicalensis*. *Pharm. Biol.* **2018**, *56*, 465–484. [CrossRef]
6. Chanchal, D.K.; Singh, K.; Bhushan, B.; Chaudhary, J.S.; Kumar, S.; Varma, A.K.; Agnihotri, N.; Garg, A. An updated review of chinese skullcap (*Scutellaria baicalensis*): Emphasis on phytochemical constituents and pharmacological attributes. *Pharmacol. Res.-Mod. Chin. Med.* **2023**, *9*, 100326. [CrossRef]
7. Gao, Z.; Huang, K.; Yang, X.; Xu, H. Free radical scavenging and antioxidant activities of flavonoids extracted from the radix of *Scutellaria baicalensis* Georgi. *Biochim. Biophys. Acta BBA-Gen. Subj.* **1999**, *1472*, 643–650. [CrossRef] [PubMed]
8. Elkin, Y.N.; Kulesh, N.I.; Stepanova, A.Y.; Solovieva, A.I.; Kargin, V.M.; Manyakhin, A.Y. Methylated flavones of the hairy root culture *Scutellaria baicalensis*. *J. Plant Physiol.* **2018**, *231*, 277–280. [CrossRef] [PubMed]
9. Zhao, Z.; Nian, M.; Qiao, H.; Yang, X.; Wu, S.; Zheng, X. Review of bioactivity and structure-activity relationship on baicalein (5,6,7-trihydroxyflavone) and wogonin (5,7-dihydroxy-8-methoxyflavone) derivatives: Structural modifications inspired from flavonoids in *Scutellaria baicalensis*. *Eur. J. Med. Chem.* **2022**, *243*, 114733. [CrossRef]
10. Liao, H.; Ye, J.; Gao, L.; Liu, Y. the main bioactive compounds of *Scutellaria baicalensis* Georgi. for alleviation of inflammatory cytokines: A comprehensive review. *Biomed. Pharmacother.* **2021**, *133*, 110917. [CrossRef]
11. Hu, Z.; Guan, Y.; Hu, W.; Xu, Z.; Ishfaq, M. An overview of pharmacological activities of baicalin and its aglycone baicalein: New insights into molecular mechanisms and signaling pathways. *Iran. J. Basic Med. Sci.* **2022**, *25*, 14–26. [CrossRef] [PubMed]
12. Chmiel, M.; Stompor-Gorący, M. Promising role of the Scutellaria baicalensis root hydroxyflavone–baicalein in the prevention and treatment of human diseases. *Int. J. Mol. Sci.* **2023**, *24*, 4732. [CrossRef]
13. Wang, C.-Z.; Calway, T.D.; Wen, X.-D.; Smith, J.; Yu, C.; Wang, Y.; Mehendale, S.R.; Yuan, C.-S. Hydrophobic Flavonoids from *Scutellaria baicalensis* induce colorectal cancer cell apoptosis through a mitochondrial-mediated pathway. *Int. J. Oncol.* **2013**, *42*, 1018–1026. [CrossRef] [PubMed]
14. Shin, H.S.; Bae, M.-J.; Choi, D.W.; Shon, D.-H. Skullcap (*Scutellaria baicalensis*) extract and its active compound, wogonin, inhibit ovalbumin-induced TH2-mediated response. *Molecules* **2014**, *19*, 2536–2545. [CrossRef] [PubMed]
15. Tao, Y.; Zhan, S.; Wang, Y.; Zhou, G.; Liang, H.; Chen, X.; Shen, H. Baicalin, the major component of traditional Chinese medicine Scutellaria baicalensis induces colon cancer cell apoptosis through inhibition of oncomirnas. *Sci. Rep.* **2018**, *8*, 14477. [CrossRef] [PubMed]
16. Wang, W.; Xia, T.; Yu, X. Wogonin suppresses inflammatory response and maintains intestinal barrier function via TLR4-MyD88-TAK1-mediated NF-κB pathway in vitro. *Inflamm. Res.* **2015**, *64*, 423–431. [CrossRef] [PubMed]
17. Jelić, D.; Lower-Nedza, A.D.; Brantner, A.H.; Blažeković, B.; Bian, B.; Yang, J.; Brajša, K.; Vladimir-Knežević, S. Baicalin and baicalein inhibit Src tyrosine kinase and production of IL-6. *J. Chem.* **2016**, *2016*, 2510621. [CrossRef]
18. Wen, Y.; Wang, Y.; Zhao, C.; Zhao, B.; Wang, J. The pharmacological efficacy of baicalin in inflammatory diseases. *Int. J. Mol. Sci.* **2023**, *24*, 9317. [CrossRef]
19. Lei, L.; Zhao, J.; Liu, X.-Q.; Chen, J.; Qi, X.-M.; Xia, L.-L.; Wu, Y.-G. Wogonin alleviates kidney tubular epithelial injury in diabetic nephropathy by inhibiting PI3K/Akt/NF-κB signaling pathways. *Drug Des. Devel. Ther.* **2021**, *15*, 3131–3150. [CrossRef]
20. Long, Y.; Xiang, Y.; Liu, S.; Zhang, Y.; Wan, J.; Yang, Q.; Cui, M.; Ci, Z.; Li, N.; Peng, W. Baicalin liposome alleviates lipopolysaccharide-induced acute lung injury in mice via inhibiting TLR4/JNK/ERK/NF-κB pathway. *Mediators Inflamm.* **2020**, *2020*, 8414062. [CrossRef]
21. Pu, W.; Bai, R.; Zhou, K.; Peng, Y.; Zhang, M.; Hottiger, M.O.; Li, W.; Gao, X.; Sun, L. Baicalein attenuates pancreatic inflammatory injury through regulating MAPK, STAT 3 and NF-κB activation. *Int. Immunopharmacol.* **2019**, *72*, 204–210. [CrossRef] [PubMed]
22. Miao, Y.; Ishfaq, M.; Liu, Y.; Wu, Z.; Wang, J.; Li, R.; Qian, F.; Ding, L.; Li, J. Baicalin Attenuates endometritis in a rabbit model induced by infection with *Escherichia coli* and *Staphylococcus aureus* via NF-κB and JNK signaling pathways. *Domest. Anim. Endocrinol.* **2021**, *74*, 106508. [CrossRef] [PubMed]
23. Ji, W.; Liang, K.; An, R.; Wang, X. Baicalin protects against ethanol-induced chronic gastritis in rats by inhibiting Akt/NF-κB pathway. *Life. Sci.* **2019**, *239*, 117064. [CrossRef] [PubMed]
24. Xiang, L.; Gao, Y.; Chen, S.; Sun, J.; Wu, J.; Meng, X. Therapeutic potential of *Scutellaria baicalensis* Georgi in lung cancer therapy. *Phytomedicine* **2022**, *95*, 153727. [CrossRef]
25. Ahmadi, A.; Mortazavi, Z.; Mehri, S.; Hosseinzadeh, H. Protective and therapeutic effects of Scutellaria baicalensis and its main active ingredients baicalin and baicalein against natural toxicities and physical hazards: A review of mechanism. *DARU. J. Pharm. Sci.* **2022**, *30*, 351–366. [CrossRef]
26. Yin, H.; Huang, L.; Ouyang, T.; Chen, L. Baicalein improves liver inflammation in diabetic Db/Db mice by regulating HMGB1/TLR4/NF-κB signaling pathway. *Int. Immunopharmacol.* **2018**, *55*, 55–62. [CrossRef]
27. Li, W.; Ma, F.; Zhang, L.; Huang, Y.; Li, X.; Zhang, A.; Hou, C.; Zhu, Y.; Zhu, Y. S-propargyl-cysteine exerts a novel protective effect on methionine and choline deficient diet-induced fatty liver via Akt/Nrf2/HO-1 pathway. *Oxid. Med. Cell. Longev.* **2016**, *2016*, 4690857. [CrossRef]
28. Zhang, G.; Li, C.; Niu, Y.; Yu, Q.; Chen, Y.; Liu, E. Osteoprotective effect of radix scutellariae in female hindlimb-suspended sprague-dawley rats and the osteogenic differentiation effect of its major constituent. *Molecules* **2017**, *22*, 1044. [CrossRef]

29. Arjmandi, B.H.; Ormsbee, L.T.; Elam, M.L.; Campbell, S.C.; Rahnama, N.; Payton, M.E.; Brummel-Smith, K.; Daggy, B.P. A combination of *Scutellaria baicalensis* and *Acacia catechu* extracts for short-term symptomatic relief of joint discomfort associated with osteoarthritis of the knee. *J. Med. Food* **2014**, *17*, 707–713. [CrossRef]
30. Khan, N.M.; Haseeb, A.; Ansari, M.Y.; Haqqi, T.M. A wogonin-rich-fraction of *Scutellaria baicalensis* root extract exerts chondroprotective effects by suppressing IL-1β-induced activation of AP-1 in human OA chondrocytes. *Sci. Rep.* **2017**, *7*, 43789. [CrossRef]
31. Zhang, K.; Lu, J.; Mori, T.; Smith-Powell, L.; Synold, T.W.; Chen, S.; Wen, W. Baicalin increases VEGF expression and angiogenesis by activating the ERRα/PGC-1α pathway. *Cardiovasc. Res.* **2011**, *89*, 426–435. [CrossRef] [PubMed]
32. Xu, J.; Zhang, Y.; Xiao, Y.; Ma, S.; Liu, Q.; Dang, S.; Jin, M.; Shi, Y.; Wan, B.; Zhang, Y. Inhibition of 12/15-lipoxygenase by baicalein induces microglia PPARβ/δ: A potential therapeutic role for CNS autoimmune disease. *Cell. Death. Dis.* **2013**, *4*, e569. [CrossRef] [PubMed]
33. Song, L.; Zhong, P.; Zhu, X.; Zhou, R.; Gao, M.; Lan, Q.; Chen, J.; Chen, Y.; Zhao, W. The anti-rotavirus effect of baicalin via the gluconeogenesis-related p-JNK–PDK1–AKT–SIK2 signaling pathway. *Eur. J. Pharmacol.* **2021**, *897*, 173927. [CrossRef] [PubMed]
34. Long, L.D.; Tung, N.H.; Thom, N.T.; Choi, G.J.; Hoang, V.D.; Dang, L.Q. Constituents and inhibitory effect on human pathogenic bacteria of the roots of *Scutellaria baicalensis*. *Vietnam. J. Sci. Technol.* **2019**, *57*, 7–14. [CrossRef]
35. Duan, C.; Matsumura, S.; Kariya, N.; Nishimura, M.; Shimono, T. In vitro antibacterial activities of *Scutellaria baicalensis Georgi* against cariogenic bacterial. *Pediatr. Dent. J.* **2007**, *17*, 58–64. [CrossRef]
36. Jang, J.-S.; Kim, J.-H.; Kwon, M.-J. Antibacterial activity of *Scutellaria baicalensis* extract against antibiotic resistant bacteria. *Korean J. Food Nutr.* **2011**, *24*, 708–712. [CrossRef]
37. Wang, L.; Zhang, D.; Wang, N.; Li, S.; Tan, H.-Y.; Feng, Y. Polyphenols of chinese skullcap roots: From chemical profiles to anticancer effects. *RSC Adv.* **2019**, *9*, 25518–25532. [CrossRef]
38. Zhang, Y.; Wang, H.; Liu, Y.; Wang, C.; Wang, J.; Long, C.; Guo, W.; Sun, X. Baicalein inhibits growth of epstein-barr virus-positive nasopharyngeal carcinoma by repressing the activity of EBNA1 Q-promoter. *Biomed. Pharmacother.* **2018**, *102*, 1003–1014. [CrossRef]
39. Yu, X.; Yang, Y.; Li, Y.; Cao, Y.; Tang, L.; Chen, F.; Xia, J. Baicalein inhibits cervical cancer progression via downregulating long noncoding RNA BDLNR and its downstream PI3 K/Akt pathway. *Int. J. Biochem. Cell Biol.* **2018**, *94*, 107–118. [CrossRef]
40. Bie, B.; Sun, J.; Guo, Y.; Li, J.; Jiang, W.; Yang, J.; Huang, C.; Li, Z. Baicalein: A review of its anti-cancer effects and mechanisms in hepatocellular carcinoma. *Biomed. Pharmacother.* **2017**, *93*, 1285–1291. [CrossRef]
41. Cai, J.; Hu, Q.; He, Z.; Chen, X.; Wang, J.; Yin, X.; Ma, X.; Zeng, J. *Scutellaria baicalensis* Georgi and their natural flavonoid compounds in the treatment of ovarian cancer: A review. *Molecules* **2023**, *28*, 5082. [CrossRef] [PubMed]
42. Yao, R.; Yang, F.; Liu, J.; Jiao, Q.; Yu, H.; Nie, X.; Li, H.; Wang, X.; Xue, F. Therapeutic drug combinations against COVID-19 obtained by employing a collaborative filtering method. *Heliyon* **2023**, *9*, e14023. [CrossRef]
43. Ma, Y.; Zhou, X.; Zhang, F.; Huang, C.; Yang, H.; Chen, W.; Tao, X. The effect of *Scutellaria baicalensis* and its active ingredients on major depressive disorder: A systematic review and meta-analysis of literature in pre-clinical research. *Front. Pharmacol.* **2024**, *15*, 1313871. [CrossRef] [PubMed]
44. EghbaliFeriz, S.; Taleghani, A.; Tayarani-Najaran, Z. Central nervous system diseases and *Scutellaria*: A review of current mechanism studies. *Biomed. Pharmacother.* **2018**, *102*, 185–195. [CrossRef]
45. Song, J.; Li, M.; Kang, N.; Jin, W.; Xiao, Y.; Li, Z.; Qi, Q.; Zhang, J.; Duan, Y.; Feng, X.; et al. Baicalein ameliorates cognitive impairment of vascular dementia rats via suppressing neuroinflammation and regulating intestinal microbiota. *Brain Res. Bull.* **2024**, *208*, 110888. [CrossRef] [PubMed]
46. Yimam, M.; Burnett, B.P.; Brownell, L.; Jia, Q. Clinical and preclinical cognitive function improvement after oral treatment of a botanical composition composed of extracts from *Scutellaria baicalensis* and *Acacia catechu*. *Behav. Neurol.* **2016**, *2016*, 7240802. [CrossRef]
47. Baba, M.; Subramanian, G.; Chand, J.; Wahedi, U.; Potlapati, V.; Jayanthi, K.; Azeemuddin, M.; Emran, T.; Nainu, F. Investigation of *Scutellaria baicalensis* for potential neuroprotective effect on the treatment of parkinson's disease. *Biointerface Res. Appl. Chem.* **2024**, *14*, 1–15. [CrossRef]
48. Peng, Y.; Zhou, C. Network pharmacology and molecular docking identify the potential mechanism and therapeutic role of *Scutellaria baicalensis* in alzheimer's disease. *Drug Des. Devel. Ther.* **2024**, *18*, 1199–1219. [CrossRef]
49. Ni, H.; Wu, Z.; Muhammad, I.; Lu, Z.; Li, J. Optimization of baicalin water extraction process from *Scutellaria baicalensis* (a traditional Chinese medicine) by using orthogonal test and HPLC. *Rev. Bras. De Farmacogn.* **2018**, *28*, 151–155. [CrossRef]
50. Choi, W.; Kwon, H.-S.; Lee, H.Y. Enhancement of anti-skin inflammatory activities of Scutellaria baicalensis using an alkaline reduced water extraction process. *Food Sci. Biotechnol.* **2014**, *23*, 1859–1866. [CrossRef]
51. Park, J.-H.; Kim, R.-Y.; Park, E. Antioxidant and glucosidase inhibitory activities of different solvent extracts of skullcap (*Scutellaria baicalensis*). *Food Sci. Biotechnol.* **2011**, *20*, 1107–1112. [CrossRef]
52. Oracz, J.; Kowalski, S.; Żyżelewicz, D.; Kowalska, G.; Gumul, D.; Kulbat-Warycha, K.; Rosicka-Kaczmarek, J.; Brzozowska, A.; Grzegorczyk, A.; Areczuk, A. The influence of microwave-assisted extraction on the phenolic compound profile and biological activities of extracts from selected *Scutellaria* species. *Molecules* **2023**, *28*, 3877. [CrossRef] [PubMed]
53. Lu, Y.; Joerger, R.; Wu, C. Study of the chemical composition and antimicrobial activities of ethanolic extracts from roots of *Scutellaria baicalensis* Georgi. *J. Agric. Food Chem.* **2011**, *59*, 10934–11094. [CrossRef] [PubMed]

54. Oomen, W.W.; Begines, P.; Mustafa, N.R.; Wilson, E.G.; Verpoorte, R.; Choi, Y.H. Natural deep eutectic solvent extraction of flavonoids of *Scutellaria baicalensis* as a replacement for conventional organic solvents. *Molecules* **2020**, *25*, 617. [CrossRef] [PubMed]
55. Lin, M.-C.; Tsai, M.-J.; Wen, K.-C. Supercritical fluid extraction of flavonoids from scutellariae radix. *J. Chromatogr. A* **1999**, *830*, 387–395. [CrossRef]
56. Olech, M.; Kasprzak, K.; Wójtowicz, A.; Oniszczuk, T.; Nowak, R.; Waksmundzka-Hajnos, M.; Combrzyński, M.; Gancarz, M.; Kowalska, I.; Krajewska, A.; et al. Polyphenol composition and antioxidant potential of instant gruels enriched with *Lycium barbarum* L. fruit. *Molecules* **2020**, *25*, 4538. [CrossRef] [PubMed]
57. Ouamnina, A.; Alahyane, A.; Elateri, I.; Boutasknit, A.; Abderrazik, M. Relationship between phenolic compounds and antioxidant activity of some moroccan date palm fruit varieties (*Phoenix dactylifera* L.): A two-year study. *Plant* **2024**, *13*, 1119. [CrossRef]
58. Gokhale, M.; Khanna, A.; Gautam, D. Ameliorative effect of *Oroxylum indicum* [L.] Vent. metabolites on tobacco extract induced cell damage in human lymphocytes. *Int. J. Biol. Pharm. Res.* **2015**, *6*, 860–866.
59. Gao, B.; Zhu, H.; Liu, Z.; He, X.; Sun, J.; Li, Y.; Wu, X.; Pehrsson, P.; Zhang, Y.; Yu, L. Chemical compositions of *Scutellaria baicalensis* Georgi. (huangqin) extracts and their effects on ACE2 binding of SARS-CoV-2 spike protein, ACE2 activity, and free radicals. *Int. J. Mol. Sci.* **2024**, *25*, 2045. [CrossRef]
60. Aliev, G.; Kaminsky, Y.G.; Bragin, V.; Kosenko, E.A.; Klochkov, S.G.; Bachurin, S.O.; Benberin, V.V. Flavones from the root of *Scutellaria baicalensis* Georgi–drugs of the future in neurodegeneration and neuroprotection? In *Systems Biology of Free Radicals and Antioxidants*; Laher, I., Ed.; Springer: Berlin/Heidelberg, Germany, 2014; pp. 2305–2323, ISBN 978-3-642-30018-9.
61. Salehi, B.; Quispe, C.; Sharifi-Rad, J.; Cruz-Martins, N.; Nigam, M.; Mishra, A.P.; Konovalov, D.A.; Orobinskaya, V.; Abu-Reidah, I.M.; Zam, W.; et al. Phytosterols: From preclinical evidence to potential clinical applications. *Front. Pharmacol.* **2021**, *11*, 599959. [CrossRef]
62. Lv, H.; Chen, S.; Xu, X.; Zhu, M.; Zhao, W.; Liu, K.; Liu, K. Isolation of linoleic acid from sambucus williamsii seed oil extracted by high pressure fluid and its antioxidant, antiglycemic, hypolipidemic activities. *Int. J. Food Eng.* **2015**, *11*, 383–391. [CrossRef]
63. Fedorova, I.; Hashimoto, A.; Fecik, R.A.; Hedrick, M.P.; Hanus, L.O.; Boger, D.L.; Rice, K.C.; Basile, A.S. Behavioral evidence for the interaction of oleamide with multiple neurotransmitter systems. *J. Pharmacol. Exp. Ther.* **2001**, *299*, 332–342.
64. Tamanna, N.; Mahmood, N. Food processing and maillard reaction products: Effect on human health and nutrition. *Int. J. Food Sci.* **2015**, *2015*, 526762. [CrossRef] [PubMed]
65. Bakhiya, N.; Monien, B.; Frank, H.; Seidel, A.; Glatt, H. Renal organic anion transporters OAT1 and OAT3 mediate the cellular accumulation of 5-sulfooxymethylfurfural, a reactive, nephrotoxic metabolite of the maillard product 5-hydroxymethylfurfural. *Biochem. Pharmacol.* **2009**, *78*, 414–419. [CrossRef] [PubMed]
66. Li, Y.-X. In vitro antioxidant activity of 5-HMF isolated from marine red alga laurencia undulata in free-radical-mediated oxidative systems. *J. Microbiol. Biotechnol.* **2009**, *19*, 1319–1327. [CrossRef] [PubMed]
67. Zhao, L.; Chen, J.; Su, J.; Li, L.; Hu, S.; Li, B.; Zhang, X.; Xu, Z.; Chen, T. In vitro antioxidant and antiproliferative activities of 5-hydroxymethylfurfural. *J. Agric. Food Chem.* **2013**, *61*, 10604–10611. [CrossRef]
68. Kedare, S.B.; Singh, R.P. Genesis and development of DPPH method of antioxidant assay. *J. Food Sci. Technol.* **2011**, *48*, 412–422. [CrossRef]
69. Prior, R.L.; Wu, X.; Schaich, K. Standardized methods for the determination of antioxidant capacity and phenolics in foods and dietary supplements. *J. Agric. Food Chem.* **2005**, *53*, 4290–4302. [CrossRef]
70. Babushok, V.I.; Linstrom, P.J.; Zenkevich, I.G. Retention indices for frequently reported compounds of plant essential oils. *J. Phys. Chem. Ref. Data* **2011**, *40*, 1–47. [CrossRef]

molecules

MDPI

Article

Inhibitory Effect of *Sorbus aucuparia* Extracts on the *Fusarium proliferatum* and *F. culmorum* Growth and Mycotoxin Biosynthesis

Sylwia Ryszczyńska [1], Natalia Gumulak-Wołoszyn [2], Monika Urbaniak [3], Łukasz Stępień [3], Marcin Bryła [4], Magdalena Twarużek [5] and Agnieszka Waśkiewicz [1,*]

1 Department of Chemistry, Faculty of Forestry and Wood Technology, Poznań University of Life Sciences, Wojska Polskiego 75, 60-625 Poznań, Poland; sylwia.ryszczynska@up.poznan.pl

2 Department of Forest Ecosystem Protection, Faculty of Forestry, University of Agriculture in Kraków, Aleja 29 Listopada 46, 31-425 Kraków, Poland; natalia.gumulak@student.urk.edu.pl

3 Plant-Pathogen Interaction Team, Institute of Plant Genetics, Polish Academy of Sciences, Strzeszyńska 34, 60-479 Poznań, Poland; murb@igr.poznan.pl (M.U.); lste@igr.poznan.pl (Ł.S.)

4 Department of Food Safety and Chemical Analysis, Prof. Waclaw Dabrowski Institute of Agricultural and Food Biotechnology—State Research Institute, Rakowiecka 36, 02-532 Warsaw, Poland; marcin.bryla@ibprs.pl

5 Department of Physiology and Toxicology, Faculty of Biological Sciences, Kazimierz Wielki University, Chodkiewicza 30, 85-064 Bydgoszcz, Poland; twarmag@ukw.edu.pl

* Correspondence: agnieszka.waskiewicz@up.poznan.pl

Abstract: Fungal infections are among the most common diseases of crop plants. Various species of the *Fusarium* spp. are naturally prevalent and globally cause the qualitative and quantitative losses of farming commodities, mainly cereals, fruits, and vegetables. In addition, *Fusarium* spp. can synthesize toxic secondary metabolites—mycotoxins under high temperature and humidity conditions. Among the strategies against *Fusarium* spp. incidence and mycotoxins biosynthesis, the application of biological control, specifically natural plant extracts, has proved to be one of the solutions as an alternative to chemical treatments. Notably, rowanberries taken from *Sorbus aucuparia* are a rich source of phytochemicals, such as vitamins, carotenoids, flavonoids, and phenolic acids, as well as minerals, including iron, potassium, and magnesium, making them promising candidates for biological control strategies. The study aimed to investigate the effect of rowanberry extracts obtained by supercritical fluid extraction (SFE) under different conditions on the growth of *Fusarium* (*F. culmorum* and *F. proliferatum*) and mycotoxin biosynthesis. The results showed that various extracts had different effects on *Fusarium* growth as well as ergosterol content and mycotoxin biosynthesis. These findings suggest that rowanberry extracts obtained by the SFE method could be a natural alternative to synthetic fungicides for eradicating *Fusarium* pathogens in crops, particularly cereal grains. However, more research is necessary to evaluate their efficacy against other *Fusarium* species and in vivo applications.

Keywords: *Sorbus aucuparia*; *Fusarium* spp.; ergosterol; mycotoxins; supercritical fluid extraction; chromatographic analysis

1. Introduction

Sorbus aucuparia is a deciduous tree species native to most of Europe and parts of Asia and Africa, commonly cultivated as an ornamental plant, which has a few climate requirements, as it can occur in both high mountains with low temperatures and the hot south. It is a species whose fruits are edible but neglected because of their natural bitter taste and the necessity of processing them before consumption to make them palatable. In some regions, particularly in Eastern Europe and Scandinavia, it is used as an ingredient in food products such as jams or syrups [1]. Rowan is rich in organic acids, including ascorbic acid, and phenolic compounds, which have shown a wide range of biological properties [2,3].

Substances extracted from rowan fruits have proven to exhibit diuretic, anti-inflammatory, vasoconstrictor, and anti-diabetic properties [4]. The rowan tree is an important biocenotic species that provides food for many birds, e.g., song thrush (*Turdus philomelos*), Eurasian bullfinch (*Pyrrhula pyrrhula*), and mistle thrush (*Turdus viscivorus*) [5]. It is also a food ingredient for mammals, such as the European pine marten (*Martes martes*) [6], roe deer (*Capreolus capreolus*) [7], and red foxes (*Vulpes vulpes*) [8], establishing the importance of mammals as dispersers of rowan seeds [8]. Due to the constant rowanberry seed multiplication, high variability is observed for most morphological and biochemical traits of wild fruits. Therefore, it is important to learn about the characteristics of specific individuals and to introduce breeding for valuable fruits [9].

Supercritical fluid extraction is an advanced extraction method utilizing supercritical fluids, typically carbon dioxide (CO_2), as a solvent. During the SFE process, the supercritical CO_2 is passed through the material containing the desired compounds. The solvating properties of supercritical CO_2 can be tuned by adjusting pressure and temperature, allowing selective extraction of specific compounds, which enables precise targeting of desired substances while leaving unwanted components behind. Compared to traditional solvent extraction, SFE operates at lower temperatures, preserving sensitive compounds, and uses non-toxic CO_2, making it environmentally friendly and leaving no solvent residues [10]. Using SFE for *Sorbus aucuparia* fruits holds promise for obtaining bioactive compounds, allowing for selective extraction of beneficial components while maintaining the fruit's nutritional integrity. The resulting extract may be a rich source of bioactive compounds for food, pharmaceutical, and biotechnological industries. Moreover, due to the variety of natural compounds, rowan fruit extracts prepared by the SFE method may have antimicrobial properties, which can be used to inhibit the growth of mycopathogens.

Up to now, a few articles have been published on the antimicrobial properties of *Sorbus aucuparia* fruit extracts. The antimicrobial effect is mainly attributed to the high content of sorbic acid and biphenyl phytoalexins, particularly aucuparin in rowanberries [11–13]. Most of the previous research concerns the effect of extracts obtained from *Sorbus aucuparia* fruits on bacteria growth [14,15]. In the research of Liepiņa et al. (2013), the highest inhibition on *Bacillus cereus* had ethanolic and aqueous extracts obtained from fresh rowanberries, which reached a percentage of inhibition equal to 10% and 9.7%, respectively [16]. A very strong antibacterial effect of acetone-derived phenolic-rich fractions of rowanberry extract against *B. cereus* was found [17], as well as for methanolic and aqueous extracts, which achieved an inhibition percentage above 75% [18]. Antibacterial properties of *Sorbus aucuparia* fruit extracts have also been observed for *Staphylococcus aureus, Escherichia coli, Pseudomonas aeruginosa, Salmonella enterica, Bacillus subtilis, Serratia marcescens*, and *Campylobacter jejuni* [16–18]. Regarding the antifungal activity of *Sorbus* extracts, only a few studies have been conducted, mainly for *Candida albicans*, and their results are inconclusive [16,17,19–23].

Fungal infections stand as some of the most destructive agricultural diseases worldwide. Within the *Fusarium* spp., *F. culmorum* and *F. proliferatum* are widespread, especially in Europe, and cause substantial losses in cereals, cereal-based items, fruits, and vegetables [24–26]. Crop yield reduction caused by fusariosis reached from 10 to 40% [27]. These *Fusarium* species synthesize mycotoxins under favorable conditions, which are high temperature, humidity, and moisture, occurring pre-harvest, post-harvest, during processing, or even in storage [28]. Additionally, most mycotoxins produced by *Fusarium* spp. exhibit heat stability and raise serious health concerns in humans and farm animals, leading to mutagenic, teratogenic, neurotoxic, and carcinogenic effects [29]. *Fusarium* spp. are associated with diseases in people with localized or invasive infections who have a history of immunosuppression [30]. These infections are difficult to treat mainly due to drug-resistant isolates [31]. Due to their characteristics, *Fusarium* fungi become resistant to fungicide exposure very quickly [32]. The crucial thing, therefore, is the implementation of appropriate methods or fungicides [33].

This work aims to investigate the antifungal properties of *Sorbus aucuparia* fruit extracts obtained by supercritical fluid extraction under various temperature and pressure conditions. The antifungal activity of rowanberry fruit extracts was assessed against two *Fusarium* strains: *F. proliferatum* and *F. culmorum*. The effect of the extract treatments on mycelial growth and mycotoxin biosynthesis was assessed by qualitative and quantitative chromatographic analysis of ergosterol and secondary fungal metabolites.

2. Results

2.1. Preparation of the Sorbus aucuparia Fruit Extracts by Supercritical Fluid Extraction Method

Rowanberry extracts were obtained by the SFE method according to the further described procedure in four variants differing in extraction parameters such as temperature and pressure (Table 1).

Table 1. List of the prepared extract variants (E1–E4), including extraction conditions and extraction yield.

Extract Variant	Extraction Conditions	Extraction Yield [%]
E1	40 °C, 300 bar	17.71 ± 0.10^{b}
E2	40 °C, 200 bar	16.93 ± 0.10^{c}
E3	70 °C, 300 bar	18.05 ± 0.06^{a}
E4	70 °C, 200 bar	16.99 ± 0.08^{c}

All values are means of three replicates ± standard deviation. The values assigned with the superscripts of different letters are significantly different (Tukey's HSD test, significant at $p < 0.05$).

The extraction yield ranged from 16.93 to 18.05%. The most efficient extraction conditions were demonstrated at 70 °C and 300 bar (variant E3). It has also been shown that higher pressure in the extraction process had a significant impact on the increase in extraction yield (extracts E3 or E1, $p < 0.05$), while lower pressure, regardless of the temperature value, resulted in a similar extraction yield.

2.2. Total Phenolic Content of the Methanolic and Aqueous Extracts

To evaluate the bioactivity of the prepared extracts, the total phenolic content (TPC) of the methanolic variants, as well as the aqueous variants obtained from the methanolic ones, were measured, and the results were expressed as mg of gallic acid (GAE) per 1 g of extract dw (Figure 1).

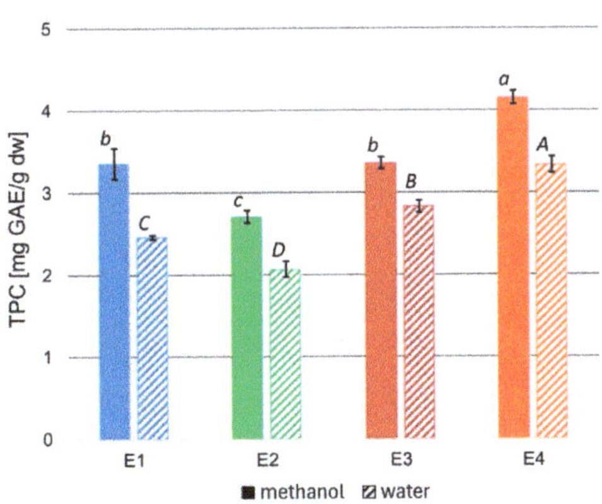

Figure 1. TPC in methanolic and aqueous rowanberry extract variants E1–E4. All values are means of three replicates; error bars represent standard deviation. The values assigned with the superscripts of different letters are significantly different (Tukey's HSD test, significant at $p < 0.05$).

The TPC of the rowanberry methanolic extracts ranged from 2.71 to 4.15 mg GAE/g dw. However, aqueous variants had significantly lower TPC than corresponding methanolic ones ($p < 0.05$), reaching values from 2.07 to 3.33 mg GAE/g dw. The highest TPC values characterized the extract variant E4, regardless of the solvent, while the lowest—extract variant E2.

2.3. The Inhibitory Effect of Sorbus aucuparia Fruit Extracts on Fusarium spp. Growth

This study focused on the antifungal activity of rowanberry extracts obtained by the SFE against *Fusarium* spp. Therefore, the effect of the obtained aqueous extract variants (E1–E4) on the growth of *F. proliferatum* and *F. culmorum* was evaluated, and the results are presented in Figure 2.

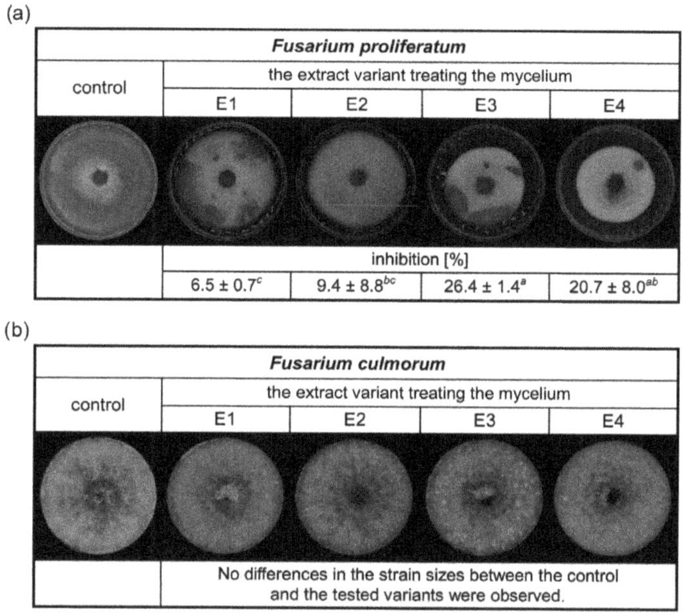

Figure 2. The inhibitory effects of different rowanberry extracts (E1–E4) on *F. proliferatum* (**a**) and *F. culmorum* (**b**) mycelia growth in the medium. All values are means of three replicates ± standard deviation. The values assigned with the superscripts of different letters are significantly different (Tukey's HSD test, significant at $p < 0.05$).

It was observed that the prepared extracts had an inhibitory effect on the growth of *F. proliferatum* (Figure 2a). The lowest inhibition of the extracts was 6.5%. The most significant antifungal activity against *F. proliferatum* was exerted by the extract variant E3, with a reduction in mycelial growth of 26.4%. In the case of the inhibition study of rowanberry extracts on the *F. culmorum* growth, no differences between the control and the tested variants were observed (Figure 2b).

2.4. The Effect of Sorbus aucuparia Fruit Extracts on Ergosterol Content

The antifungal activity of the obtained extracts on the *F. proliferatum* and *F. culmorum* growth was investigated based on the analysis of the ergosterol (ERG) content in the control and tested samples (Figure 3).

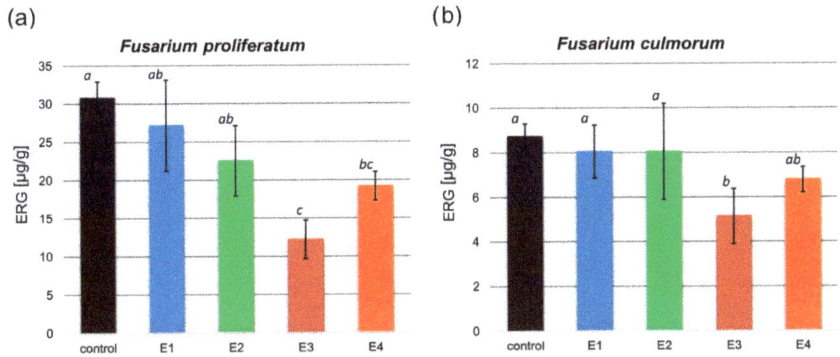

Figure 3. Ergosterol (ERG) content [µg/g] in the control and tested samples treated with the obtained extracts (E1–E4) after inoculation with *F. proliferatum* (**a**) and *F. culmorum* (**b**). All values are means of three replicates; error bars represent standard deviation. The values assigned with the superscripts of different letters are significantly different (Tukey's HSD test, significant at $p < 0.05$).

In both examined *Fusarium* spp. treated with the rowanberry extracts, ERG content decreased compared to the control group. A significant difference in ERG concentration was observed after treating *F. proliferatum* with each of the obtained extract variants, but only the E3 and E4 extract variants had a significant effect on *F. culmorum* ($p < 0.05$). The ERG content of the tested samples was in the range between 12.18 and 27.15 µg/g (30.79 µg/g control) for *F. proliferatum* (Figure 3a) and 5.13–8.04 µg/g (8.73 µg/g control) for *F. culmorum* (Figure 3b). The most significant decrease in ergosterol concentration ($p < 0.05$) was observed in samples treated with the extract variant E3.

The ERG reduction (Table 2) in samples with the addition of *Sorbus aucuparia* fruit extracts ranged from 12.41 to 60.67% for *F. proliferatum* and from 7.70 to 41.62% for *F. culmorum*, depending on the extraction variants. The greatest reduction in ergosterol content compared to the control group was observed after treatment of both *Fusarium* spp. with extract variant E3.

Table 2. The percentage of ergosterol (ERG) reduction for *Fusarium* spp. samples treated with different variants of rowanberry extracts (E1–E4).

Extract Variant	ERG Content Reduction [%]	
	Fusarium proliferatum	*Fusarium culmorum*
E1	12.41 ± 14.07^{c}	8.01 ± 11.41^{b}
E2	27.30 ± 10.29^{bc}	7.70 ± 5.62^{b}
E3	60.67 ± 5.89^{a}	41.62 ± 11.48^{a}
E4	37.85 ± 1.92^{ab}	22.42 ± 4.03^{ab}

All values are means of three replicates ± standard deviation. The values assigned with the superscripts of different letters are significantly different (Tukey's HSD test, significant at $p < 0.05$).

2.5. Mycotoxin Identification

In this study, the multi-mycotoxins method was used to analyze the effect of rowanberry extracts on mycotoxin biosynthesis by *Fusarium* spp. For *F. proliferatum*, the content of fumonisins—B_1 (FB_1), B_2 (FB_2), and B_3 (FB_3), and beauvericin (BEA) were determined. However, in the case of *F. culmorum*, among the 12 analyzed mycotoxins—deoxynivalenol (DON), 3- and 15-acetyldeoxynivalenol (3- and 15-AcDON), deoxynivalenol-3-glucoside (DON-3G), nivalenol (NIV), nivalenol-3-glucoside (NIV-3G), fusarenon X (FUSX), zearalenone (ZEN), zearalenone-14-sulfate (ZEN-14S), zearalenone-14-glucoside (ZEN-14G), α-zearalenol (α-ZOL), and β-zearalenol (β-ZOL)—only six were identified and quantified: DON, 3- and 15-AcDON, ZEN, ZEN-14S, and α-ZOL (Figure 4).

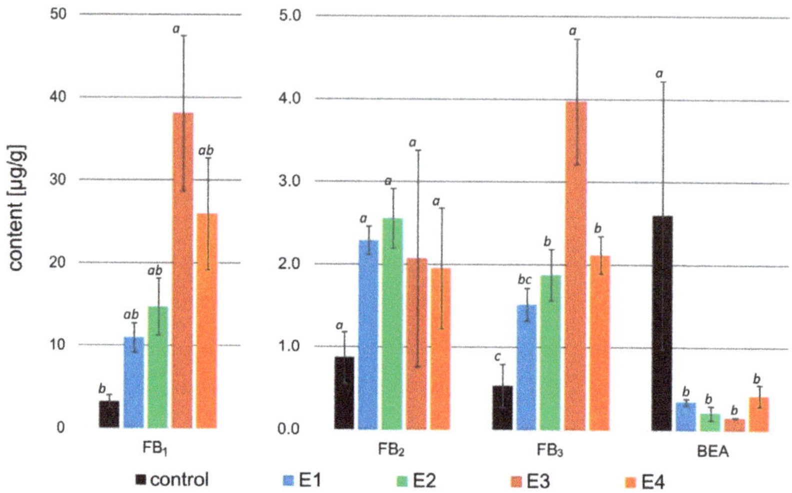

Figure 4. Mycotoxin content [µg/g] in the samples treated with rowanberry extracts (four variants, E1–E4) after inoculation with *Fusarium proliferatum*. FB_1—fumonisin B_1, FB_2—fumonisin B_2, FB_3—fumonisin B_3, BEA—beauvericin. All values are means of three replicates; error bars represent standard deviation. The values assigned with the superscripts of different letters are significantly different (Tukey's HSD test, significant at $p < 0.05$).

The application of rowanberry extracts to *Fusarium proliferatum* resulted in an intriguing effect on mycotoxin biosynthesis. A significant decrease in BEA content in the samples treated with the rowanberry extracts was observed, ranging from 0.15 to 0.42 to µg/g compared to the control group—2.60 µg/g ($p < 0.05$). However, fumonisins were biosynthesized by *F. proliferatum* more prominently in the samples exposed to the prepared extracts. The highest concentration of FB_1 (38.08 µg/g) and FB_3 (3.97 µg/g) in the prepared samples was connected with the addition of extract variant E3 (control: 3.23 µg/g, and 0.53 µg/g, respectively). However, the highest content of FB_2 (2.55 µg/g) was observed after the addition of the extract variant E2 (control: 0.87 µg/g), whereas the differences in the production of this fumonisin by each extract were not statistically significant ($p > 0.05$). The total content of the fumonisins in the tested samples with *F. proliferatum* treated with the rowanberry extracts was: 14.72, 19.09, 44.11, and 29.98 µg/g for E1, E2, E3, and E4 variants, respectively (control: 4.63 µg/g). The most significant difference in the content of fumonisins was observed between the control sample and the sample treated with the E3 extract variant ($p < 0.05$).

In the case of the *Fusarium proliferatum*, the percentage of BEA reduction by prepared fruit extracts reached the highest value of 92.1% after treatment with variant E3, whereas this result is statistically similar to other extract variants ($p > 0.05$, Table 3). As previously mentioned, no reduction was observed in the content of fumonisins—in fact, their biosynthesis exhibited an increase compared to the control group. In the case of variant E3, a significant increase in the production of FB_1 (1017.9%) and FB_3 (716.1%) was observed ($p < 0.05$). However, the increase in FB_2 production was at a similar level (123.1–208.5%), regardless of the used extract variant.

Instead, the biosynthesis of all mycotoxins by *F. culmorum* after treatment with the rowanberry extracts was significantly lower compared to the control group ($p < 0.05$ for all determined mycotoxins; Figure 5).

Table 3. The percentage of mycotoxin increase or decrease in tested samples of *F. proliferatum* treated with different variants of rowanberry extracts (E1–E4).

Extract Variant	Variation of Mycotoxin Content [%]			
	Increase			Decrease
	FB_1	FB_2	FB_3	BEA
E1	242.9 ± 41.4 [a]	180.5 ± 72.4 [a]	216.3 ± 93.8 [a]	82.5 ± 11.3 [a]
E2	354.2 ± 54.4 [a]	208.5 ± 61.3 [a]	286.5 ± 105.5 [a]	90.4 ± 4.9 [a]
E3	1017.9 ± 481.1 [b]	123.1 ± 65.6 [a]	716.1 ± 216.1 [b]	92.1 ± 5.6 [a]
E4	704.1 ± 75.6 [ab]	126.2 ± 63.0 [a]	344.5 ± 142.1 [ab]	80.6 ± 8.8 [a]

FB_1—fumonisin B_1, FB_2—fumonisin B_2, FB_3—fumonisin B_3, BEA—beauvericin. All values are means of three replicates ± standard deviation. The values assigned with the superscripts of different letters are significantly different (Tukey's HSD test, significant at $p < 0.05$).

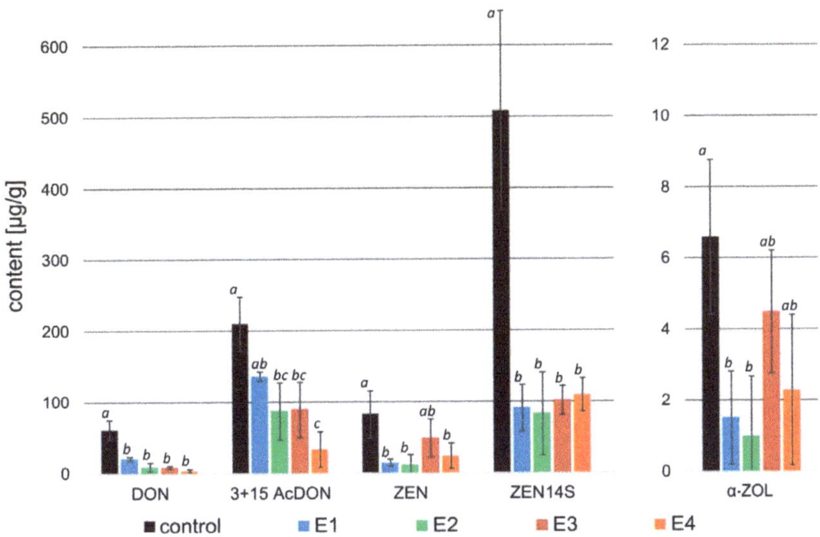

Figure 5. Mycotoxin content [µg/g] in the samples treated with rowanberry extracts (four variants, E1–E4) after inoculation with *Fusarium culmorum*. DON—deoxynivalenol, 3 + 15 AcDON—3- and 15-acetyldeoxynivalenol, ZEN—zearalenone, ZEN14S—zearalenone-14-sulfate, α-ZOL—α-zearalenol. All values are means of three replicates; error bars represent standard deviation. The values assigned with the superscripts of different letters are significantly different (Tukey's HSD test, significant at $p < 0.05$).

In the case of *F. culmorum*, the addition of all extract variants resulted in a significant decrease in the DON (3.14–20.21 µg/g), as well as 3- and 15-AcDON (32.45–136.12 µg/g) content compared to the control group (60.30 and 209.36 µg/g, respectively). The greatest decrease in the concentration of these mycotoxins was observed after the treatment of *F. culmorum* with the extract variant E4. However, in the case of DON, the differences in the effects of individual extracts were not significant ($p > 0.05$), while a significant difference was observed in the case of 3- and 15-AcDON content after treatment with E4 ($p < 0.05$). Moreover, all extract variants decreased in ZEN (11.10–48.01 µg/g), ZEN-14S (82.32–108.61 µg/g), and α-ZOL (0.98–4.48 µg/g) concentrations in comparison to the control (81.75, 508.53, and 6.58 µg/g, respectively). The highest decrease in these mycotoxin contents was caused by the addition of the extract variant E2 to the mycelium, whereas in the case of ZEN and α-ZOL, there were no significant differences between the activity of extract variants E1 and E2 ($p > 0.05$), but for the concentration of ZEN-14S, a significant difference caused by the extract variant E2 was observed ($p < 0.05$).

The application of rowanberry extracts on *F. culmorum* resulted in a reduction in all detected mycotoxins by at least 32.7%, whereas the different extract variants inhibited mycotoxin production in various grades (Table 4). A similar reduction in mycotoxins ($p < 0.05$) was observed regardless of the extract variant for DON: 86.6–95.1% (exception: E1: 66.1%), ZEN: 74.7–89.1% (exception E3: 43.6%), and ZEN-14S: 78.7–82.5%. The extract variant E4 reduced the synthesis of 3- and 15-AcDON by *F. culmorum* the most (85.5%), but the extract variant E2 reduced the synthesis of α-ZOL by 89.2% ($p < 0.05$).

Table 4. The percentage of mycotoxin reduction in tested samples of *F. culmorum* treated with different variants of rowanberry extracts (E1–E4).

Extract Variant	Mycotoxin Content Reduction [%]				
	DON	**3- and 15-AcDON**	**ZEN**	**ZEN-14S**	**α-ZOL**
E1	66.1 ± 3.1 ^{b}	33.8 ± 10.3 ^{b}	83.2 ± 1.2 ^{a}	82.5 ± 3.2 ^{a}	79.0 ± 18.9 ^{ab}
E2	86.8 ± 6.1 ^{a}	60.0 ± 12.6 ^{ab}	89.1 ± 10.1 ^{a}	82.3 ± 12.1 ^{a}	89.2 ± 18.7 ^{a}
E3	86.6 ± 2.3 ^{a}	59.0 ± 11.6 ^{ab}	43.6 ± 12.4 ^{b}	79.8 ± 2.3 ^{a}	32.7 ± 9.3 ^{b}
E4	95.1 ± 2.3 ^{a}	85.5 ± 9.2 ^{a}	74.7 ± 12.8 ^{a}	78.1 ± 4.1 ^{a}	69.2 ± 26.6 ^{ab}

DON—deoxynivalenol, 3- and 15-AcDON—3- and 15-acetyldeoxynivalenol, ZEN—zearalenone, ZEN14S—zearalenone-14-sulfate, α-ZOL—α-zearalenol. All values are means of three replicates ± standard deviation. The values assigned with the superscripts of different letters are significantly different (Tukey's HSD test, significant at $p < 0.05$).

3. Discussion

Despite the considerable advancements in modern agriculture, fungal infections affecting agricultural produce, notably fusariosis, persist as a substantial threat [34]. Consequently, there is a critical need to develop innovative methodologies against pathogenic fungi. A promising approach involves utilizing biological methods to inhibit fungal development in crops, e.g., through the use of plant extracts possessing antimicrobial properties [35]. One such natural candidate is rowanberry, known for its richness in organic acids as well as phenolic and flavonoid compounds [14]. However, the imperative involves the judicious selection of an appropriate extraction method and conditions to obtain the most effective extract.

Supercritical fluid extraction is an innovative and promising method offering significant advantages in the extraction of natural compounds while aligning with the growing demand for environmentally sustainable and high-quality extraction [36]. This approach is currently widely used for the extraction of essential oils from, e.g., *Lavendula genus*, *Pimpinella anisum*, or *Jasminum sambac* [37–39]. In the case of *Sorbus aucuparia*, the SFE technique was used to extract pomace, seeds, or waste to obtain oil extracts [40–42]. The highest yield of the lipophilic fraction was received by Bobinaite et al. (2020) at 60 °C and 450 bar in 180 min [40]. However, Ivakhnov et al. (2019) reached the highest yield of the oil fraction, equal to 9.02 wt % (weight percent) at 85 °C and 329.3 bar in 72 min [42]. Testing modified extraction (CO_2 expanded ethanol combined with sonication) with different co-solvents for receiving lipids from berry seeds, including rowanberries, showed that the optimal operating conditions were as follows: ethanol, 52 °C, 100 bar, and 7 min [41]. However, no research has been carried out on other (aqueous) fractions of extracts obtained, especially from *Sorbus aucuparia* fruits. Due to divergent data on extraction parameters, we decided to prepare four different extract variants using low and high temperatures (40 and 70 °C) at intermediate pressure values (200 and 300 bar). We used methanol as a co-solvent, which is a commonly used solvent in numerous extractions. To prepare the richest extracts, the extraction time was set to 180 min. The highest extraction yield was obtained when high temperature and pressure were used (extract variant E3), while the lowest yield was caused by the use of 200 bar, without a significant influence of temperature (E2 and E4). The extraction yield ranged from 16.90 to 18.05% (169.60–184.40 g/kg) and was similar to the yields of the typical extractions of rowanberry pomace using methanol, e.g., 20.75% [43].

In general, at a specific temperature, an increase in pressure increases the density of the solvent and the solubility of targeted compounds, promoting an increase in extraction yield. Under constant pressure, the temperature increase decreases the solvent density, reducing its solvating power but improving the vapor pressure, which increases the analyte solubility, thus increasing extraction efficiency [44]. The obtained results are consistent with these data.

To assess the quality of the obtained rowanberry extracts, the total phenolic contents were measured (see Figure 1). To date, the highest TPC values of the defatted methanol–water rowanberry fruit extract (1:1, *v*/*v*) were equal to 26.03 mg GAE/g dw and of the water residue—14.27 mg GAE/g dw [45]. However, most reports on methanolic extracts of *Sorbus aucuparia* fruits reported lower TPC values in a range of 2.53–10.05 mg GAE/g, depending on rowanberries genotypes, 4.35–8.19 mg GAE/g for different cultivars, and between 2.68 and 3.35 mg GAE/g [13,46–48]. The extract variant E4 prepared at 70 °C and 200 bar was characterized by the highest TPC (4.15 and 3.33 mg GAE/g dw for methanol and water, respectively). In comparison, the extract variant E2 obtained at 40 °C and the same pressure had the lowest TPC (2.71 and 2.07 mg GAE/g dw for methanol and water, respectively). These results suggest that higher temperatures may promote the release of phenolic compounds from the tested material due to the improvement of their solubility or acceleration of diffusion connected with the increasing temperature [49]. Aqueous solutions prepared from the corresponding methanolic extracts had significantly lower TPC values, which can be explained based on the different solubility of phenolic compounds in water and methanol. Nevertheless, the content of phenolic compounds in aqueous extracts was substantial, and due to their potential to provide a biocompatible environment given their applications as fungicides and ensuring the safety of both plants and their consumers, we chose them for microbiological experiments.

Previously published research on rowan fruit extracts has primarily concentrated on their antibacterial properties, with limited studies on their antifungal effects, which have yielded inconclusive results. No inhibitory effect on *Candida albicans* growth was observed after treatment with the aqueous and ethanolic, as well as acetone-derived phenolic-rich, rowanberry extracts [16,17]. However, Maliuvanchuk et al. (2023) determined an inhibition zone of ethanolic fruit extract obtained by the percolation method on *C. albicans* equal to 5.81 mm [23]. Further, an antifungal effect was not observed against *Rhizopus stolonifer* treated with ethanolic–aqueous rowanberry extract [19]. Importantly, several reports present promising results regarding the growth inhibition of various fungal species by extracts obtained from the other *Sorbus* spp. The extracts prepared from *S. sibirica* fruits were able to inhibit the growth of *Aspergillus niger* (fungistatic diameter: 1.51 mm) [50]. Moreover, antifungal properties of *S. cashmiriana* whole plant extracts were tested on various fungal species, and the most efficient inhibition was obtained against *Aspergillus flavus* (68%) and the lowest against *Rhizoctonia solani* (24%) [20]. However, acidified ethanolic extracts obtained from *S. domestica* fruits did not show antifungal activity against *C. albicans* and *Saccharomyces cerevisiae* [21]. In our experiment, the addition of rowanberry extracts inhibited the growth of *Fusarium proliferatum* from 6.5 to 26.4%, but for *F. culmorum*, growth inhibition was not observed (see Figure 2). However, to determine the actual antifungal activity of the obtained extracts, we measured ergosterol content in the treated mycelia. ERG, which is a sterol compound found in the cell membranes of fungi that serves as a crucial component for maintaining the structural integrity and fluidity of fungal cell membranes, is used as a specific biomarker to quantify fungal biomass [22]. A reduction in ERG content was observed for all samples, but extract variant E3 was the most effective (ERG reduction: 60.67% for *F. proliferatum* and 41.62% for *F. culmorum*). It is noteworthy that the extraction yield of this particular variant was the highest, potentially indicating an increased concentration of bioactive compounds, which exhibit antifungal properties, but this is not directly related to the total content of phenolic compounds. The extracts obtained at low temperature (40 °C) had significantly lower antifungal properties, but still, their decrease in ERG content was not lower than 7.70% against *F. culmorum* and 12.41% against

F. proliferatum. In the case of *F. proliferatum*, ERG concentration reduction corresponded to the observed inhibition of mycelium growth. However, the reduction in ERG content with the absence of differences in strain sizes between the control and the tested samples of *F. culmorum* may be caused by the specific structure of the mycelium, for which it was not possible to observe a size decrease.

The promising effects of the microbiological studies induced us to analyze secondary metabolites of *Fusarium* spp. *F. proliferatum* produced beauvericin and fumonisins from the B group. All tested rowanberry extracts significantly decreased the production of BEA, reaching reduction values from 80.6 to 92.1% (without significant differences in the degree of reduction between individual extracts, $p > 0.05$). Interestingly, in the case of fumonisins, there was an enormous increase in their biosynthesis in all samples after extract addition (see Figure 4). So far, this effect has been only described in a few reports [51,52]. This phenomenon can be interpreted as part of a stress response mechanism caused by the external factor, where the fungi change metabolic priorities [53]. Under stress caused by the presence of rowanberry extracts, *F. proliferatum* reallocates resources to prioritize the synthesis of compounds that enhance their survival, like fumonisins, which disrupt sphingolipid metabolism in host organisms [54]. Moreover, the significant increase in fumonisin production may divert metabolic resources away from beauvericin biosynthesis, or the regulatory pathways controlling BEA synthesis might be directly inhibited by the stress, further reducing its production. Furthermore, under stress conditions, fungi can release alkaline proteases, which, further, are able to break peptide bonds presented in the BEA structure [55]. Instead, *Fusarium culmorum* synthesized B-trichothecenes (DON, 3- and 15-AcDON) and zearalenone with derivatives (ZEN, ZEN-14S, α-ZOL). The simultaneous presence of free and modified forms can be attributed to the intricate interplay between the fungus and its host environment. Modified mycotoxins may originate from the metabolic pathways of infected plants or be synthesized de novo by the fungus, leading to their coexistence alongside their free forms [56]. The obtained results showed that *F. culmorum* produced ZEN-14S in high concentration (control: 508.53 μg/g) and α-ZOL in low concentration (control: 6.58 μg/g). The application of rowanberry fruit extracts resulted in a notable decrease ($p < 0.05$) in the biosynthesis of all tested mycotoxins, with reductions not less than 32.7%.

4. Materials and Methods

4.1. Plant Material

The plant material consisted of ripe fruits of the rowan tree *Sorbus aucuparia* obtained from trees growing in Poznań and Zagórów (Central Poland). The fruits were collected directly from the tree, then lyophilized and ground in liquid nitrogen. The material was stored frozen at −20 °C.

4.2. Fungal Material

Fusarium proliferatum (PEA 1) and *Fusarium culmorum* (KF 846) were isolated from pea seeds and wheat kernels, respectively. They were identified by molecular techniques [35,57] and preserved at the Institute of Plant Genetics, Polish Academy of Science, Poznań, Poland.

4.3. Reagents

Carbon dioxide CO_2 (99.9995%, Air Products, Warszawa, Poland), methanol (99.5%, Chempur, Piekary Śląskie, Poland), and deionized water (Milipore, Burlington, MA, USA) were used to prepare rowanberry extracts. Potato dextrose agar (PDA) used for the in vitro experiment was supplied by Oxoid, Basingstoke, UK. Sodium hydroxide (≥98.8%, POCH, Gliwice, Poland), hydrochloric acid (≥37%, Sigma Aldrich, Taufkirchen, Germany), methanol (99.99%, POCH, Gliwice, Poland), and pentane (99%, POCH, Gliwice, Poland) were used to extract ergosterol or mycotoxins. For the HPLC measurements, the appropriate solvents with LC grade were used. Analytical standards purchased as ready-to-use solutions from Romer Labs (Tulln, Austria) included ERG, FB_1, FB_2, FB_3, BEA, DON, DON-

3G, 3-AcDON, 15-AcDON, NIV, NIV-3G, FUSX, ZEN, and ZEN-14G (100 μg/mL). The α- and β-ZOL concentrations were 10 μg/mL. ZEN-14S (100 μg/mL) was purchased from Aokin (Berlin, Germany).

4.4. Supercritical Fluid Extraction

The extraction was performed utilizing the MV-10ASFE extractor (Waters, Manchester, MA, USA), which included a CO_2 cylinder, cooling system, fluid delivery module, column oven, back pressure regulator, heat exchanger, and fraction collection module, as well as ChromScope v1.20 software (Waters, Manchester, MA, USA). The extraction of 25 g of rowanberries conducted under various temperature (40 and 70 °C) and pressure (200 and 300 bar) conditions enabled the preparation of 4 different extract variants (E1–E4). The CO_2 flow rate was 4 mL/min, and the methanol (co-solvent) flow was 1 mL/min. Each experimental run took 180 min, with the first dynamic time of 15 min, the static time of 30 min, and the second dynamic time of 135 min. The extraction yield was calculated using the following formula:

$$\text{extraction yield} = \frac{m_a - m_b}{m_a} \times 100\% \quad (1)$$

where m_a is the mass of *Sorbus aucuparia* fruits taken for extraction, and m_b is the mass of the remaining rowanberry residues after extraction.

The initial methanolic extracts obtained from *Sorbus aucuparia* fruits were transferred to deionized water, the remaining methanol was removed, and the aqueous extracts were concentrated using an evaporator. Finally, from the 1 g of rowanberries, 1 mL of each aqueous extract variant was prepared.

4.5. Determination of Total Phenolic Content

The total phenolic content (TPC) was determined using the modified Folin–Ciocâlteu method [58], as follows: 0.1 mL of gallic acid solutions with concentrations of 0.01, 0.05, 0.10, 0.15, 0.20, and 0.25 mg/mL, respectively, were mixed with 0.25 mL of Folin–Ciocâlteu's reagent, and after 3 min, 3 mL of 10% calcium carbonate solution was added. After 40 min in the darkness, the absorbances of the prepared calibrating solutions were measured at 765 nm using a UV–Vis Varian Cary 300 spectrophotometer (Thermo Fisher Scientific, Waltham, MA, USA). Then, the absorbance of appropriately diluted methanolic and aqueous extracts of rowanberries prepared analogously to the standard samples was measured similarly. TPC was determined based on the obtained calibration curve ($y = 4.018x$, $R^2 = 0.9999$), and the results were expressed as mg of gallic acid per 1 g of extract dw.

4.6. Study on the Effect of Sorbus aucuparia Fruit Extracts on Fusarium Growth

The antifungal effect of rowanberries was estimated using a modified method of Uwineza et al. (2022) [56]: 15 mL of a mixture of PDA medium with 10% aqueous rowanberry extract was inoculated with a 5 mm mycelium of *F. proliferatum* (PEA 1) or *F. culmorum* (KF 846) on each Petri dish. The control group was PDA medium containing the investigated fungal strains without extracts. All samples were incubated at 25 °C in the darkness for 10 days in three replications. After 10 days of the experiment, the radial mycelial growth was measured, and the antifungal properties of each extract variant were calculated based on the following formula:

$$\text{mycelium growth inhibition} = \frac{D_c - D_e}{D_c} \times 100\% \quad (2)$$

where D_c is the average diameter of the control group (PDA medium with *F. proliferatum* or *F. culmorum*), and D_e is the average diameter of tested samples (*Fusarium* spp. with PDA medium treated by rowanberry extracts).

4.7. Determination of Ergosterol Content

To 0.1 g of dried mycelium samples, 2 mL of methanol and 0.5 mL of 2 M sodium hydroxide were added. Samples were microwave-heated 3 times at 370 W for 20 s. After cooling, the samples were neutralized with 1 mL of 1 M hydrochloric acid and 2 mL of methanol. The suspensions were extracted three times with 4 mL of n-pentane, and then the solvent was air-evaporated. The dried residues were dissolved in 1 mL of methanol, filtered through 15 mm syringe filters with a 0.20 µm pore diameter (Chromafil, Macherey-Nagel, Duren, Germany), and analyzed by the UPLC-PDA technique according to Uwineza et al. (2022) [56]. The ERG content was given in µg per g of sample, and the ERG reduction percentage was calculated by the given formula:

$$\text{ERG reduction} = \frac{E_c - E_e}{E_c} \times 100\% \quad (3)$$

where E_c is the ergosterol content in the control group (PDA medium with *F. proliferatum* or *F. culmorum*), and E_e is the ergosterol content in the test samples (PDA medium with *Fusarium* spp. treated with rowanberry extracts).

4.8. Analysis of Mycotoxins

Mycotoxins were extracted from the dried mycelium samples (0.335–0.506 g) by adding 7 mL of methanol and stirring for 24 h. Then, the samples were centrifuged at 5000 rpm for 10 min, and the supernatants were filtered using a disc filter with a 0.20 µm pore diameter (Chromafil, Macherey-Nagel, Duren, Germany). Mycotoxin detection and determination were performed with UHPLCHESI-MS/MS using a non-porous C18 Cortecs chromatographic column (100 mm × 2.1 mm × 1.6 µm, Waters, Manchester, MA, USA). The mobile phase consisted of water–methanol 90:10 (A) and methanol–water 90:10 (B); both phases had 5 mM ammonium formate and 0.2% formic acid. The separation parameters and operation of the chromatograph are given in detail in a previous publication [56]. The percentage of mycotoxin biosynthesis reduction by *F. proliferatum* and *F. culmorum* was determined using the following formula:

$$\text{mycotoxin reduction} = \frac{M_c - M_e}{M_c} \times 100\% \quad (4)$$

where M_c is the mycotoxin content in the control group (PDA medium with *F. proliferatum* or *F. culmorum*), and M_e is the mycotoxin content in the tested samples (PDA medium with *Fusarium* spp. treated with rowanberry extracts).

4.9. Statistical Analysis

Experimental data were statistically evaluated using the Statistica 14 software package (TIBCO Software Inc., Palo Alto, CA, USA). A one-way analysis of variance (ANOVA) was used to calculate the means and standard deviations and to assess the significance of the differences in the effects of individual extracts. Subsequently, the post hoc Tukey's honest significant difference (HSD) test was used for paired comparisons ($p = 0.05$).

5. Conclusions

This study aimed to determine the influence of *Sorbus aucuparia* fruit extracts obtained by a supercritical fluid extraction method under different conditions on *Fusarium proliferatum* and *F. culmorum* growth and mycotoxin biosynthesis. It was observed that the obtained extracts inhibited the growth of only *F. proliferatum*. However, the decrease in ergosterol concentrations showed that the growth of both *Fusarium* species was reduced, but this effect was more evident for *F. proliferatum* than *F. culmorum*. Rowanberry extract obtained at 70 °C and 300 bars showed the most significant antifungal activity, with an ERG reduction of 60.67% for *F. proliferatum* and 41.62% for *F. culmorum*. Despite a significant decrease in ERG content in both *Fusarium* spp., which indicated fungi reduction, there was a diversification in their biosynthesis of secondary metabolites. All tested extract variants reduced the production of BEA by *F. proliferatum*. However, fumonisins showed substantial increases in their biosynthesis compared to the control group. This phenomenon can be attributed to the fungal survival mechanism that stimulates mycotoxin production under specific stress conditions. Instead, extracts obtained from rowan fruits consistently decreased the production of all tested mycotoxins by *F. culmorum*. All prepared extract variants similarly reduced the content of DON, ZEN, and ZEN-14S. However, the most significant reduction in 3- and 15-AcDON content occurred when samples were treated with the extract prepared at 40 °C and 200 bar, and in the case of α-ZOL, with the extract prepared at 70 °C and 200 bar. The obtained results are important for practically assessing the effectiveness of *Sorbus aucuparia* fruit extracts against *Fusarium* pathogens in cereal crops. Future studies on the gene expression of the tested *Fusarium* spp. are planned to understand the mechanisms of their growth inhibition and the different mycotoxin biosynthesis effects. The current research shows great potential for advancing sustainable agricultural practices, addressing the challenges of fungal diseases in cereal crops, and enhancing food safety.

Author Contributions: Conceptualization, S.R. and A.W.; methodology, S.R., N.G.-W., M.U., M.T. and A.W.; software, A.W.; validation, S.R., Ł.S. and A.W.; formal analysis, S.R. and A.W.; investigation, S.R., N.G.-W., M.U., M.B. and A.W.; resources, A.W., M.B., M.T. and Ł.S.; data curation, A.W.; writing—original draft preparation, S.R.; writing—review and editing, S.R., N.G.-W., M.U., Ł.S. and A.W.; visualization, S.R. and A.W.; supervision, A.W.; project administration, A.W. All authors have read and agreed to the published version of the manuscript.

Funding: This research was founded by Polish Minister of Science and Education, under the program "Regional Initiative of Excellence" in 2019–2022 (Grant No. 008/RID/2018/19).

Institutional Review Board Statement: Not applicable.

Informed Consent Statement: Not applicable.

Data Availability Statement: Dataset available on request from the authors.

Conflicts of Interest: The authors declare no conflicts of interest.

References

1. Moskalets, V.; Hulko, B.; Rozhko, I.; Moroz, V.; Ivankiv, M. Morpho-Physiological Characteristics of Plants and Biochemical Parameters of Rowan Berries, Common Rowan, and Domestic Rowan Grown in the Conditions of the Northern Forest-Steppe of Ukraine. *Sci. Horiz.* **2023**, *26*, 78–92. [CrossRef]
2. Huang, S.; Yang, N.; Liu, Y.; Gao, J.; Huang, T.; Hu, L.; Zhao, J.; Li, Y.; Li, C.; Zhang, X. Grape Seed Proanthocyanidins Inhibit Colon Cancer-Induced Angiogenesis through Suppressing the Expression of VEGF and Angl. *Int. J. Mol. Med.* **2012**, *30*, 1410–1416. [CrossRef] [PubMed]
3. Baby, B.; Antony, P.; Vijayan, R. Antioxidant and Anticancer Properties of Berries. *Crit. Rev. Food Sci. Nutr.* **2018**, *58*, 2491–2507. [CrossRef]
4. Shikov, A.N.; Pozharitskaya, O.N.; Makarov, V.G.; Wagner, H.; Verpoorte, R.; Heinrich, M. Medicinal Plants of the Russian Pharmacopoeia; Their History and Applications. *J. Ethnopharmacol.* **2014**, *154*, 481–536. [CrossRef]
5. Raspé, O.; Findlay, C.; Jacquemart, A. *Sorbus aucuparia* L. *J. Ecol.* **2000**, *88*, 910–930. [CrossRef]
6. Jędrzejewski, W.; Zalewski, A.; Jędrzejewska, B. Foraging by Pine Marten *Martes Martes* in Relation to Food Resources in Białowieża National Park, Poland. *Acta Theriol.* **1993**, *38*, 405–426. [CrossRef]
7. Krasnov, V.; Shelest, Z.; Boiko, S.; Gulik, I.; Sieniawski, W. The Diet of the Roe Deer (*Capreolus capreolus*) in the Forest Ecosystems of Zhytomirske Polesie of the Ukraine. *For. Res. Pap.* **2015**, *76*, 184–190. [CrossRef]
8. Guitián, J.; Munilla, I. Responses of Mammal Dispersers to Fruit Availability: Rowan (*Sorbus aucuparia*) and Carnivores in Mountain Habitats of Northern Spain. *Acta Oecologica* **2010**, *36*, 242–247. [CrossRef]
9. Bozhuyuk, M.R.; Ercisli, S.; Ayed, R.B.; Jurikova, T.; Fidan, H.; Ilhan, G.; Ozkan, G.; Sagbas, H.I. Compositional Diversity in Fruits of Rowanberry (*Sorbus aucuparia* L.) Genotypes Originating from Seeds. *Genetika* **2020**, *52*, 55–65. [CrossRef]
10. Herrero, M.; Mendiola, J.A.; Cifuentes, A.; Ibáñez, E. Supercritical Fluid Extraction: Recent Advances and Applications. *J. Chromatogr. A* **2010**, *1217*, 2495–2511. [CrossRef]
11. Brunner, U. Some Antifungal Properties of Sorbic Acid Extracted from Berries of Rowan (*Sorbus aucuparia*). *J. Biol. Educ.* **1985**, *19*, 41–47. [CrossRef]
12. Kokubun, T.; Harborne, J.B.; Eagles, J.; Waterman, P.G. Antifungal Biphenyl Compounds Are the Phytoalexins of the Sapwood of *Sorbus aucuparia*. *Phytochemistry* **1995**, *40*, 57–59. [CrossRef]

13. Qiu, X.; Lei, C.; Huang, L.; Li, X.; Hao, H.; Du, Z.; Wang, H.; Ye, H.; Beerhues, L.; Liu, B. Endogenous Hydrogen Peroxide Is a Key Factor in the Yeast Extract-Induced Activation of Biphenyl Biosynthesis in Cell Cultures of *Sorbus aucuparia*. *Planta* **2012**, *235*, 217–223. [CrossRef]
14. Arvinte, O.M.; Senila, L.; Becze, A.; Amariei, S. Rowanberry—A Source of Bioactive Compounds and Their Biopharmaceutical Properties. *Plants* **2023**, *12*, 3225. [CrossRef]
15. Sołtys, A.; Galanty, A.; Podolak, I. *Ethnopharmacologically Important but Underestimated Genus Sorbus: A Comprehensive Review*; Springer: Dordrecht, The Netherlands, 2020; Volume 19, ISBN 0123456789.
16. Liepiņa, I.; Nikolajeva, V.; Jākobsone, I. Antimicrobial Activity of Extracts from Fruits of *Aronia melanocarpa* and *Sorbus aucuparia*. *Environ. Exp. Biol.* **2013**, *11*, 195–199.
17. Nohynek, L.J.; Alakomi, H.L.; Kähkönen, M.P.; Heinonen, M.; Helander, I.M.; Oksman-Caldentey, K.M.; Puupponen-Pimiä, R.H. Berry Phenolics: Antimicrobial Properties and Mechanisms of Action against Severe Human Pathogens. *Nutr. Cancer* **2006**, *54*, 18–32. [CrossRef] [PubMed]
18. Krisch, J.; Galgóczy, L.; Tölgyesì, M.; Papp, T.; Vágvölgyi, C. Effect of Fruit Juices and Pomace Extracts on the Growth of Gram-Positive and Gram-Negative Bacteria. *Acta Biol. Szeged.* **2008**, *52*, 267–270.
19. Bazarnova, Y.G.; Ivanchenko, O.B. Investigation of the Composition of Biologically Active Substances in Extracts of Wild Plants. *Vopr. Pitan.* **2016**, *85*, 100–107.
20. Khan, S.; Kazmi, M.H.; Fatima, I.; Malik, A.; Inamullah, F.; Farheen, S.; Abbas, T. Cashmirins A and B, New Antifungal and Urease Inhibitory Prenylated Coumarins from *Sorbus cashmiriana*. *Braz. J. Pharm. Sci.* **2022**, *58*, e21493. [CrossRef]
21. Sagdic, O.; Polat, B.; Yetim, H. Bioactivities of Some Wild Fruits Grown in Turkey. *Erwerbs-Obstbau* **2022**, *64*, 299–305. [CrossRef]
22. Hu, Y.; Zhang, J.; Kong, W.; Zhao, G.; Yang, M. Mechanisms of Antifungal and Anti-Aflatoxigenic Properties of Essential Oil Derived from Turmeric (*Curcuma longa* L.) on *Aspergillus flavus*. *Food Chem.* **2017**, *220*, 1–8. [CrossRef] [PubMed]
23. Maliuvanchuk, S.; Grytsyk, A.; Melnyk, M.; Kutsyk, R.; Yurkiv, K.; Raal, A.; Koshovyi, O. *Sorbus Aucuparia* L. Fruit Extract and Its Cosmetics—As Promising Agents for Prophylactic and Treatment of Pyodermitis: Phytochemical and Microbiological Research. *Open Agric. J.* **2023**, *17*, e18743315268063. [CrossRef]
24. Shi, W.; Tan, Y.; Wang, S.; Gardiner, D.M.; De Saeger, S.; Liao, Y.; Wang, C.; Fan, Y.; Wang, Z.; Wu, A. Mycotoxigenic Potentials of *Fusarium* Species in Various Culture Matrices Revealed by Mycotoxin Profiling. *Toxins* **2017**, *9*, 6. [CrossRef] [PubMed]
25. El Chami, J.; El Chami, E.; Tarnawa, Á.; Kassai, K.M.; Kende, Z.; Jolánkai, M. Effect of *Fusarium* Infection on Wheat Quality Parameters. *Cereal Res. Commun.* **2023**, *51*, 179–187. [CrossRef]
26. Tava, V.; Prigitano, A.; Cortesi, P.; Esposto, M.C.; Pasquali, M. *Fusarium Musae* from Diseased Bananas and Human Patients: Susceptibility to Fungicides Used in Clinical and Agricultural Settings. *J. Fungi* **2021**, *7*, 784. [CrossRef] [PubMed]
27. Pinotti, L.; Ottoboni, M.; Giromini, C.; Dell'Orto, V.; Cheli, F. Mycotoxin Contamination in the EU Feed Supply Chain: A Focus on Cereal Byproducts. *Toxins* **2016**, *8*, 45. [CrossRef]
28. Bottalico, A.; Perrone, G. Toxigenic *Fusarium* Species and Mycotoxins Associated with Head Blight in Small-Grain Cereals in Europe. *Eur. J. Plant Pathol.* **2002**, *108*, 611–624. [CrossRef]
29. Saadullah, A.A.M. Studies on Teratogenic and Maternal Effects of Trichothecene (TCT) Extracted from *Fusarium* and *trichoderma* Culture on Pregnant Albino Mice. *Bionatura* **2023**, *8*, 1–4. [CrossRef]
30. Dabas, Y.; Bakhshi, S.; Xess, I. Fatal Cases of Bloodstream Infection by *Fusarium solani* and Review of Published Literature. *Mycopathologia* **2016**, *181*, 291–296. [CrossRef]
31. Al-Hatmi, A.M.S.; Bonifaz, A.; Ranque, S.; de Hoog, G.S.; Verweij, P.E.; Meis, J.F. Current Antifungal Treatment of Fusariosis. *Int. J. Antimicrob. Agents* **2018**, *51*, 326–332. [CrossRef]
32. Ribas e Ribas, A.D.; Spolti, P.; Del Ponte, E.M.; Donato, K.Z.; Schrekker, H.; Fuentefria, A.M. Is the Emergence of Fungal Resistance to Medical Triazoles Related to Their Use in the Agroecosystems? A Mini Review. *Braz. J. Microbiol.* **2016**, *47*, 793–799. [CrossRef] [PubMed]
33. Batista, B.G.; de Chaves, M.A.; Reginatto, P.; Saraiva, O.J.; Fuentefria, A.M. Human Fusariosis: An Emerging Infection That Is Difficult to Treat. *Rev. Soc. Bras. Med. Trop.* **2020**, *53*, e20200013. [CrossRef] [PubMed]
34. Peng, Y.; Li, S.J.; Yan, J.; Tang, Y.; Cheng, J.P.; Gao, A.J.; Yao, X.; Ruan, J.J.; Xu, B.L. Research Progress on Phytopathogenic Fungi and Their Role as Biocontrol Agents. *Front. Microbiol.* **2021**, *12*, 670135. [CrossRef] [PubMed]
35. Uwineza, P.A.; Urbaniak, M.; Stępień, Ł.; Gramza-Michałowska, A.; Waśkiewicz, A. *Lamium album* Flower Extracts: A Novel Approach for Controlling *Fusarium* Growth and Mycotoxin Biosynthesis. *Toxins* **2023**, *15*, 651. [CrossRef]
36. Reverchon, E.; De Marco, I. Supercritical Fluid Extraction and Fractionation of Natural Matter. *J. Supercrit. Fluids* **2006**, *38*, 146–166. [CrossRef]
37. Vairinhos, J.; Miguel, M.G. Essential Oils of Spontaneous Species of the Genus *Lavandula* from Portugal: A Brief Review. *Z. Naturforsch. C* **2020**, *75*, 233–245. [CrossRef]
38. Rodrigues, V.M.; Rosa, P.T.V.; Marques, M.O.M.; Petenate, A.J.; Meireles, M.A.A. Supercritical Extraction of Essential Oil from Aniseed (*Pimpinella anisum* L.) Using CO_2: Solubility, Kinetics, and Composition Data. *J. Agric. Food Chem.* **2003**, *51*, 1518–1523. [CrossRef]
39. Ye, Q.; Jin, X.; Wei, S.; Zheng, G.; Li, X. Effect of Subcritical Fluid Extraction on the High Quality of Headspace Oil from *Jasminum sambac* (L.) Aiton. *J. AOAC Int.* **2016**, *99*, 725–729. [CrossRef]

40. Bobinaitė, R.; Kraujalis, P.; Tamkutė, L.; Urbonavičienė, D.; Viškelis, P.; Venskutonis, P.R. Recovery of Bioactive Substances from Rowanberry Pomace by Consecutive Extraction with Supercritical Carbon Dioxide and Pressurized Solvents. *J. Ind. Eng. Chem.* **2020**, *85*, 152–160. [CrossRef]
41. Al-Hamimi, S.; Turner, C. A Fast and Green Extraction Method for Berry Seed Lipid Extraction Using CO_2 Expanded Ethanol Combined with Sonication. *Eur. J. Lipid Sci. Technol.* **2020**, *122*, 1900283. [CrossRef]
42. Ivakhnov, A.D.; Sadkova, K.S.; Sobashnikova, A.S.; Skrebets, T.E. Optimization of Oil Extraction from Rowanberry Waste in Alcoholic Beverage Production. *Russ. J. Phys. Chem. B* **2019**, *13*, 1135–1138. [CrossRef]
43. Bobinaitė, R.; Grootaert, C.; Van Camp, J.; Šarkinas, A.; Liaudanskas, M.; Žvikas, V.; Viškelis, P.; Rimantas Venskutonis, P. Chemical Composition, Antioxidant, Antimicrobial and Antiproliferative Activities of the Extracts Isolated from the Pomace of Rowanberry (*Sorbus aucuparia* L.). *Food Res. Int.* **2020**, *136*, 109310. [CrossRef]
44. Uwineza, P.A.; Waśkiewicz, A. Recent Advances in Supercritical Fluid Extraction of Natural Bioactive Compounds from Natural Plant Materials. *Molecules* **2020**, *25*, 3847. [CrossRef]
45. Rutkowska, M.; Kolodziejczyk-Czepas, J.; Owczarek, A.; Zakrzewska, A.; Magiera, A.; Olszewska, M.A. Novel Insight into Biological Activity and Phytochemical Composition of *Sorbus aucuparia* L. Fruits: Fractionated Extracts as Inhibitors of Protein Glycation and Oxidative/Nitrative Damage of Human Plasma Components. *Food Res. Int.* **2021**, *147*, 110526. [CrossRef]
46. Sarv, V.; Venskutonis, P.R.; Rätsep, R.; Aluvee, A.; Kazernavičiūtė, R.; Bhat, R. Antioxidants Characterization of the Fruit, Juice, and Pomace of Sweet Rowanberry (*Sorbus aucuparia* L.) Cultivated in Estonia. *Antioxidants* **2021**, *10*, 1779. [CrossRef] [PubMed]
47. Mlcek, J.; Rop, O.; Jurikova, T.; Sochor, J.; Fisera, M.; Balla, S.; Baron, M.; Hrabe, J. Bioactive Compounds in Sweet Rowanberry Fruits of Interspecific Rowan Crosses. *Cent. Eur. J. Biol.* **2014**, *9*, 1078–1086. [CrossRef]
48. Wolf, J.; Göttingerová, M.; Kaplan, J.; Kiss, T.; Venuta, R.; Nečas, T. Determination of the Pomological and Nutritional Properties of Selected Plum Cultivars and Minor Fruit Species. *Hortic. Sci.* **2020**, *47*, 181–193. [CrossRef]
49. Antony, A.; Farid, M. Effect of Temperatures on Polyphenols during Extraction. *Appl. Sci.* **2022**, *12*, 2107. [CrossRef]
50. Wei, J.; Shi, J.; Gao, J.; Zhou, Z.; Fan, J. Biological Activities of Extract Prepared from *Sorbus sibirica* Fruit. *J. Chem. Pharm. Res.* **2014**, *6*, 1369–1372.
51. Ferrochio, L.; Cendoya, E.; Farnochi, M.C.; Massad, W.; Ramirez, M.L. Evaluation of Ability of Ferulic Acid to Control Growth and Fumonisin Production of *Fusarium verticillioides* and *Fusarium proliferatum* on Maize Based Media. *Int. J. Food Microbiol.* **2013**, *167*, 215–220. [CrossRef]
52. Bodoira, R.; Velez, A.; Maestri, D.; Herrera, J. Bioactive Compounds Obtained from Oilseed By-Products with Subcritical Fluids: Effects on *Fusarium verticillioides* Growth. *Waste Biomass Valorization* **2020**, *11*, 5913–5924. [CrossRef]
53. Reynoso, M.M.; Torres, A.M.; Ramirez, M.L.; Rodríguez, M.I.; Chulze, S.N.; Magan, N. Efficacy of Antioxidant Mixtures on Growth, Fumonisin Production and Hydrolytic Enzyme Production by *Fusarium verticillioides* and *F. proliferatum* in Vitro on Maize-Based Media. *Mycol. Res.* **2002**, *106*, 1093–1099. [CrossRef]
54. Reynoso, M.M.; Torres, A.M.; Chulze, S.N. Fusaproliferin, Beauvericin and Fumonisin Production by Different Mating Populations among the Gibberella Fujikuroi Complex Isolated from Maize. *Mycol. Res.* **2004**, *108*, 154–160. [CrossRef] [PubMed]
55. Perincherry, L.; Lalak-Kańczugowska, J.; Stępień, Ł. Fusarium-Produced Mycotoxins in Plant-Pathogen Interactions. *Toxins* **2019**, *11*, 664. [CrossRef]
56. Uwineza, P.A.; Urbaniak, M.; Bryła, M.; Stepien, Ł.; Modrzewska, M.; Waśkiewicz, A. In Vitro Effects of Lemon Balm Extracts in Reducing the Growth and Mycotoxins Biosynthesis of *Fusarium culmorum* and *F. proliferatum*. *Toxins* **2022**, *14*, 355. [CrossRef]
57. Perincherry, L.; Urbaniak, M.; Pawłowicz, I.; Kotowska, K.; Waśkiewicz, A.; Stępień, Ł. Dynamics of *Fusarium* Mycotoxins and Lytic Enzymes during Pea Plants' Infection. *Int. J. Mol. Sci.* **2021**, *22*, 9888. [CrossRef]
58. Kupina, S.; Fields, C.; Roman, M.C.; Brunelle, S.L. Determination of Total Phenolic Content Using the Folin-C Assay: Single-Laboratory Validation. *J. AOAC Int.* **2019**, *102*, 320–321. [CrossRef]

molecules

MDPI

Communication

Caudiquinol: A Meroterpenoid with an Intact C20 Geranylgeranyl Chain Isolated from *Garcinia caudiculata*

Maya Valmiki [1], Stephen Ping Teo [2], Pedro Ernesto de Resende [3], Simon Gibbons [4] and A. Ganesan [1,*]

[1] School of Pharmacy, University of East Anglia, Norwich Research Park, Norwich NR4 7TJ, UK; m.valmiki@uea.ac.uk
[2] Forest Department Sarawak Headquarters, Medan Raya, Petra Jaya, Kuching 93050, Sarawak, Malaysia; stephetp@sarawak.gov.my
[3] The John Innes Centre, Norwich Research Park, Norwich NR4 7UH, UK; pedro.de-resende@jic.ac.uk
[4] Natural and Medical Sciences Research Center, University of Nizwa, PC 616, Birkat Al-Mauz, Nizwa P.O. Box 33, Oman; simon@unizwa.edu.om
* Correspondence: a.ganesan@uea.ac.uk

Abstract: The tropical *Garcinia* genus of flowering plants is a prolific producer of aromatic natural products including polyphenols, flavonoids, and xanthones. In this study, we report the first phytochemical investigation of *Garcinia caudiculata* Ridl. from the island of Borneo. Fractionation, purification, and structure elucidation by MS and NMR resulted in the discovery of two meroterpenoids. One was a benzofuranone lactone previously isolated from *Iryanthera grandis* and *Rhus chinensis*, and the second was a new hydroquinone methyl ester that we named caudiquinol. Both natural products are rare examples of plant meroterpenoids with an intact geranylgeranyl chain.

Keywords: natural products; meroterpenoids; hydroquinones; *Garcinia* species

1. Introduction

The *Garcinia* genus of saptrees within the Clusiaceae family comprises several hundred species of flowering shrubs and trees that are widely distributed in tropical regions around the world [1]. In addition to producing edible fruit, such as the mangosteen from *Garcinia mangostana*, *Garcinia* species are a prolific source of biologically active secondary metabolites, including polyphenols, flavonoids, and xanthones [2–4]. Within this genus, Ridley classified the bunau tree found on the island of Borneo as a new species, *G. caudiculata*, nearly a century ago [5]. However, no phytochemical investigations have appeared until the present work, where we report two meroterpenoids with a geranylgeranyl sidechain: the benzofuranone lactone **1** (Figure 1), previously isolated from two other plant genera, and a new quinol **2** that was given the name caudiquinol.

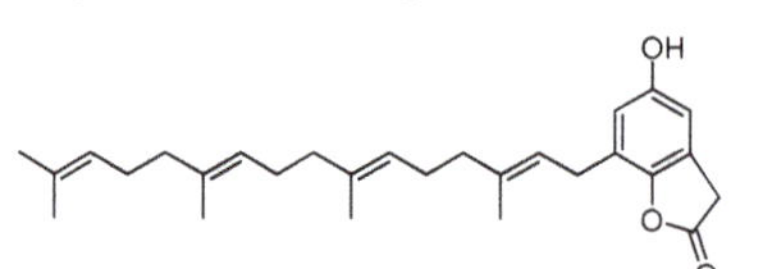

Benzofuranone (**1**)

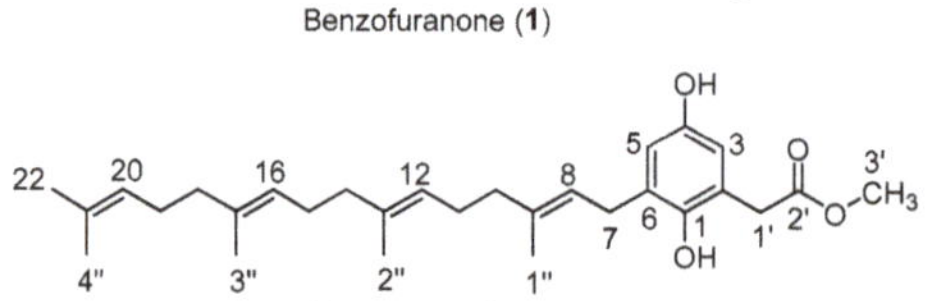

Caudiquinol (**2**)

Figure 1. Benzofuranone (**1**) and caudiquinol (**2**), geranylgeranyl meroterpenoids isolated from *Garcinia caudiculata*.

2. Results

Leaves of *G. caudiculata* Ridl. were collected from Lundu, Sarawak, Malaysia, and air-dried before being ground into a powder and extracted with dichloromethane. In a minimum inhibitory concentration (MIC) antibacterial assay, the crude extract was active against methicillin-susceptible *Staphylococcus aureus* (MSSA) 25923 at a level of 128 µg/mL. A portion of this extract of 5 g was subjected to vacuum liquid chromatography (VLC), eluting with a gradient of hexane/ethyl acetate (100:0 to 0:100) to provide 16 fractions. Upon evaporation, the most abundant yellow fractions 10 and 11, each containing ~0.5 g of residue, were selected for further purification. By recycling preparative HPLC, we ultimately obtained 7 mg of **1** from fraction 10 and 6 mg of **2** from fraction 11. Based on the spectroscopic and mass spectrometric data (Supporting Information), we assigned **1** as a benzofuranone lactone (Figure 1) with a C20 geranylgeranyl sidechain. This lactone was first identified in *Iryanthera grandis* of the Myristicaceae family [6] and later, in *Rhus chinensis* of the Anacardiaceae family [7]. It was the subject of a recent total synthesis due to its anti-HIV activity [8].

Compound **2** was isolated as a yellow oil with IR absorptions at 3387 and 1714 cm^{-1} suggesting the presence of OH and C=O functional groups. The ^1H and ^{13}C NMR chemical shifts of **2** (Table 1) indicated a carbonyl group at δ_C 173.9, a tetrasubstituted aromatic benzene ring with two proton signals in a meta relationship at δ_H 6.50 and 6.41 (J = 3 Hz), and an unsaturated terpenoid chain with four double bonds and five methyl groups at δ_C 16.2, 16.2, 16.3, 17.8, and 25.6. All these features were common to both **1** and **2**. However, **2** uniquely contained a singlet at δ_H 3.66 (3H) that correlated with a signal at δ_C 52.5. Furthermore, the pseudo-molecular ion of m/z 455.316 observed in the positive-mode ESI MS of **2** was higher than that of **1** by 32 Da. We concluded that the two natural products differed by the addition of a methoxy group. Since the geranylgeranyl moiety and the two aromatic protons within **1** were preserved in **2**, we deduced that the methoxy group was attached as either a phenolic ether or as an ester of the ring-opened lactone. In the HMBC spectrum (Supporting Information), an absence of correlations between the methoxy group and the aromatic ring ruled out the ether structures. Meanwhile, a ^3J coupling observed between the methyl group and the carbonyl (Figure 2) enabled us to conclusively elucidate **2** as the methyl ester that we named caudiquinol.

Table 1. ^1H NMR (500 MHz) and ^{13}C NMR (126 MHz) data for compound **2** in $CDCl_3$.

Position	δ_C, Type	δ_H, Type
1	147.0, C	
2	121.5, C	
3	115.0, CH	6.50, d
4	149.0, C	
5	115.9, CH	6.41, d
6	131.3, C	
7	29.2, CH_2	3.27, d
8	121.6, CH	5.20–5.25, m
9	138.0, C	
10	37.2, CH_2	1.90–2.02, m
11	26.5, CH_2	1.90–2.02, m
12	123.9, CH	5.00–5.09, m
13	135.3, C	
14	39.69, CH_2	1.90–2.02, m
15	26.6, CH_2	1.90–2.02, m
16	124.2, CH	5.00–5.09, m
17	135.0, C	
18	39.73, CH_2	1.90–2.02, m
19	26.7, CH_2	1.90–2.02, m
20	124.4, CH	5.00–5.09, m

Table 1. *Cont.*

Position	δ_C, Type	δ_H, Type
21	130.6, C	
22	25.6, CH_3	1.54, s
1′	39.75, CH_2	3.53, s
2′	173.9, C	
3′	52.5, CH_3	3.66, s
1″	16.3, CH_3	1.66, s
2″	16.2, CH_3	1.61, s
3‴	16.2, CH_3	1.54, s
4″	17.8, CH_3	1.54, s

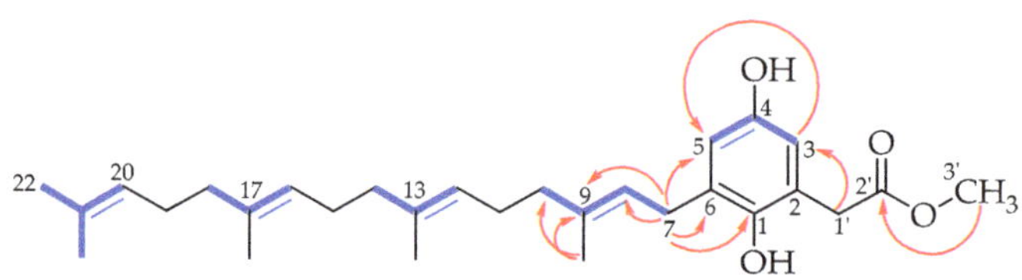

Figure 2. Observed COSY (blue; bold) and HMBC (red arrows) correlations in caudiquinol.

3. Discussion

The lactone versus methyl ester relationship between meroterpenoids **1** and **2** was precedented by a shorter C10 geranyl sidechain by the pair of natural products denudalide (**3**) and denudaquinol (**4**) (Figure 3) isolated from fruits of *Magnolia denudata* of the Magnoliaceae family [9]. Although denudalide could give rise to denudaquinol, in principle, by methanolic hydrolysis, the authors could not demonstrate this conversion in the laboratory. Similarly, given our mild HPLC conditions (aq MeOH; pH 7; rt), we believe that caudiquinol is an authentic natural product.

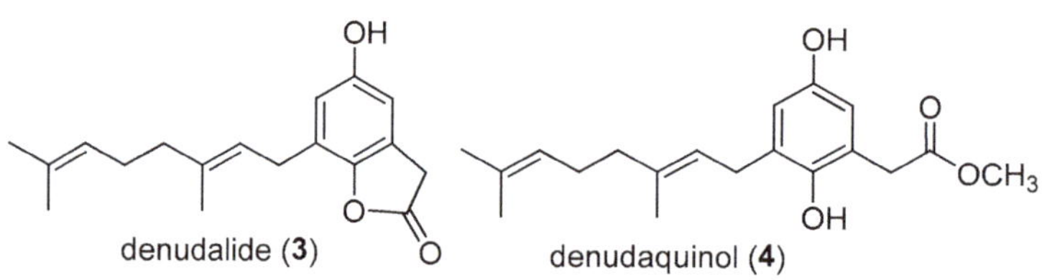

Figure 3. The geranyl meroterpenoids denudalide (**3**) and denudaquinol (**4**).

Meroterpenoids that contain units larger than the simple C5 prenyl (dimethylallyl) group typically undergo further transformations, such as oxidation or cyclization, whereas **1–4** feature unmodified C10 or C20 sidechains. The discovery of **1–4** from four different tree families suggests a common biosynthetic pathway to such meroterpenoids within the plant kingdom, and that congeners with an intermediate C15 sidechain are also likely to be found in nature. Furthermore, in addition to **1** and **2**, we are aware of only four other meroterpenoids (**5–8** (Figure 4)) of plant origin with an intact C20 geranylgeranyl unit [7,10–12].

Purified **1** and **2** were inactive against seven Gram-positive bacterial strains assayed (MSSA 25923, methicillin-resistant *S. aureus* (MRSA) 13373, SA XU212, SA 1199B, SA RN4220, *Enterococcus faecalis* 12967, and *E. faecalis* 51299) with MIC values of >250 μM or against the A549 lung cancer cell line at 100 μM. We did not have access to the HIV virus or the SFME cell line against which **1**, **3**, and **4** were reported to be active. We conclude the original antibiotic activity of the crude extract arises from other components within the mixture.

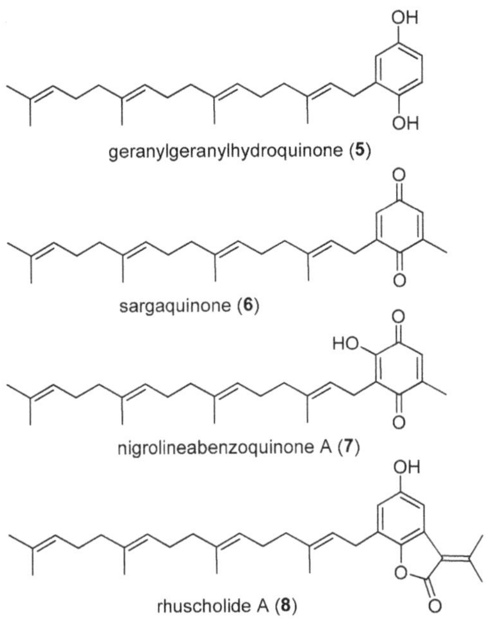

Figure 4. Plant meroterpenoids, other than **1** and **2**, with an intact geranylgeranyl sidechain.

4. Materials and Methods

General experimental procedures: Vacuum liquid chromatography (VLC) was performed using dry silica gel 60 $PF_{254+366}$ (Merck, London, UK). LC-QToF-MS/MS data were acquired using an Agilent (Santa Clara, CA, USA) 6546 Quadrupole/Time-of-Flight (Q-ToF) mass spectrometer with 1290 UHPLC, equipped with a Phenomenex Kintex C_{18} column (100 × 2.1 mm, 2.6 μm, 100 Å) using deionized H_2O/MeCN (95:5 to 5:95 gradient with 0.1% HCO_2H over 5 min 50 s) eluent mixture. Preparative HPLC was performed using a recycling LaboACE LC-5060 series HPLC instrument fitted with a C18 column (20 × 500 mm, 10 μm, 120 Å) (JAI, Tokyo, Japan) and a flow rate of 10 mL/min. One- and two-dimensional (1D and 2D) NMR spectra were recorded with a 500 MHz spectrometer (Bruker, Billerica, MA, USA) using a chloroform-d solvent. The spectra were processed using the MestReNova 14.1 software. UV–visible absorption spectra were recorded with a Perkin Elmer (Shelton, CT, USA) UV/Vis Lambda 365 spectrophotometer. IR absorbance spectra were recorded with a Perkin Elmer FT-IR System Spectrum BX.

Plant material: Leaves of *Garcinia caudiculata* Ridl. were collected at Lundu, Sarawak, Malaysia (1°37′15″ N, 109°45′57″ E). The samples were taxonomically identified by one of the authors, Stephen Ping Teo, and deposited as a voucher specimen STP86 at the Forest Herbarium (SAR), the Forest Department Sarawak. The leaves were air-dried and ground into a fine powder before storage.

Extraction and isolation: The dry, powdered leaves (100 g) were extracted by macerating them with CH_2Cl_2 at room temperature (1 L × 3 times for 24 h each). The extracts were filtered, and the supernatant was concentrated under reduced pressure at 40 °C to obtain the combined crude extract (10 g). Half of the crude extract (5 g) was separated using VLC via silica gel into 16 fractions using a mixture of two solvents (hexane and ethyl acetate) of increasing polarity. Of these, fractions 10 (554 mg) and 11 (459 mg), eluted with 50% and 30% hexane, respectively, were purified by preparative HPLC, using 1 mL volume injections. Fraction 10 was injected at 10 mg/mL and was eluted using 100% MeOH to yield compound **1** (7.4 mg). Fraction 11 was injected at 12 mg/mL and eluted using a gradient of deionized H_2O/MeOH of 20%:80% for 10 min followed by a linear gradient reaching 100% MeOH at 15 min to yield caudiquinol **2** (6.0 mg).

Methyl 3-[(2E, 6E, 10E, 14E)-3,7,11,15-tetramethyl-2,6,10,14-hexadecatetraen-1-yl]-2,5-dihydroxybenzeneacetate (caudiquinol **2***)*. 6.0 mg; yellow oil; UV λ_{max} (MeOH): 220, 232, and 294 nm; IR: 3387, 2914, 1714, and 1435 cm^{-1}; m/z 455.316 [M + H]$^+$ (calcd. for $C_{29}H_{43}O_4$; 455.316; Δ = 0 ppm); ^{1}H NMR (500 MHz): δ 6.50 (d, *J* = 3.1 Hz, 1H), 6.41 (d, *J* = 3.1 Hz, 1H), 5.20–5.25 (m, 1H), 5.00–5.09 (m, 3H), 3.66 (s, 3H), 3.53 (s, 2H), 3.27 (d, *J* = 7.0 Hz, 2H), 1.90–2.02 (m, 12H), 1.66 (s, 3H), 1.61 (s, 3H), and 1.53 (s, 9H). ^{13}C NMR (126 MHz): δ 173.9, 149.0, 147.0, 138.0, 135.3, 135.0, 131.3, 130.6, 124.4, 124.2, 123.9, 121.5, 121.6, 115.9, 115.0, 52.5, 39.75, 39.73, 39.69, 37.2, 29.2, 26.7, 26.6, 26.5, 25.6, 17.8, 16.3, 16.20, and 16.15.

Supplementary Materials: The following supporting information can be downloaded at https://www.mdpi.com/article/10.3390/molecules29153613/s1. Characterization data for **1** and **2** comprising NMR, MS, IR, and UV spectra.

Author Contributions: Conceptualization, S.P.T., S.G. and A.G.; methodology, all; investigation, M.V.; resources, S.P.T., S.G. and A.G.; data curation, M.V.; writing—original draft preparation, A.G.; writing—review and editing, M.V., S.P.T., P.E.d.R., S.G. and A.G.; visualization, M.V.; supervision, P.E.d.R., S.G. and A.G.; project administration, S.P.T., P.E.d.R., S.G. and A.G.; funding acquisition, S.P.T., S.G. and A.G. All authors have read and agreed to the published version of the manuscript.

Funding: This research received no external funding. We thank the University of East Anglia for awarding a studentship to M.V.

Institutional Review Board Statement: Not applicable.

Informed Consent Statement: Not applicable.

Data Availability Statement: The data are contained within this article and the Supplementary Materials.

Conflicts of Interest: The authors declare no conflicts of interest.

References

1. *Garcinia* L. Plants of the World Online. Available online: https://powo.science.kew.org/taxon/urn:lsid:ipni.org:names:19345-1 (accessed on 24 May 2024).
2. Espirito Santo, B.L.S.D.; Santana, L.F.; Kato Junior, W.H.; de Araújo, F.O.; Bogo, D.; Freitas, K.C.; Guimarães, R.C.A.; Hiane, P.A.; Pott, A.; Filiú, W.F.O.; et al. Medicinal Potential of *Garcinia* Species and Their Compounds. *Molecules* **2020**, *25*, 4513. [CrossRef] [PubMed]
3. Brito, L.C.; Marques, A.M.; da Camillo, F.C.; Figueiredo, M.R. *Garcinia* Spp: Products and by-products with potential pharmacological application in cancer. *Food Biosci.* **2022**, *50*, 102110. [CrossRef]
4. Nchiozem-Ngnitedem, V.-A.; Mukavi, J.; Omosa, L.K.; Kuete, V. Phytochemistry and antibacterial potential of the genus *Garcinia*. In *Advances in Botanical Research*; Kuete, V., Ed.; Academic Press: London, UK, 2023; Volume 107, pp. 105–175.
5. Ridley, H.N. Additions to the Flora of Borneo and Other Malay Islands: VI. *Bull. Misc. Inf. (Royal Gardens Kew)* **1938**, *1938*, 110–123. [CrossRef]
6. Vieira, P.C.; Gottlieb, O.R.; Gottlieb, H.E. Tocotrienols from *Iryanthera grandis*. *Phytochemistry* **1983**, *22*, 2281–2286. [CrossRef]
7. Gu, Q.; Wang, R.R.; Zhang, X.M.; Wang, Y.H.; Zheng, Y.T.; Zhou, J.; Chen, J.J. A New Benzofuranone and Anti-HIV Constituents from the Stems of *Rhus chinensis*. *Planta Med.* **2007**, *73*, 279–282. [CrossRef] [PubMed]
8. Li, T.Z.; Geng, C.A.; Chen, J.J. First total synthesis of rhuscholide A, glabralide B and denudalide. *Tetrahedron Lett.* **2019**, *60*, 151059. [CrossRef]
9. Noshita, T.; Kiyota, H.; Kidachi, Y.; Ryoyama, K.; Funayama, S.; Hanada, K.; Murayama, T. New Cytotoxic Phenolic Derivatives from Matured Fruits of *Magnolia denudata*. *Biosci. Biotechnol. Biochem.* **2009**, *73*, 726–728. [CrossRef] [PubMed]
10. Reynolds, G.W.; Rodriguez, E. Prenylated hydroquinones: Contact allergens from trichomes of *Phacelia minor* and *P. parryi*. *Phytochemistry* **1981**, *20*, 1365–1366. [CrossRef]
11. Voutquenne, L.; Lavaud, C.; Massiot, G.; Sevenet, T.; Hadi, H.A. Cytotoxic polyisoprenes and glycosides of long-chain fatty alcohols from *Dimocarpus fumatus*. *Phytochemistry* **1999**, *50*, 63–69. [CrossRef] [PubMed]
12. Rukachaisirikul, V.; Kamkaew, M.; Sukavisit, D.; Phongpaichit, S.; Sawangchote, P.; Taylor, W.C. Antibacterial Xanthones from the Leaves of *Garcinia nigrolineata*. *J. Nat. Prod.* **2003**, *66*, 1531–1535. [CrossRef] [PubMed]

molecules

MDPI

Communication

Secobutanolides Isolated from *Lindera obtusiloba* Stem and Their Anti-Inflammatory Activity

Hye Jin Yang [1,2], Young-Sang Koh [3], MinKyun Na [1,*] and Wei Li [2,*]

1 College of Pharmacy, Chungnam National University, Daejeon 305-764, Republic of Korea; hjyang@kiom.re.kr
2 Korean Medicine (KM) Application Center, Korea Institute of Oriental Medicine, Daegu 41062, Republic of Korea
3 School of Medicine and Brain Korea 21 PLUS Program, Institute of Medical Science, Jeju National University, Jeju 690-756, Republic of Korea; yskoh7@jejunu.ac.kr
* Correspondence: mkna@cnu.ac.kr (M.N.); liwei1986@kiom.re.kr (W.L.)

Abstract: In this study, a new secobutanolide, named secosubamolide B (**3**), along with three previously known secobutanolides (**1**, **2**, and **4**), were successfully isolated from a methanol extract of the stem of *Lindera obtusiloba*. The chemical structures of these compounds were elucidated through the analysis of spectroscopic data, and then compared with the existing literature to confirm their identities. Furthermore, the anti-inflammatory effect of these isolated compounds on bone marrow-derived dendritic cells stimulated by lipopolysaccharide (LPS) was evaluated. Compounds **1–3** showed the significant suppression of LPS-triggered IL-6 and IL-12 p40 production, with IC_{50} values between 1.8 and 24.1 μM. These findings may provide a scientific foundation for developing anti-inflammatory agents from *L. obtusiloba*.

Keywords: *Lindera obtusiloba*; Lauraceae; secobutanolide; anti-inflammatory activity

1. Introduction

Traditional Oriental medicine presents an abundant source of potential therapeutic agents that have yet to be fully explored by Western medicine. In Korea, decoctions prepared from the wood and bark of *Lindera obtusiloba* (Lauraceae) have been traditionally utilized for their anti-inflammatory properties, as well as for enhancing blood circulation and treating skin conditions such as dermatitis [1–3]. Additionally, herbal infusions made from *L. obtusiloba* have been historically used in the management of chronic liver diseases. The plant's broad medicinal applications are largely due to its diverse range of bioactive compounds. Among the bioactive components identified in the leaves of *L. obtusiloba* are lignans and butanolides, which have been shown to possess significant anti-tumor, anti-allergic, and anti-inflammatory effects. Further research has also highlighted the plant's antioxidant, hepatoprotective, and anti-coagulant properties, underscoring its considerable therapeutic potential [4–6]. In particular, secobutanolides from genus Lindera have shown promising anticancer effects against HepG2, P-388, A549 and HT-29 cancer cell lines, and also shown collagen-induced platelet aggregation inhibitory activity [7–9].

In our recent study, we set out to isolate and identify a novel butanolide that could expand the known spectrum of active compounds derived from *L. obtusiloba*. Through our investigation, we successfully isolated and elucidated the chemical structure of a new secobutanolide, in addition to three previously known secobutanolides, from a methanol extract of the plant's stem. Moreover, we explored the anti-inflammatory potential of these isolated compounds by assessing their effects on lipopolysaccharide (LPS)-stimulated bone marrow-derived dendritic cells (BMDCs).

Lipopolysaccharide (LPS), a predominant component of the outer membrane of gram-negative bacteria, is known for its ability to suppress inflammatory responses in vivo. However, LPS is also capable of inducing the release of various inflammatory cytokines in different cell types, leading to a robust acute inflammatory response to pathogens. Among the cytokines released, interleukin (IL)-6, IL-12, and tumor necrosis factor (TNF)-α are particularly important as key mediators of inflammation. IL-6 is especially noteworthy due to its dual role in promoting and regulating inflammation, making it a central player in many inflammatory diseases. IL-12 is crucial in the initiation and regulation of the cellular immune response, with its p40 and p80 subunits exhibiting distinct biological activities, independent of IL-12 and IL-23. Dendritic cells (DCs) are essential in orchestrating immune responses, primarily through their ability to modulate T-cell functions [10–13]. To uncover the anti-inflammatory components of *L. obtusiloba*, we tested the isolated compounds for their ability to modulate the LPS-induced expression of the pro-inflammatory cytokines IL-6 and IL-12 p40 in BMDCs. This investigation has contributed to a deeper understanding of the anti-inflammatory potential of compounds derived from *L. obtusiloba*, further supporting its use in traditional medicine and its potential use in therapeutic applications.

2. Results and Discussion

2.1. Isolation and Structural Characterization

Four secobutanolides were successfully isolated from a MeOH extract of *L. obtusiloba* stems. These compounds were identified as secomahubanolide (**1**) [14], secosubamolide A (**2**) [15], secosubamolide B (**3**), and secoisolitsealiicolide B (**4**) [16] (Figure 1). Notably, secosubamolide B (**3**) is a newly discovered compound, and all four compounds were identified in *L. obtusiloba* for the first time. This study represents the first thorough chemical investigation of secobutanolides in *L. obtusiloba*.

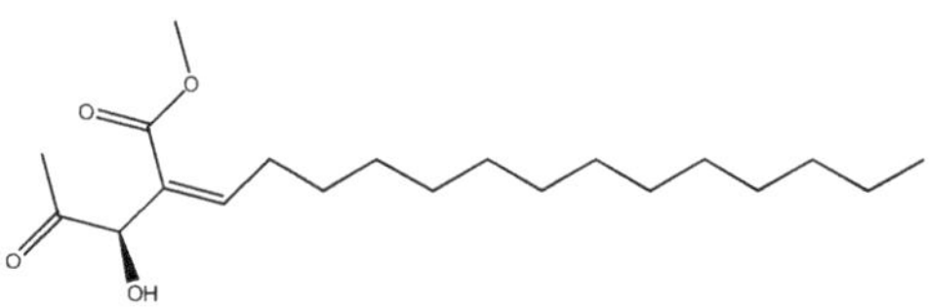

1

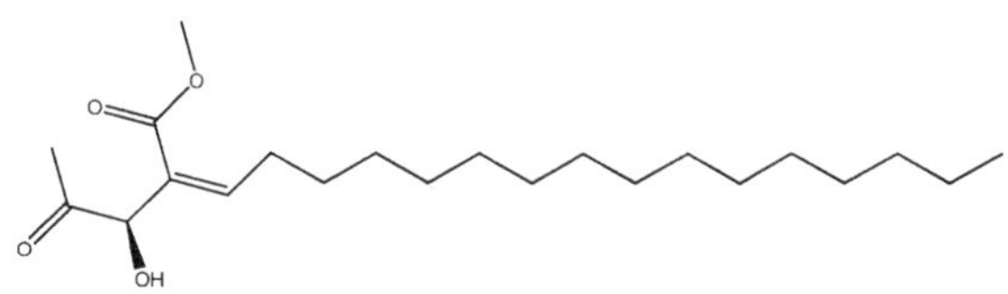

2

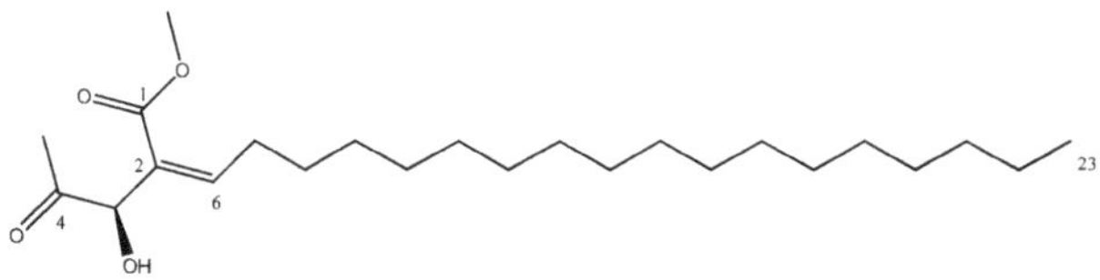

3

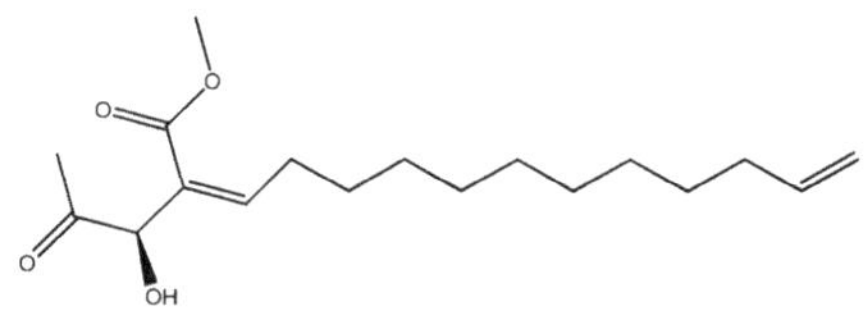

4

Figure 1. Structures of compounds **1–4** from the stem of *L. obtusiloba*.

Compound **1** was isolated as a yellow oil. High-resolution electrospray ionization time-of-flight mass spectrometry (HR-ESI-TOF-MS) established its molecular formula as $C_{20}H_{36}O_4$, with an observed m/z of 341.2690 ($[M + H]^+$). The ^{1}H-NMR spectrum of compound **1** (Table 1) displayed three characteristic signals: an olefinic proton at δ_H 7.08 (1H, t, J = 7.6 Hz), an oxymethine proton at δ_H 4.90 (1H, d, J = 5.0 Hz), and a methoxy group at δ_H 3.73 (3H, s). Correspondingly, the ^{13}C-NMR spectrum (Table 1) revealed eight signals, including quaternary carbon atoms of a double bond at δ_C 129.7 (C-2) and 149.1 (C-6), a ketone group at δ_C 206.4 (C-4), and a carboxyl group at δ_C 166.6 (C-1), consistent with the structure of secomahubanolide, as reported by Cheng in 2005. These data indicated the presence of a secobutanolide skeleton. Compound **1** exhibited the *E*-form geometry of the trisubstituted double bond [δ_H 7.08 (1H, t, J = 7.6 Hz)] and demonstrated optical activity with $[\alpha]_D^{25}$: $-13.1°$ (c 0.32, $CHCl_3$), confirming a 3*R* configuration [15,16]. Based on these observations, the structure of compound **1** was identified as secomahubanolide.

Table 1. ^{1}H and ^{13}C NMR spectroscopic data for compounds **1–3** in $CDCl_3$.

	1		**2**		**3**	
	δ_H ᵃ (*J*/Hz)	δ_C ᵇ	δ_H ᵃ (*J*/Hz)	δ_C ᵇ	δ_H ᵃ (*J*/Hz)	δ_C ᵇ
1	-	166.6	-	166.2	-	166.6
2	-	129.7	-	132.2	-	129.7
3	4.90 (d, 5.0)	73.3	4.84 (d, 5.0)	72.2	4.89 (d, 5.0)	73.3
4	-	206.4	-	208.8	-	206.5
5	2.15 (s)	24.71	2.11 (s)	26.0	2.14 (s)	25.8
6	7.08 (t, 7.4)	149.1	6.91 (t, 7.2)	147.2	7.07 (t, 7.2)	149.2
7	2.35 (q, 7.4)	28.6	2.28 (q, 7.2)	28.0	2.34 (q, 7.2)	27.9
8	1.51 (m)	28.8	1.41 (m)	28.4	1.41 (m)	28.1
9	1.25 (m)	29.2	1.24 (m)	28.5–29.0	1.24 (m)	29.0–29.5
10	1.25 (m)	29.2	1.24 (m)	28.5–29.0	1.24 (m)	29.0–29.5
11	1.25 (m)	29.3	1.24 (m)	28.5–29.0	1.24 (m)	29.0–29.5
12	1.25 (m)	29.9	1.24 (m)	28.5–29.0	1.24 (m)	29.0–29.5
13	1.25 (m)	29.4	1.24 (m)	28.5–29.0	1.24 (m)	29.0–29.5
14	1.25 (m)	29.4	1.24 (m)	28.5–29.0	1.24 (m)	29.0–29.5
15	1.25 (m)	29.4	1.24 (m)	28.5–29.0	1.24 (m)	29.0–29.5
16	1.25 (m)	29.2	1.24 (m)	28.5–29.0	1.24 (m)	29.0–29.5
17	1.25 (m)	31.8	1.24 (m)	28.5–29.0	1.24 (m)	29.0–29.5
18	1.25 (m)	22.5	1.24 (m)	28.5–29.0	1.24 (m)	29.0–29.5
19	0.87 (t, 6.8)	13.9	1.24 (m)	31.5	1.24 (m)	29.0–29.5
20			1.24 (m)	22.5	1.24 (m)	29.0–29.5
21			0.86 (t, 6.7)	13.9	1.24 (m)	31.8
22					1.24 (m)	22.6
23					0.87 (t, 6.2)	13.9
1-OCH_3	3.73 (s)	51.9	3.62 (s)	51.8	3.72 (s)	51.9

J values (Hz) are given in parentheses. ᵃ 600 MHz. ᵇ 150 MHz.

Compound **2** was obtained as a yellow oil. The molecular formula was established as $C_{22}H_{40}O_4$ by high-resolution electrospray ionization time-of-flight mass spectrometry (HR-ESI-TOF-MS; *m*/*z* 369.2996 ([M + H]$^+$)). Both the ^{1}H-NMR spectrum [δ_H 6.91 (1H, t, *J* = 7.2 Hz), δ_H 4.84 (1H, d, *J* = 5.0 Hz), and δ_H 3.62 (3H, s)] and ^{13}C-NMR spectrum [δ_C 132.2 (C-2), 147.2 (C-6), 166.2 (C-1), and 208.8 (C-4)] were similar to that of secomahubanolide (**1**) (Table 1). Compound **2** exhibited the *E*-form geometry of the trisubstituted double bond, as indicated by the ^{1}H NMR signal at δ_H 6.91 (1H, t, *J* = 7.6 Hz). Its optical rotation was measured at $[\alpha]_D^{25}$: −11.7° (c 0.15, $CHCl_3$), suggesting a 3*R* configuration [15,16]. Additionally, compound **2** featured two extra methylene units at δ_H 1.24 in its side chain compared to secomahubanolide (**1**). Based on these characteristics, the structure of compound **2** was determined to be secosubamolide A.

Compound **3** was also isolated as a yellow oil, with its molecular formula established as $C_{24}H_{44}O_4$ through high-resolution electrospray ionization time-of-flight mass spectrometry (HR-ESI-TOF-MS), yielding an *m*/*z* of 397.3319 ([M + H]$^+$). The HR-MS spectrum clearly shows that compound **3** has two more methylene groups than compound **2**, just as compound **2** has two more methylene groups than compound **1**. The HR-MS data for all three compounds exhibit a consistent increasing trend. Therefore, we can conclude that compound **3** is compound **2** with two additional methylene groups on the side chain, and this inference is further confirmed by the NMR spectra. The ^{1}H NMR (Table 1) spectrum revealed signals at δ_H 7.07 (1H, t, *J* = 7.2 Hz), δ_H 4.89 (1H, d, *J* = 5.0 Hz), and δ_H 3.72 (3H, s), while the ^{13}C NMR spectrum showed resonances at δ_C 132.0 (C-2), 146.8 (C-6), 166.2 (C-1), and 208.9 (C-4), closely resembling those of secomahubanolide (**1**) and secosubamolide A (**2**). Compound **3** maintained the *E*-form geometry of the trisubstituted double bond, as seen in the signal at δ_H 6.82 (1H, t, *J* = 7.6 Hz), and exhibited an optical rotation of $[\alpha]_D^{25}$: −21.7° (c 0.52, $CHCl_3$), confirming the 3*R* configuration [15,16]. Moreover, compound **3** displayed two additional methylene units at δ_H 1.24 in its side chain compared to secosubamolide A (**2**).

Based on these observations, the structure of compound **3** was identified as secosubamolide B.

2.2. Anti-Inflammatory Activity of Compounds **1–3**

Prior to evaluating the anti-inflammatory potential of secobutanolides isolated from the methanol extract of *L. obtusiloba*, we initially examined the effects of compounds **1–3** (at a 50 μM concentration) on BMDC cell viability using an MTT colorimetric assay. The results showed that these compounds did not exhibit any cytotoxicity at the concentrations tested (data not shown).

Following this, the effects of compounds **1–3** on the production of IL-6 and IL-12 p40 were assessed at various concentrations ranging from 1 to 50 μM. In these experiments, 4-(4-Fluorophenyl)-2-(4-methylsulfinylphenyl)-5-(4-pyridyl)-1H-imidazole (SB203580) was utilized as a positive control [14]. SB203580 demonstrated a marked ability to inhibit the production of both IL-6 and IL-12 p40, with calculated IC_{50} values of 3.5 μM and 5.0 μM, respectively (Figure 2). The tested secobutanolides, identified as compounds **1–3**, also showed significant inhibitory effects on IL-6 production, with IC_{50} values determined to be 1.8 μM, 3.4 μM, and 24.1 μM, respectively. Similarly, these compounds exerted strong inhibitory activity against IL-12 p40 production, with IC_{50} values of 3.1 μM, 2.9 μM, and 16.3 μM, as shown in Table 2.

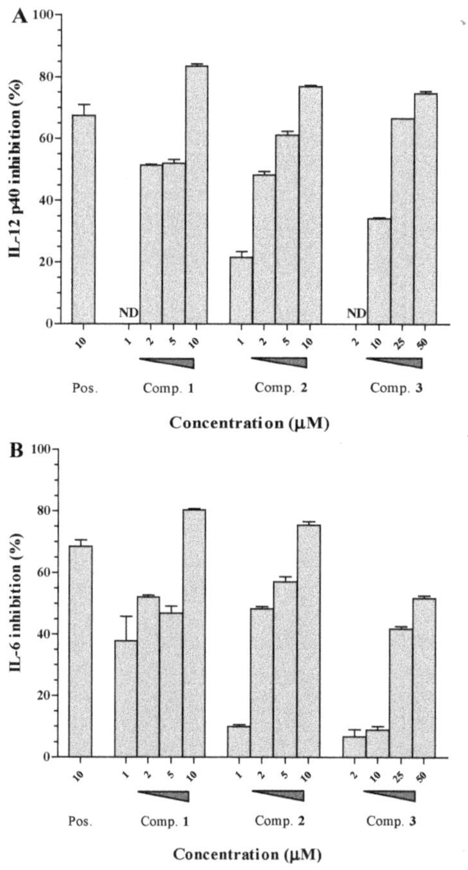

Figure 2. Effect of compounds **1–3** on the production of IL-12 p40 (**A**) and IL-6 (**B**) in LPS-stimulated BMDCs at concentrations of 1.0–50.0 μM. The results are shown as inhibition rates (%) relative to the levels observed in vehicle-treated DCs. Pos: SB203580. ND = Not Detected.

Table 2. Anti-inflammatory effects of compounds **1–3**.

Compound	IC_{50} (μM) [a]	
	IL-12 p40	IL-6
1	3.1 ± 0.4	1.8 ± 0.1
2	2.9 ± 0.6	3.4 ± 0.2
3	16.3 ± 0.6	24.1 ± 1.2
SB203580 [b]	5.0 ± 0.1	3.5 ± 0.2

The [a] IC_{50} values for selected compounds are given in column IL-12 p40 and IL-6. [b] Positive control.

This study not only highlights the robust anti-inflammatory properties of these secobutanolides, but also provides important insights into their structure–activity relationships. Notably, the inhibitory effects were significantly amplified by the presence of unsaturated aliphatic substituents on the acylglycerol framework. Moreover, the activity of these compounds appeared to increase proportionally with the length of the fatty acid side chains, indicating that these structural characteristics are critical for their bioactivity. These findings could be pivotal in informing the design and synthesis of new anti-inflammatory agents derived from the secobutanolide structure, offering a promising direction for future drug development efforts.

3. Materials and Methods

3.1. General Information

Column chromatography: silica gel (Kieselgel 60, 70−230, and 230−400 mesh, Merck, Darmstadt, Germany); YMC RP-18 resins, and thin-layer chromatography (TLC): silica-gel 60 F_{254} and RP-18 F_{254}S plates (both 0.25 mm, Merck, Darmstadt, Germany); optical rotation: Jasco DIP-370 automatic polarimeter (Easton, MD, USA); NMR spectra: JEOL ECA 600 spectrometer (Tokyo, Japan); High-resolution electrospray ionization mass spectra (HR-MS): Agilent 6530 Accurate-Mass Q-TOF LC/MS system (Santa Clara, CA, USA).

3.2. Plant Material

In 2001, dried stems of *L. obtusiloba* were provided by Bomyeong Herbal Market (Seoul, Republic of Korea), Korea, and the voucher specimen (KM-000101) was deposited with the Korean Medicine (KM) Application Center, Korea Institute of Oriental Medicine, Korea.

3.3. Extraction and Isolation

The air-dried and chopped stems of *L. obtusiloba* (3.0 kg) were extracted three times with MeOH for 10 h in a water bath. The MeOH extract obtained was evaporated to dryness in a rotary evaporator, affording crude extract (185.3 g, 6.1% yield). The extract was dissolved in distilled water (1.5 L), and 1.2 L of the resulting solution was consecutively partitioned using $CHCl_3$, EtOAc, and *n*-BuOH; $CHCl_3$ (48.5 g), EtOAc (20.4 g), *n*-BuOH (36.8g), and H_2O fractions were obtained.

A portion of the $CHCl_3$ layer (48.5 g) was subjected to column chromatography using silica gel column chromatography and eluted with a stepwise gradient of EtOAc:MeOH (20:1–10:1–5:1–2:1–1:1, *v*/*v*) to yield three fractions (1a–c). Fraction 1a was chromatographed on a silica gel column using a gradient of *n*-Hexane:EtOAc (8:1–3:1 *v*/*v*) to afford subfractions (2a–c). Fraction 2b was further chromatographed on a silica gel chromatography column with $CHCl_3$:MeOH (90:1–70:1 *v*/*v*) to afford subfractions (3a–b). Fraction 3d was further chromatographed on a silica gel chromatography column with $CHCl_3$:MeOH (60:1 *v*/*v*) to yield compound **1** (12.1 mg). Fraction 3b was chromatographed on a silica gel column *n*-Hexane:acetone (8:1–7:1 *v*/*v*) to give compounds **2** (24.1 mg), **3** (2.4 mg), and **4** (49.9 mg).

Secomahubanolide (**1**): yellow oil; UV (MeCN): λ_{max} 215 nm; HR-ESI-TOF-MS *m*/*z* 341.2690 $[M + H]^+$; $[\alpha]_D^{25}$: −13.1° (c = 0.32, $CHCl_3$); IR (KBr, cm^{-1}) 3450, 1735, 1710 cm^{-1}; ^{1}H-NMR (600 MHz) and ^{13}C-NMR (150 MHz): see Table 1.

Secosubamolide A (**2**): yellow oil; UV (MeCN): λ_{max} 215 nm; HR-ESI-TOF-MS *m/z* 369.2996 [M + H]$^+$; $[\alpha]_D^{25}$: −11.7° (c = 0.15, $CHCl_3$); IR (KBr, cm^{-1}) 3450, 1735, 1710 cm^{-1}; ^{1}H-NMR (600 MHz) and ^{13}C-NMR (150 MHz): see Table 1.

Secosubamolide B (**3**): yellow oil; UV (MeCN): λ_{max} 215 nm; HR-ESI-TOF-MS *m/z* 397.3319 [M + H]$^+$; $[\alpha]_D^{25}$: −21.7° (c = 0.52, $CHCl_3$); IR (KBr, cm^{-1}) 3450, 1735, 1710 cm^{-1}; ^{1}H-NMR (600 MHz) and ^{13}C-NMR (150 MHz): see Table 1. HR-MS, UV, and NMR data of compound **3** can be found in the Supplementary Materials.

Secoisolitsealiicolide B (**4**): yellow oil; UV (MeOH): λ_{max} 213 nm; ^{1}H-NMR (400 MHz, $CDCl_3$, δ_H): 7.08 (1H, t, *J* = 7.5, H-6), 5.81 (1H, ddt, *J* = 17.5, 10.6, 6.8 Hz, H-16), 4.99 (1H, dd, *J* = 17.2, 2.1 Hz, H-17a), 4.93 (1H, dd, *J* = 10.3, 1.4 Hz, H-17b), 4.90 (1H, s, H-3), 4.03 (1H, s, 3-OH), 3.73 (3H, s, 1-OCH_3), 2.35 (2H, quartet, *J* = 7.6 Hz, H-7), 2.16 (3H, s, H-5), 2.04 (2H, quartet, *J* = 6.9 Hz, H-15), 1.28 (2H, br s, H-9), 1.28 (2H, br s, H-10), 1.28 (2H, br s, H-11), 1.28 (2H, br s, H-12), 1.28 (2H, br s, H-13), 1.28 (2H, br s, H-14); ^{13}C-NMR (100 MHz, $CDCl_3$, δ_C): 206.5 (C-4), 166.6 (C-1), 149.2 (C-6), 139.2 (C-16), 129.8 (C-2), 114.2 (C-17), 73.4 (C-3), 52.0 (1-OCH_3), 33.7 (C-15), 28.8 (C-7), 28.6 (C-8), 28.7–29.5 (C-9), 28.7–29.5 (C-10), 28.7–29.5 (C-11), 28.7–29.5 (C-12), 28.7–29.5 (C-13), 28.7–29.5 (C-14), 24.9 (C-5).

3.4. Cell Culture

The BMDCs were derived from wild-type C57BL/6 mice. Briefly, bone marrow cells were cultured in RPMI 1640 medium containing 10% heat-inactivated fetal bovine serum (FBS), 50 μM of 2-ME, and 2 mM of glutamine supplemented with 3% J558L hybridoma cell culture supernatant containing granulocyte macrophage colony-stimulating factor for dendritic cell generation. The culture medium was replaced with fresh medium every second day. On day 6 of culture, non-adherent cells and loosely adherent DC aggregates were harvested, washed and resuspended in RPMI 1640 supplemented with 5% FBS. All animal procedures were approved and performed according to the guidelines of the Institutional Animal Care and Use Committee of Jeju National University (#2010−0028).

3.5. Cytokine Production Measurements

The effect of isolated compounds on cytokine production by LPS-stimulated cells was determined using ELISA (BD PharMingen, San Diego, CA, USA). The final concentration of chemical solvents did not exceed 0.1%. BMDCs were cultured in DMEM medium containing macrophage colony-stimulating factor. On day 6 of incubation, BMDMs and BMDCs were harvested and seeded in 48-well plates at a density of 1×10^5 cells/0.5 mL, and then treated with the PTE for 1 h before stimulation with LPS (10 ng/mL). At the end of the incubation period, the IL-12 p40 and IL-6 levels in the medium were determined by ELISA. All experiments were performed in triplicate.

3.6. Statistical Analysis

The student's t-test and one-way ANOVA were used to determine the statistical significance of the differences between values for a variety of experimental and control groups. Data were expressed as the mean ± SD (at least in triplicate). The statistical significance was determined by one-way analysis of variance followed by Dunnett's multiple comparison tests, $p < 0.05$.

4. Conclusions

In this study, four compounds, including a newly discovered secobutanolide (compound **3**), were successfully isolated from a methanol extract of *Lindera obtusiloba*. The inhibitory effects of the isolated compounds (**1–3**) were assessed in terms of their ability to suppress the production of cytokines IL-12p40 and IL-6, which are stimulated by lipopolysaccharide (LPS) in bone marrow-derived dendritic cells (BMDCs). Notably, this research represents the first investigation into the secobutanolide components of *L. obtusiloba* and their potential anti-inflammatory properties. The compounds isolated in this study were identified from the plant for the first time, indicating their novelty. These findings suggest that secobutanolides are active constituents of *L. obtusiloba* and may serve as a scientific foundation for developing new anti-inflammatory agents derived from this plant.

Supplementary Materials: The following supporting information can be downloaded at: https://www.mdpi.com/article/10.3390/molecules29184292/s1, Figure S1. HR-ESI-MS spectrum of compound 3; Figure S2. UV spectrum of compound 3 MeCN; Figure S3. ^{1}H-NMR spectrum of compound 3 (600 MHz $CDCl_3$); Figure S4. ^{13}C-NMR spectrum of compound 3 (150 MHz $CDCl_3$).

Author Contributions: H.J.Y.: Conceptualization, methodology, validation, formal analysis, writing—original draft preparation; Y.-S.K.: methodology, validation; M.N.: writing—review and editing, visualization, supervision, writing—original draft preparation; W.L.: resources, funding, writing—original draft preparation. All authors have read and agreed to the published version of the manuscript.

Funding: This research was supported by a grant (KSN2411011) awarded to the Korean Institute of Oriental Medicine by the Ministry of Education, Science and Technology (MEST), Republic of Korea.

Institutional Review Board Statement: Not applicable.

Data Availability Statement: The original contributions presented in the study are included in the article/Supplementary Materials, further inquiries can be directed to the corresponding author/s.

Conflicts of Interest: The authors declare no competing financial interests.

References

1. Kwon, D.J.; Kim, J.K.; Bae, Y.S. Essential oils from leaves and twigs of Lindera obtusiloba. *J. Korean For. Soc.* **2007**, *96*, 65–69.
2. Seo, K.H.; Baek, M.Y.; Lee, D.Y.; Cho, J.G. Isolation of Flavonoids and Lignans from the Stem Wood of Lindera Obtusioba Blume. *J. Appl. Biol. Chem.* **2011**, *54*, 178–183. [CrossRef]
3. Kwon, H.C.; Choi, S.U.; Lee, J.O.; Bae, K.H.; Zee, O.P.; Lee, K.R. Two new lignans from Lindera obtusiloba blume. *Arch. Pharm. Res.* **1999**, *22*, 417–422. [CrossRef] [PubMed]
4. Lee, J.O.; Oak, M.H.; Jung, S.H.; Park, D.H.; Auger, C.; Kim, K.R.; Lee, S.W.; Schini-Kerth, V.B. An ethanolic extract of Lindera obtusiloba stems causes NO-mediated endothelium-dependent relaxations in rat aortic rings and prevents angiotensin II-induced hypertension and endothelial dysfunction in rats. *Naunyn. Schmiedebergs. Arch. Pharmacol.* **2011**, *383*, 635–645. [CrossRef] [PubMed]
5. Lee, K.Y.; Kim, S.H.; Jeong, E.J.; Park, J.H.; Kim, S.H.; Kim, Y.C.; Sung, S.H. New secoisolariciresinol derivatives from Lindera obtusiloba stems and their neuroprotective activities. *Planta Med.* **2010**, *76*, 294–297. [CrossRef] [PubMed]
6. Bang, C.Y.; Choung, S.Y. Antioxidant and whitening effects of Lindera obtusiloba BL. 70% EtOH extract. *Planta Med.* **2009**, *75*, PI27. [CrossRef]
7. Kwon, H.C.; Baek, N.I.; Choi, S.U.; Lee, K.R. New cytotoxic butanolides from Lindera obtusiloba BLUME. *Chem. Pharm. Bull.* **2000**, *48*, 614–616. [CrossRef] [PubMed]
8. Tsai, I.L.; Hung, C.H.; Duh, C.Y.; Chen, I.S. Cytotoxic butanolides and secobutanolides from the stem wood of Formosan Lindera communis. *Planta Med.* **2002**, *68*, 142–145. [CrossRef] [PubMed]
9. Lin, C.T.; Chu, F.H.; Chang, S.T.; Chueh, P.J.; Su, Y.C.; Wu, K.T.; Wang, S.Y. Secoaggregatalactone-A from Lindera aggregata induces apoptosis in human hepatoma hep G2 cells. *Planta Med.* **2007**, 1548−1553. [CrossRef] [PubMed]
10. Weinstein, D.L.; O'Neill, B.L.; Metcalf, E.S. Salmonella typhi stimulation of human intestinal epithelial cells induces secretion of epithelial cell-derived interleukin-6. *Infect. Immun.* **1997**, *65*, 395–404. [CrossRef] [PubMed]
11. Mannon, P.J.; Fuss, I.J.; Mayer, L.; Elson, C.O.; Sandborn, W.J.; Present, D.; Dolin, B.; Goodman, N.; Groden, C.; Hornung, R.L.; et al. Anti-interleukin-12 antibody for active Crohn's disease. *N. Engl. J. Med.* **2004**, *351*, 2069–2079. [CrossRef] [PubMed]
12. Cooper, A.M.; Khader, S.A. IL-12p40: An inherently agonistic cytokine. *Trends Immunol.* **2007**, *28*, 33–38. [CrossRef] [PubMed]
13. Tung, N.H.; Quang, T.H.; Son, J.H.; Koo, J.E.; Hong, H.J.; Koh, Y.S.; Song, G.Y.; Kim, Y.H. Inhibitory effect of ginsenosides from steamed ginseng-leaves and flowers on the LPS-stimulated IL-12 production in bone marrow-derived dendritic cells. *Arch. Pharm. Res.* **2011**, *34*, 681–685. [CrossRef] [PubMed]
14. Cheng, M.J.; Tsai, I.L.; Lee, S.J.; Jayaprakasam, B.; Chen, I.S. Steryl epoxide, secobutanolide and butanolides from the stem wood of Machilus zuihoensis. *Phytochemistry* **2005**, *66*, 1180–1185. [CrossRef] [PubMed]
15. da Costa-Silva, T.A.; Conserva, G.A.A.; Galisteo, A.J., Jr.; Tempone, A.G.; Lago, J.H.G. Antileishmanial activity and immunomodulatory effect of secosubamolide, a butanolide isolated from Nectandra oppositifolia (Lauraceae). *J. Venom. Anim. Toxins. Incl. Trop. Dis.* **2019**, e20190008. [CrossRef] [PubMed]
16. Cheng, M.J.; Wang, T.A.; Lee, S.J.; Chen, I.S. A new butanolide and a new secobutanolide from Litsea lii var. nunkao-tahangensis. *Nat. Prod. Res.* **2010**, *24*, 647–656. [CrossRef] [PubMed]

molecules

MDPI

Review

The Mediterranean Species *Calendula officinalis* and *Foeniculum vulgare* as Valuable Source of Bioactive Compounds

Filomena Monica Vella [1,*], Domenico Pignone [2] and Bruna Laratta [1,*]

[1] National Research Council (CNR), Institute of Biosciences and BioResources (IBBR), Via P. Castellino 111, 80131 Naples, Italy
[2] Institute for Veterinary and Agri-Food Bioethics (IBV-A), 00054 Fiumicino, Italy; direttore@istitutoibva.it
* Correspondence: filomenamonica.vella@cnr.it (F.M.V.); bruna.laratta@cnr.it (B.L.)

Abstract: Research studies on plant secondary metabolites have increased over the last decades as a consequence of the growing consumer demand for natural products in pharmaceutics and therapeutics, as well as in perfumery and cosmetics. In this perspective, many Mediterranean plant species could be an appreciated source of bioactive compounds with pharmacological and health-promoting properties, including antioxidant, antimicrobial, antiviral, anti-inflammatory, and antitumor ones. *Calendula officinalis* and *Foeniculum vulgare* are commercially important plants of the Mediterranean flora, with great therapeutic use in the treatment of many disorders since ancient times, and are now listed in several world pharmacopoeias and drug agencies. The present review offers an overview of the main phytochemicals, phenols, terpenes, and alkaloids, biosynthesized in *C. officinalis* and *F. vulgare*, both species endemic to the Mediterranean region. Further, all current knowledge and scientific data on taxonomic classification, botanical description, traditional uses, pharmacological studies, and potential toxicity of both species were reported. The principal aim of this review is to point out the prospective use of *C. officinalis* and *F. vulgare* as valuable reservoirs of beneficial plant-derived products with interesting biological properties, also providing suggestions and future challenges for the full exploitation of these two Mediterranean species for human life improvement.

Keywords: Mediterranean; *Calendula officinalis; Foeniculum vulgare*; phenols; terpenes; alkaloids; bioactivity

1. Secondary Metabolites, Biochemistry, and Biological Activity

The recent societal shift toward a sustainable lifestyle has resulted in increased demand for products derived from natural sources. Consequently, the consumption of plant secondary metabolites (SMs) is growing in interest for consumers and also for companies due to their promising biological activities, including antioxidant, antimicrobial, antiviral, and antitumor activities [1–4]. SMs have been found to have applications in many fields, such as supplements, nutraceuticals, pigments, cosmetics, bio-pesticides, herbicides, and bio-diesel due to their therapeutic and useful effects [3,4]. SMs, which are generally recognized as safe (GRAS), can be employed as alternatives to their synthetic counterparts, avoiding the undesirable toxic effects of chemicals on human wellness [2].

From an ecological point of view, unlike primary metabolites (such as sugars, amino acids, and nucleotides), SMs are not directly involved in essential plant functions like growth and reproduction though they play a crucial role in their long-term survival. Their role is multifaceted, encompassing plant communication, photo-protection, pest and parasite defense, and pollinator attraction [5]. Thus, the concentration of SMs varies seasonally and daily, and their production is influenced by numerous biotic and abiotic factors [1,4]. This literature review will demonstrate that these phytochemicals have a vast range of biological roles that are strictly determined by their chemical structures and are species-specific and organ-specific compounds [1,4].

The chemical classification typically divides SMs into three main groups depending on their biosynthetic pathways: phenols, terpenes, and alkaloids [1,4].

1.1. Phenols

One of the largest and most complex biomolecules among phytochemicals is phenolic compounds. Based on the number of aromatic rings, carbon atoms, and hydroxyl groups, they are divided into different sub-classes: phenolic acids, flavonoids, and non-flavonoids compounds, the latter including stilbenes, lignans, and tannins [6,7]. Polyphenols are widely distributed in all plant organs. In particular, phenolic acids are generally found in seeds, leaves, roots, and stems, instead flavonoids are prominently in aerial parts, and tannins in roots, bark, and seeds [1].

Phenolic compounds act in plants mainly as defense and protective molecules against biotic and abiotic stresses [8]. To overcome the photo-oxidative stress, they act by maintaining the redox balance of plant cells and avoiding the generation of ROS (Reactive Oxygen Species) or quenching them. In addition, phenolic compounds can act as good UV screeners by absorbing the shortest solar wavelengths and reducing the highly energetic ones [9,10].

In detail, phenolic acids are divided into benzoic acid derivates (C_6-C_1; i.e., gallic acid, vanillic acid, syringic acid) and hydroxycinnamic derivates (C_6-C_3; i.e., caffeic acid, ferulic acid, coumaric acid) with promising therapeutic properties. In fact, for their antidepressants, anti-hypertensives, anti-inflammatory, neuroprotective, anti-hyperglycemic, anti-cancer, and antidiarrheal properties, they are considered versatile dietary components naturally present in all fruits and vegetables [11,12].

Flavonoids (C_6-C_3-C_6) are mainly grouped into seven subclasses based on modifications to their basic skeletons: flavones (luteolin, apigenin), flavanols (hesperitin, naringenin, and eriodictyol), flavanones (naringin, hesperidin, eriocitrin), flavonols (quercetin, galangin, kaempferol, and myricetin), isoflavones (genistein, daidzein, and glycitein), and anthocyanins (cyanidin, delphinidin, malvidin) [13]. In plants, flavonoids are mainly found in the form of glycosides and are considered important molecules for the human diet, suggested as being active ingredients in food supplements and nutraceuticals, in the cosmetic field, and as natural dyes [13]. In fact, they are renowned for their several biological activities as antioxidant, anti-inflammatory, anti-cancer, and antihypertensive, and as circulation-improving and hypolipidemic agents [13].

Among the non-flavonoid compounds, tannins are classified as hydrolyzable (gallotannins and ellagitannins) and condensed (proanthocyanidins). They are mostly used in the veterinary field as anthelmintic and antimicrobial agents, as well as in the leather industry for their tanning properties [14].

1.2. Terpenes

Terpenes belong to the largest family of natural products. They are also known as isoprenoids since they originate from isoprene, a five-carbon atom compound, whose units are arranged in various structural patterns. Therefore, they are extremely diverse in structure, function, and properties, accounting for more than 50,000 known molecules. From a chemical point of view, they are classified according to the number of isoprene units into monoterpenes $(C_5H_8)_2$, sesquiterpenes $(C_5H_8)_3$, diterpenes $(C_5H_8)_4$, sesterterpenes $(C_5H_8)_5$, triterpenes $(C_5H_8)_6$, tetraterpenes $(C_5H_8)_8$, and so on [15,16]. Monoterpenes and sesquiterpenes are common components of essential oils and are responsible for the odorous properties of these compounds. Triterpenes are derived from the squalene biosynthetic pathway, through cyclization and various modifications to produce the diverse triterpene compounds. They contain numerous methyl groups that can be oxidized into alcohols, aldehydes, and acids, leading to various biologically active molecules. This group includes phytogenic bio-surfactants, historically utilized for their soap-like properties, and steroids, with cholesterol being the most significant representative. Finally, tetraterpenes are a class of terpenes composed of eight isoprene units, totaling 40 carbon atoms. They include carotenoids, like β-carotene, which are vital for photosynthesis and provide pigmentation in plants. Tetraterpenes exhibit antioxidant properties and serve as precursors for vitamin A synthesis [15,16].

In plants, the primary function of terpenes is to act as signaling molecules. Their emissions are linked to biotic and abiotic stresses, such as vital cycles, extreme temperature, radiation, drought, fire, air pollution, or herbivore attack [17]. The emission of monoterpenes and sesquiterpenes allows the plant to reduce ROS-induced damage and to improve ozone and thermal tolerance [18].

The growing interest in the potential application of terpenes can be attributed to their broad range of biological properties, including cancer chemoprevention, antimicrobial, antiviral, analgesic, anti-inflammatory, antifungal, and anti-parasitic activities [15,19]. Due to their numerous bioactivities, these SMs are demanded in several industrial sectors such as pharmaceuticals, food, cosmetics, perfumery, aromatherapy, and agricultural, and can be used as drugs, food supplements, flavors, fragrances, and bio-pesticides [1,20].

1.3. Alkaloids

Alkaloids are nitrogen-containing organic molecules that are very abundant in plants. So far, over 10,000 SMs have been classified as alkaloids from numerous families with a varied distribution in plant organs according to the phase of the life cycle [21].

Their name comes from the "alkali-like" nature of the nitrogen atoms present in their structure. These molecules include amine-type elements in the structure, usually in a heterocyclic ring. They are synthesized through various metabolic pathways, involving amino acids as precursors. In chemical classification, the alkaloids are categorized into three groups: true alkaloids, proto-alkaloids, and pseudo-alkaloids [22]. Specifically, true alkaloids have heterocyclic rings with nitrogen and are derived from amino acids; proto-alkaloids do not have heterocyclic rings with nitrogen and derive from amino acids; pseudo-alkaloids have heterocyclic rings with nitrogen and are derived from terpenoids or purines [22]. These SMs have wide-ranging biological activities, including analgesic, antimalarial, and stimulant properties, making them valuable in pharmacology and medicine. In plants, their biosynthesis is promoted as a consequence of abiotic and biotic stresses [23]. Therefore, they act as natural toxins for different organisms, defending plants from pathogens and preventing herbivore grazing. In apparent contrast, some alkaloids are fundamental for plant–pollinator interactions, thus favoring seed dispersion and plant reproduction [23]. For millennia, alkaloids have been used in all cultures as medicines, poisons, and drugs, and they are still important nowadays [22]. In fact, molecules such as stimulant alkaloids in coffee, tea, cacao, and nicotine in tobacco are consumed worldwide. Molecules with hallucinogenic, narcotic, or analgesic properties, such as morphine and atropine, have found applications in medicine [22]. Therefore, alkaloids are used as preparation for sedatives, stimulants, muscle relaxants, tranquilizers, and anesthetics, but also in therapy as antimalarial, antimicrobial, anti-diabetic, anti-cancer, anti-HIV, and antioxidants. Nevertheless, alkaloids are often abused being distributed as illegal drugs such as cocaine, heroin, and opium [22].

2. *Asteraceae* and *Apiaceae* Families

The families of *Asteraceae* and *Apiaceae* among Mediterranean plant species have been used since ancient times in folk medicine for the treatment of illnesses and pain relief [24–27]. They are also considered valuable reservoirs of botanical flavors and fragrances, utilized in foods and cosmetics as supplements and additives [24–27]. Today, these families have a cosmopolitan distribution and are easily adaptable around the world.

The *Asteraceae* (*Compositae*) is the most abundant flowering plant family in many European countries: it consists of approximately 25,000 species and 1700 genera [24,27]. Demonstrating a high level of adaptability, species of this family are distributed worldwide, to different habitats and climatic conditions, except in Antarctica. This family includes a number of well-known food species, such as chicory, sunflower, and lettuce, as well as a number of medicinal plants, such as wormwood, chamomile, marigold, and dandelion [24,27].

Another major and popular family of flowering plants is the *Apiaceae* (*Umbelliferae*), which encompasses almost 400 genera and about 4000 species across the globe. The *Apiaceae* family mainly consists of aromatic plants, commonly used as food, spice, and ornamental plants, as well as for medical purposes, in perfumery, and in the pharmaceutical and cosmetic industries. The most economically important crops and herbs belonging to this family are celery, carrot, parsley, coriander, cumin, fennel, anise, and dill [25,26].

A phylogenetic relationship between *Asteraceae* and *Apiaceae* has been hypnotized based on some studies on phytochemicals, which identify molecules with similar structures [28,29]. In particular, several sesquiterpene lactones, based on skeletal stereo-structural types, have been surprisingly detected in *Asteraceae* species, since they are also representative of the *Apieceae* family [29]. In truth, both families are regarded as the richest plants for sesquiterpene lactones, such as germacranolides, guaianolides, eudesmanolides, eremophilanolides, and elemanolides [28]. All these sesquiterpenes have been used as medicines, poisons, flavorings, and fragrances for millennia [30]. The discovery that sesquiterpene lactones are the most important chemicals in allergies and systemic contact dermatitis was also intriguing [31]. The therapeutic properties of the guaianolide lactone group are well documented for the treatment of inflammation and cancer [30,32]. More recently, guaianolide lactones have been reported to be useful even in treating type-2 diabetes patients [33].

The basic skeletal types of guaianolide lactones of the *Apiaceae* and *Asteraceae* families are very similar. Despite the common structure of the γ-lactone function, they differ in the stereochemistry around the lactone ring. Lactone biosynthesis also has a similar pathway as is reported in both plant families, but the difference in the spatial arrangement and chemical configuration of some protons have been remarked on [29].

The genus *Calendula* is considered one of the largest and most evolved of the *Asteraceae* family [34]. This genus encompasses both annual and perennial plants, native to the Mediterranean basin [35,36]. The most common species are *Calendula officinalis* Linn., and *Calendula arvensis* Linn., with *C. officinalis* being the most studied species for medicinal purposes and its high economic value [37]. Nevertheless, a few studies have been carried out for the other species growing in the Mediterranean basin (*Calendula stellata, Calendula suffruticosa*, and *Calendula tripterocarpa*), as reported by Arora et al. [37]. Recently, seven other accepted species (*Calendula eckerleinii, Calendula karakalensis, Calendula lanzae, Calendula maroccana, Calendula meuselii, Calendula pachysperma*, and *Calendula palaestina*) have been reported [38]. This demonstrates the need for further investigation to understand the evolution of the genus *Calendula*.

An important member of the *Apiaceae* family is the Mediterranean endemic *Foeniculum vulgare* Mill. [39,40]. Today, fennel is the most studied culinary, medicinal, aromatic, and flavoring plant [39]. As reported by Malhotra [41], three main varieties have been described for *F. vulgare*: var. *piperitum* (bitter fennel), var. *dulce* (sweet fennel), and var. *azoricum* (Florence fennel or finocchio). Moreover, two subspecies of fennel are reported: *piperitum*, whose inflorescences and tops are used to make vinegar; and *capicellaceum*, which tastes bitter and whose seeds are still used to flavor liqueurs [42,43]. Nevertheless, there are only a few studies on the two subspecies.

3. Review Methods

For this review, the following international electronic databases were queried: Scopus, Web of Science, PubMed, Medline, and Google Scholar. Only original papers written in English were considered. Keywords used to search the databases included plant names (e.g., *Calendula, C. officinalis, Foeniculum, F. vulgare*) combined with names of each class of SMs considered (phenols, terpenes, alkaloids). Moreover, ethnobotanical knowledge, culinary uses, and biological activities associated with each species were searched to identify the main uses and potential future applications for both *C. officinalis* and *F. vulgare*.

To date, a large amount of scientific information on *Calendula* and *Foeniculum* species is available in the literature. Considering only *Calendula* species, a high number of studies were conducted during 2010–2019 (representing 44% of publications); furthermore, during 2020–2022, approximately 19% of studies on this genus were published, as reported by Olennikov and Kashchenko [38]. In the case of *Foeniculum* species, 20% of the manuscripts were published from 2001 to 2005. This value increased to approximately 38% from 2006 to 2010, until it reached 39% of the articles reported from 2011 to 2013, indicating the trend toward the scientific topic [40].

The main objective of this review is to better understand the prospective use of these two Mediterranean species as valuable sources of beneficial plant natural products with potential therapeutic properties, taking into account all the information available in the literature on the uses, phytochemicals, and pharmacological studies reported on *C. officinalis* and *F. vulgare*.

4. Calendula Officinalis

4.1. Taxonomic Classification and Botanical Description

C. officinalis belongs to the huge family of *Asteraceae*, as reported in Table 1 [44]. It is usually known as common marigold. The name Calendula derives from the Latin *calendae* means "the first day of the month", referring to its blooming period [36].

Table 1. Taxonomic classification of *C. officinalis* [44].

Rank	Scientific Name
Kingdom	*Plantae*
Division	*Magnoliophyta*
Class	*Magnoliopsida*
Order	*Asterales*
Family	*Asteraceae*
Genus	*Calendula*
Species	*C. officinalis*

C. officinalis is widely cultivated in sunny locations and usually grows in a variety of soils (acidic, sandy, and clayey). Although perennial, it is commonly treated as an annual or biennial plant. In temperate areas, seeds are sown in spring and typically bloom quickly in flowers [44]. It can reach a height of 30–60 cm with a stem angular, hairy, and solid (Figure 1). Lower leaves are spatulate, with a length of 10–20 cm long. The inflorescences comprise a thick capitulum or flower head of 4–7 cm diameter, surrounded by two rows of hairy bracts. In particular, flower heads vary in color from bright yellow to orange, and the corolla is around 15–25 mm long and 3 mm wide [37,44].

Figure 1. *C. officinalis* (https://dryades.units.it/floritaly/index.php, accessed on 22 June 2024).

4.2. Ethnopharmacology and Human Food Uses

C. officinalis ranks among the oldest cultivated flowers, first described in the third century B.C.; ancient Romans and Greeks used calendula flowers in many rituals and ceremonies, to make crowns or garlands. The nickname marigold given to calendula derives from "Mary's Gold", referring to the use of the flowers in early Christian events [36]. The plant has been in cultivation and used for medicinal purposes only since the 12th century and has a long history of use. In the Middle Ages, calendula was used for hepatic disorders, poisoning, and cardiac tonicity. Doctors realized that the plant could stop bleeding and promote wound healing around the 18th century [36]. Traditionally, calendula has provided different uses, including the elaboration of food, dyes, cosmetics, and traditional remedies. Along with these uses, different plant parts of C *officinalis* have been used for medicinal purposes, above all leaves and flowers [35,37,45].

Currently, *C. officinalis* has been listed in multiple national pharmacopoeias and agencies, such as the European Pharmacopoeia, British Herbal Pharmacopoeia, and European Medicines Agency, for its well-known therapeutic applications such as antipyretic, anti-inflammatory, antiepileptic, and antimicrobial properties [46]. *C. officinalis* is used in wound healing and to treat internal inflammation, gastrointestinal ulcers, and dysmenorrhea, and as a diuretic and diaphoretic in convulsions [35,37,45]. Particularly, calendula tea is used as eyewash and to make gargles in the treatment of inflammatory conditions of the skin and mucous membranes [37]. Mother tincture of *C. officinalis* is involved in homoeopathy for the treatment of mental tension and insomnia [47].

Recently, calendula has assumed the function of an edible flower, improving the appearance, flavor, and aesthetic value of food, according to consumer's tastes [35]. Other food uses of calendula petals include the addition of yellow color to foods as a substitute for saffron, as well as in decorations of cakes, sweets, and savory toppings. Fresh flowers are utilized in lettuce, rice, fish, herb butter, yogurt, and cheese spread, while the dried flowers are often used to make tea. The leaves, on the other hand, taste spicy and are well suited to flavorings soups and salads [48]. However, the use of this species in food products is subject to the removal of the pollen, as it may trigger severe allergic reactions [48]. Moreover, dyes and tinctures can be extracted from the dried and fresh flowers to produce a range of useful colors from yellow to orange [35].

4.3. Phytochemicals

Several classes of SMs have been reported in *C. officinalis* [37,38,49] and the main phytochemicals, phenols (phenolic acids, flavonoids), terpenes (monoterpenes, sesquiterpenes, saponins, carotenoids), and alkaloids are summarized in Table 2.

Table 2. Main chemical compounds in *C. officinalis*.

Compounds	Plant Part	References
Phenols		
Phenolic acids		
Hydroxybenzoic acid, salicylic acid, protocatechuic acid, vanillic acid, syringic acid, hydroxycinnamic acid, ferulic acid, fumaric acid, chlorogenic acid, caffeic acid	flowers flowers, leaves, roots	[50] [51]
Flavonoids		
Hesperidin, catechin, miquelianin, isoquercitrin, rutin, cosmosiin, astragalin, nicotiflorin, quercetin, luteolin, amentoflavone	flowers, leaves, roots	[51]
Isorhamnetin 3-*O*-neohesperidoside, isorhamnetin 3-*O*-rhamnosylrutinoside, isorhamnetin 3-*O*-rutinoside, isorhamnetin 3-*O*-glucoside, isorhamnetin-3-*O*-(6″-acetyl)-glucoside, quercetin-3-*O*-rhamnosylrutinoside, quercetin-*O*-pentosylhexoside, quercetin 3-*O*-glucoside, quercetin 3-*O*-rutinoside, quercetin-*O*-acetyldeoxyhexosylhexoside, quercetin-3-*O*-(6″-acetyl)-glucoside, rutinoside, neohesperidoside, quercetin glucoside, kaempferol-*O*-rhamnosylrutinoside	flowers	[52–55]
Cyanidin, delphinidin, malvidin, paeonidin, pelargonidin, petunidin (glycosides)	flowers	[55]

Table 2. *Cont.*

Compounds	Plant Part	References
Terpenes		
Monoterpenes		
α-thujene, α-pinene, sabinene, β-pinene, limonene, 1,8-cineol, p-cymene, trans-β-ocimene, α-phellandrene, γ-terpenene, δ-3-carene, geraniol, bornyl acetate, sabinyl acetate	leaves, flowers	[56,57]
Sesquiterpenes		
α-cubebene, α-copaene, α-bourbonene, β-cubebene, α-gurjunene, aromadendrene, β-caryophyllene, α-ylangene, α-humulene, epi-bicyclo-sequiphellandrene, germacrene D, allo-aromadendrene, β-salinene, calarene, α-muurolene, γ-muurolene, δ-cadinene, cadina-1,4-diene, α-cadinene, nerolidol, palustrol, β-oplopenone, α-cadinol, τ-muurolol	leaves, flowers	[56,57]
Officinoside C, officinoside D	flowers	[54]
Triterpenes		
Sitosterol, stigmasterol	seedlings, leaves	[58]
3-monoesters taraxasterol, lupeol	flowers	[59,60]
Ursadiol	flowers	[61,62]
Faradiol-3-*O*-palmitate, faradiol-3-*O*-myristate, faradiol-3-*O*-laurate, arnidiol-3-*O*-palmitate, arnidiol-3-*O*-myristate, arnidiol-3-*O*-laurate, calenduladiol-3-*O*-palmitate, calenduladiol-3-*O*-myristate	flowers	[60,63]
Calendasaponins A, B, C, and D, officinosides A, and B	flowers	[54]
Calendulaglycoside A, calendulaglycoside A 6′-*O*-methyl ester, calendulaglycoside A 6′-*O*-n-butyl ester, calendulaglycoside B, calendulaglycoside B 6′-*O*-n-butyl ester, calendulaglycoside C, calendulaglycoside C 6′-*O*-methyl ester, calendulaglycoside C 6′-*O*-n-butyl ester, calendulaglycoside F 6′-*O*-butyl ester, calendulaglycoside G 6′-*O*-methyl ester	flowers	[52]
Carotenoids		
Neoxanthin, 9Z-neoxanthin, violaxanthin, luteoxanthin, auroxanthin, 9Z-violaxanthin, flavoxanthin, mutatoxanthin, 9Z-anthroxanthin, lutein, 9/9A-lutein, 13/13Z-lutein, α-cryptoxanthin, β-cryptoxanthin, z-cryptoxanthin, lycopene, α-carotene, β-carotene	flowers	[64]
(5Z, 9Z)-lycopene, (5Z, 9Z, 5′Z, 9′Z)-lycopene, (5′Z)-γ-carotene, (5′Z, 9′Z)-rubixanthin, (5Z, 9Z, 5′Z)-lycopene	flowers	[65]
Alkaloids		
Sitsirikine, vinblastine, vindoline, catharanthine, vinleurosine	flowers	[66]
Platynecine, platynecine-type	aerial parts	[34]

4.3.1. Phenols

Phenolic acids and flavonoids have been identified in *C. officinalis* mainly from inflorescences, as shown in Table 2. Both benzoic acid and hydroxycinnamic acid derivates were reported among phenolic acids (Table 2). In particular, hydroxybenzoic acid, salicylic acid, protocatechuic acid, vanillic acid, syringic acid, hydroxycinnamic acid, ferulic acid, fumaric acid, chlorogenic acid, and caffeic acid were detected [50,51]. Considering flavonoids, Ak et al. [51]

described the presence of hesperidin, catechin, miquelianin, isoquercitrin, rutin, cosmosiin, astragalin, nicotiflorin, quercetin, luteolin, and amentoflavone (Table 2). In addition, several flavonol 3-*O*-glycosides from *C. officinalis* flowers have been isolated and characterized by many authors, as reported in Table 2 [52–55]. Anthocyanins have been detected in calendula as the components of the flowers that tend to be red in color and are mainly glycosides of cyanidin, delphinidin, malvidin, paeonidin, pelargonidin, and petunidin [55].

4.3.2. Terpenes

The terpenic profile of *C. officinalis* is mainly composed of monoterpenes, sesquiterpenes, triterpenes, and tetraterpenes (carotenoids), as shown in Table 2.

Monoterpenes and sesquiterpenes are responsible for the odor of calendula flowers. The main monoterpenes isolated in *C. officinalis* are α-thujene, α-pinene, sabinene, β-pinene, limonene, 1,8-cineol, p-cymene, trans-β-ocimene, α-phellandrene, γ-terpenene, δ-3-carene, geraniol, bornyl acetate, and sabinyl acetate, as characterized by Okoh et al. [56] and Ak et al. [57]. Compounds of a sesquiterpene nature were detected in *C. officinalis* both in non-glycosidic and glycosides form. In particular, Okoh et al. [56] and Ak et al. [57] detected α-cubebene, α-copaene, α-bourbonene, β-cubebene, α-gurjunene, aromadendrene, β-caryophyllene, α-ylangene, α-humulene, epi-bicyclo-sequiphellandrene, germacrene D, allo-aromadendrene, β-salinene, calarene, α-muurolene, γ-muurolene, δ-cadinene, cadina-1,4-diene, α-cadinene, nerolidol, palustrol, β-oplopenone, α-cadinol, and τ-muurolol (Table 2). The sesquiterpene glycosides officinoside C and officinoside D are natural terpenes isolated exclusively from *C. officinalis*, in which a hydroxyl group is substituted by fucose, as characterized by Yoshikawa et al. [54].

Triterpenes in *C. officinalis* are present both in the free state and as an ester with fatty acids or alcohols, as well as in the glycosidic form, as listed in Table 2 [58–63]. Yoshikawa et al. [54] isolated four new triterpene glycosides, named calendasaponins A, B, C, and D (Table 2). Moreover, ten oleanane-type triterpene glycosides, including four new compounds, calendulaglycoside A 6′-*O*-methyl ester, calendulaglycoside A 6′-*O*-n-butyl ester, calendulaglycoside B 6′-*O*-n-butyl ester, and calendulaglycoside C 6′-*O*-n-butyl ester, were isolated from the flowers [52].

Approximately one hundred carotenoids (tetraterpenoids) in free and esterified forms have been found and identified in *C. officinalis* [64]. Owing to a wide range of petal colors, different types and amounts of carotenoids can be detected in calendula flowers. The carotenoids found in the petals were neoxanthin, 9Z-neoxanthin, violaxanthin, luteoxanthin, auroxanthin, 9Z-violaxanthin, flavoxanthin, mutatoxanthin, 9Z-anthroxanthin, lutein, 9/9A-lutein, 13/13Z-lutein, α-cryptoxanthin, β-cryptoxanthin, z-cryptoxanthin, lycopene, α-carotene, and β-carotene (Table 2). In addition, ten carotenoids were unique to orange-flowered cultivars. Among them, (5Z, 9Z)-lycopene, (5Z, 9Z, 5′Z, 9′Z)-lycopene, (5′Z)-γ-carotene, (5′Z, 9′Z)-rubixanthin, and (5Z, 9Z, 5′Z)-lycopene have been identified [65].

4.3.3. Alkaloids

Few studies on the phytochemical characterization of *C. officinalis* describe the presence of alkaloid compounds (Table 2). Alkaloids, including sitsirikine, vinblastine, vindoline, catharanthine, and vinleurosine, have been identified in *C. officinalis* in detail [66] (Hernández-Saavedra et al., 2016). Moreover, in *C. officinalis,* significant quantities of pyrrolizidine alkaloids (platynecine-type) have been observed, with a share of 41.5% [34].

4.4. Biological Activities

C. officinalis is registered as a herbal drug and several ailments have been treated with *C. officinalis* [36]. Actually, many scientific researches have established that *C. officinalis* has a wide spectrum of pharmacological effects, including having antioxidant, cardio-protective, antimicrobial, cytotoxic, anti-cancer, anti-diabetic, nootropic, anti-inflammatory, wound-healing, hepato-protective, nephro-protective, and antiviral properties [37,45], as itemized in Table 3.

Table 3. Pharmacological activities along with their phytochemical constituents in *C. officinalis*.

Pharmacological Activity	Phytochemicals	References
Cardio-protective	Phenols	[67]
	-	[68]
Antioxidant	Phenols	[69]
	-	[70]
	-	[71]
	Sesquiterpenes	[57]
Antimicrobial	-	[47]
	-	[72]
	Sesquiterpenes	[73]
Cytotoxic and anti-cancer	Triterpenes	[74]
	Triterpenes	[75]
	Triterpenes	[52]
	-	[76]
Anti-diabetic and hypoglycemic	Phenols	[51]
	Triterpenes	[54]
Nootropic	Triterpenes	[77]
Anti-inflammatory	Triterpenes	[78]
Wound-healing	Tetraterpenes	[79]
Hepato-protective and nephro-protective	Tetraterpenes	[80]
Antiviral	-	[81]
	-	[82]

Plant polyphenols such as phenolic acid and flavonoids are among the most significant natural compounds with biological properties. In particular, flavonoids of *C. officinalis* are involved in cardiovascular issues, as depicted in Table 3. Martinez [67] carried out a pre-clinical study to evaluate the effects of calendula flowers on the vascular smooth muscle of rats. A concentration-dependent relation was obtained in endothelium-denuded rat aortic rings, and the vaso-relaxant effect was attributed to the flavonoid quercetin. *C. officinalis* has also been proven to be cardio-protective against ischemic heart disease by stimulating left ventricular pressure and aortic flow, as well as by reducing myocardial infarct size and cardiomyocyte apoptosis [68]. In this research, cardio-protection appears to be achieved by modulating antioxidant and anti-inflammatory properties, but no information was provided on the calendula SMs involved.

C. officinalis flavonoids and phenolic acids showed strong radical-scavenging capacity and free radical protection. As reported by Rigane et al. [69], rutin, quercetin-3-*O*-glucoside, scopoletin-7-*O*-glucoside, isorhamnetin-3-*O*-glucoside, and gallic acid have been identified as major antioxidant phytoconstituents (Table 3). Moreover, petal and flower extracts tested for antioxidant activity by lipid peroxidation, indicated that the petal extract was more potent than the flower head [70]. Calendula exerts also anti-ROS and anti-reactive nitrogen species (RNS) activity in a concentration-dependent manner, with significant effects even at very low concentrations [71]. Moreover, Ak et al. [57] reported the sesquiterpene α-cadinol as the most abundant constituent of the essential oil with high antioxidant capacity through free radical scavenging and reducing mechanisms (Table 3).

The discovery and isolation of new bioactive compounds from medicinal plants is an immediate and pressing need due to the growing incidence of drug-resistant pathogens. With this in mind, the antimicrobial activity of calendula flowers against Gram-positive (*Escherichia coli* and *Staphylococcus aureus*), Gram-negative (*Salmonella typhae* and *Vibrio cholera*), and fungi (*Candida albicans*) was studied using different extraction solvents [47]. In particular, the ethanolic extract showed activity against *E. coli*, *V. cholera*, and *C. albicans*, whereas the methanolic extract was active only against *C. albicans*. The chloroform extract gave antimicrobial activity against all microbes, while the acetone extract was active only against *E. coli* [47]. However, the compounds or classes of molecules potentially involved in the mechanism of action linked to microbial growth inhibition were not identified in the study. The antimicrobial potential of methanol and ethanol extracts from *C. officinalis* petals was also evaluated against a panel of clinical microorganisms, including bacteria (*Bacillus subtilis, Pseudomonas aeruginosa, Bacillus cereus, E. coli, S. aureus, Klebsiella aerogenes, Enterococcus faecalis, Bacillus pumilis, Klebsiella pneumoniae*) and fungi (*C. albicans, Candida krusei, Candida glabrata, Candida parapsilosis, Aspergillus flavus, Aspergillus fumigatus, Aspergillus niger*, and *Exophiala dermatitidis*). Both extracts showed an antimicrobial activity comparable with the standard antibiotic, Fluconazole. Further clinical studies are required to examine the *C. officinalis* antimicrobial compounds [72]. In recent times, Darekar and Hate [73] investigated the antibacterial potential of chloroform extract of *C. officinalis* against *Bacillus subtilis, Klebsiella pneumonia, S. aureus*, and *Enterococcus faecalis*. The results revealed strong antibacterial activity against all tested strains. The study also aimed to identify phytochemicals with potential antibacterial activity present in *C. officinalis*. As a result, the major components of *C. officinalis* were caryophyllene (12.97%), lupeol (9.45%), stigmasterol (9.38%), and γ-sitosterol (5.07%), suggesting these terpenic biomolecules as potential calendula antibiotics (Table 3) [73].

C. officinalis has been shown to exhibit antimutagenic action. In particular, saponin-like triterpene compounds were employed in the screening of the antimutagenic activity, by using benzo-[a]pyrene, a well-known pro-mutagenic molecule [74]. *C. officinalis* flowers have shown in vitro cytotoxic activity, too. In particular, the triterpenic glycoside compounds, calendulaglycoside F6′-*O*-butyl ester and calendulaglycoside G6′-*O*-methyl ester (Table 3), resulted active against leukemia, colon cancer, and melanoma cell lines [52,75]. Recently, Cruceriu et al. [76] reported that *C. officinalis* could exert anti-cancer activity by inducing apoptosis, activating caspase 3 and caspase 7, and downregulating cyclin D1, D3, A, E, and several cyclin-dependent kinases, suggesting the prospective usage of *C. officinalis* in cancer management, particularly in cancer prevention, treatment, and palliative care for patients.

The triterpene saponins, calendasaponins A, B, C, and D, have shown a potent inhibitory effect on serum glucose levels. In fact, Yoshikawa et al. [54] demonstrated that *C. officinalis* flowers had a hypoglycemic effect, inhibitory activity of gastric emptying, and gastro-protective effect in glucose-loaded rats (Table 3). On the contrary, phenolic compounds in *C. officinalis* extracts were reported to exhibit weak inhibition against α-amylase and α-glucosidase, the main enzyme involved in decreasing postprandial hyperglycemia. In particular, the flower extract showed higher inhibition against α-amylase, followed by the leaf extract and the root extract, while the root extract was more active against α-glucosidase and flower one was the least active [51].

Nootropic activity was also reported in *C. officinalis*. Ercetin et al. [77] reported the enzyme inhibitory effects of *C. officinalis* extracts with different solvents (*n*-hexane, dichloromethane, acetone, ethyl acetate, methanol, and water) against acetylcholinesterase (AChE) and butyrylcholinesterase (BChE). The results revealed that the methanolic extracts of leaves and flowers have the highest activity against enzymes involved in cognitive metabolism, and therefore with potential to treat dementia and Alzheimer's disease as nootropic agents (Table 3).

Excellent anti-inflammatory activity was reported in *C. officinalis* (Table 3). Using in vivo pharmacological testing, it has been determined that the triterpenes fatty acid esters (lauryl, myristoyl, and palmitoyl esters of faradiol) are responsible for the anti-inflammatory effects of flower extract, as reported by Silva et al. [78]. They demonstrated that *C. officinalis* flower was much more effective for treating both acute and chronic swelling in mice. Further, the results showed that the potent anti-inflammatory response of *C. officinalis* extract may be mediated by the inhibition of pro-inflammatory cytokines (IL-6, IL-1β; TNF-α, and IFN-γ) and cyclooxygenase 2 (COX-2).

Calendula flowers may have an impact on the inflammatory process and the new tissue generation phase, as demonstrated by Nicolaus et al. [79], but the active compounds that are responsible are still a matter of debate. They found that while triterpenes may play a minor role, tetraterpenic compounds, such as carotenoids or their derivatives, may be more useful in the treatment of wound healing (Table 3).

Further, tetraterpene extracts from *C. officinalis* flower are considered responsible for the protective role against hepato-toxicity and nephro-toxicity. Preethi and Kuttan [80] suggested that these activities are due to the presence of different carotenoids, such as lutein, zeaxanthin, and lycopene (Table 3).

Finally, the antiviral activity of *C. officinalis* flowers extract has been reported, as shown in Table 3. In particular, Bogdanova et al. [81] conducted a study on *Herpes simplex*, Influenza A2, and Influenza APR8 in vitro, and found that *C. officinalis* flowers extract was an effective agent against these viruses. Afterward, Kalvatchev et al. [82] demonstrated that *C. officinalis* flowers exhibited potent anti-HIV activity. This property was attributed to the inhibition of human immunodeficiency virus type 1 (HIV-1) reverse transcriptase in a dose-dependent manner as well as to the suppression of the replication of HIV-1. Until today, no studies have been conducted in order to identify the phytochemicals of *C. officinalis* involved in the antiviral mechanism.

4.5. Safety

Calendula cream and products have shown very few allergic and side effects, approximately 2% of the patients have reacted to skin contact with calendula products [83]. Generally, in the *Asteraceae* family, the only main group of chemicals that may cause allergies and contact dermatitis is sesquiterpene lactones [31,84].

5. Foeniculum Vulgare

5.1. Taxonomic Classification and Botanical Description

The genus *F. vulgare* is a member of the family *Apiaceae* and is classified as shown in Table 4 [40]. The plant was placed in genus *Anethum* by Linnaeus, but later placed in the new genus *Foeniculum* by De Candolle [41]. The name *Foeniculum*, used by the Romans, is diminutive of the Latin *foenum*, meaning hay, given that fennel smells like hay.

Table 4. Taxonomic classification of *F. vulgare* [40].

Rank	Scientific Name
Kingdom	*Plantae*
Division	*Tracheophyta*
Class	*Magnoliopsida*
Order	*Apiales*
Family	*Apiaceae*
Genus	*Foeniculum*
Specie	*F. vulgare*

Fennel is an herbaceous and aromatic plant comprising biennial or perennial varieties [40]. *F. vulgare* is commonly cultivated in tropical and temperate regions and this herbaceous plant is grown in semi-arid or arid environments [40]. In the Italian regions, especially in the south, in stony and sub-mountains up to an altitude of 700 m and along the coasts, wild spontaneous species of *F. vulgare* are still present with perennial plants, provided with robust and fittoning roots that form a false bulb named grumolo [43]. Fennel is characterized by stems grooved and intermittent leaves. Flowers are usually bisexual with yellow umbrellas in the form of oval beads (Figure 2). Fennel diachenes have a narrow, long, and cylindrical appearance with a length of about 8 mm and a width of 3 mm, with an aromatic odor and sweet taste [41].

Over the past few decades, modern horticultural practices have favored the use of seeds derived from careful germplasm selection and/or new varieties selected through genetic improvement programs, in contrast with the old practice of self-production of seeds that had brought to notice some important fennel landraces, particularly adapted to specific regional environments. Nevertheless, it is still possible to find varieties such as the Dolce di Firenze, Nostrale di Chioggia, Romanesco, Marchigiano, Mantovano, Di Bologna, and Di Napoli, which derive from the careful selection of germplasm and fixation of desired characters [43].

Figure 2. *F. vulgare* (https://dryades.units.it/floritaly/index.php, accessed on 22 June 2024).

5.2. Ethnopharmacology and Human Food Uses

F. vulgare was renowned by the ancient Egyptians, Romans, Indians, and Chinese. In early Sanskrit writings, fennel was known as *madhurika* and its cultivation in India dates back to at least 2000 BC. To the ancient Greeks, fennel represented success and was called *marathon* because the battle of Marathon (490 BC) was fought in a fennel field [40]. Fennel was also a triumph symbol for the Romans and leaves were used to crown winners of games. Emperor Charlemagne encouraged the cultivation of fennel in Central Europe for its therapeutic properties [40]. Chewing the diachenes was believed to be important in curing stomach indisposition in the Middle Ages. In the 5th century, fennel was thought to have a sedative effect and, from the 9th century, numerous therapeutic properties were attributed to this plant [85].

F. vulgare is widely used in traditional medicine for a number of conditions and is recognized as an alternative medicine in various traditional systems of medicine like the Ayurveda, Unani, Siddha, Indian, and Iranian [40]. Different parts of the plant are employed to treat many digestive ailments [85]. It also is very useful in the treatment of diabetes, bronchitis, chronic cough, and kidney stones [39,40,85]. Due to its diuretic effect, fennel is also used to treat kidney and bladder diseases, and to relieve nausea. Further, it is applied to improve eye illnesses such as cataracts and conjunctivitis [85].

As it is a highly aromatic and flavorful herb, fennel is traditionally employed for culinary purposes. Fennel was considered a royal spice, served to kings with fruit, bread, and in fish dishes as early as the 13th century [41]. Today, all parts of the fennel plant are edible: diachenes, leaves, stalks, and false bulbs are regularly consumed in modern French and Italian cooking. Flowers and leaves are also utilized to make yellow and brown dyes [41]. Fennel diachenes are anise-like in aroma and are used as flavorings in baked products, meat and fish dishes, ice cream, alcoholic beverages, and herb mixtures [86]. The false bulb is a crisp vegetable and may be sautéed, fried, stewed, braised, grilled, or eaten raw [39].

5.3. Phytochemicals

Fennel, one of the most appreciated sweet and aromatic greens, raw or cooked, has a low energy content but is particularly rich in beneficial substances. Research led to the isolation and characterization of phytochemicals from *F. vulgare*, including phenolic acids, flavonoids, stilbenes, terpenes, and alkaloids (Table 5).

Table 5. Main chemical compounds in *F. vulgare*.

Compounds	Plant Part	References
Phenols		
Phenolic acids		
3-*O*-caffeoylquinic acid, 4-*O*-caffeoylquinic acid, 5-*O*-caffeoylquinic acid, 1,3-*O*-di-caffeoylquinic acid, 1,4-*O*-di-caffeoylquinic acid, 1,5-*O*-di-caffeoylquinic acid	fruits	[87]
Rosmarinic acid, chlorogenic acid	diachenes	[88]
Flavonoids		
Eriodictyol-7-rutinoside, quercetin-3-rutinoside	fruits	[87]
Quercetin, apigenin	diachenes	[88]
Quercetin-3-*O*-galactoside, kaempferol-3-*O*-rutinoside, kaempferol-3-*O*-glucoside, quercitin-3-*O*-glucuronide, kaempferol-3-*O*-glucuronide, isoquercitin, isorhamnetin-3-*O*-glucoside	whole plants	[89]
Stilbenes		
Foeniculoside X, Foeniculoside XI, cis-miyabenol C, trans-miyabenol C, trans-resveratrol-3-*O*-β-D-glucopyranoside, sinapoyl glucoside, syringin-4-*O*-β-glucoside, oleanolic acid, 7a-hydroxycampesterol, (3b,5a,8a,22E) 5,8-epidioxy-ergosta-6,22-dien-3-ol, 2,3-dihydropropylheptadec-5-onoate	fruits	[90]
Terpenes		
Monoterpenes		
Trans-anethole, estragole, fenchone, p-anisaldehyde, α-phellandrene, nerol, α-pinene, γ-terpinene, o-cymene, D-limonene, β-myrcene	stems diachenes whole plants leaves, diachenes fruits	[91] [92,93] [94,95] [96,97] [98]
Sesquiterpenes		
Caryophyllene, germacrene D	aerial parts	[99]
Bergamotene, β-farnesene, α-farnesene, α-curcumene	fruits	[100]
Alkaloids		
Pyrrolizidine alkaloids	fruits leaves	[101] [102,103]

5.3.1. Phenols

Fennel fruits and diachenes are characterized to be rich in phenolic compounds, in particular phenolic acids and flavonoids, as shown in Table 5. Especially, *F. vulgare* fruits have been reported to contain 3-*O*-caffeoylquinic acid, 4-*O*-caffeoylquinic acid, 5-*O*-caffeoylquinic acid, 1,3-*O*-di-caffeoylquinic acid, 1,4-*O*-di-caffeoylquinic acid, 1,5-*O*-di-caffeoylquinic acid, as phenolic acids [87]. The flavonoids like eriodictyol-7-rutinoside and quercetin-3-rutinoside have also been isolated from *F. vulgare* fruit [87]. *F. vulgare* diachenes were reported to contain rosmarinic and chlorogenic acids as major phenolic acids (14.9% and 6.8%), and quercetin and apigenin as major flavonoids (17.1% and 12.5%), as demonstrated by Roby et al. [88]. As listed in Table 5, flavonoids quercetin-3-*O*-galactoside, kaempferol-3-*O*-rutinoside, kaempferol-3-*O*-glucoside, quercitin-3-*O*-glucuronide, kaempferol-3-*O*-glucuronide, isoquercitin and isorhamnetin-3-*O*-glucoside have also been reported to occur in *F. vulgare* [89].

Two new phenolic compounds, identified as diglucoside stilbene trimers (named Foeniculoside X and Foeniculoside XI) have also been isolated from *F. vulgare* fruits together with cis-miyabenol C, trans-miyabenol C, trans-resveratrol-3-*O*-β-D-glucopyranoside, sinapyl glucoside, syringin-4-*O*-β-glucoside, oleanolic acid, 7a-hydroxycampesterol, (3b,5a,8a,22E) 5,8-epidioxy-ergosta-6,22-dien-3-ol, and 2,3-dihydropropylheptadec-5-onoate, as reported in Table 5 [90].

5.3.2. Terpenes

The characteristic anise odor of *F. vulgare* is mainly due to the monoterpenes and sesquiterpenes that mainly constitute this essential oil. Fennel has been reported to contain about 80 different monoterpenic compounds, the major ones being trans-anethole, fenchone, estragole (methyl-chavicol), p-anisaldehyde, and α-phellandrene, nerol, α-pinene, γ-terpinene, o-cymene, D-limonene, and β-myrcene, as shown in Table 5 [91–98]. The relative concentration of these compounds varies considerably depending on the phenological phase and geographical origin of the plant [91]. Further, the terpenic composition of *F. vulgare* exhibits considerable chemo-diversity depending upon the method of extraction and the accumulation of these compounds is different in each fennel part (roots, stem, diachenes, flowers, and fruits), as reported by Diaz-Maroto et al. [91].

Sesquiterpenes compounds present in *F. vulgare* are listed in Table 5. In particular, caryophyllene, germacrene D, bergamotene, β-farnesene, α-farnesene, α-curcumene were identified [99,100].

5.3.3. Alkaloids

Fennel fruits were reported to contain alkaloids. In fact, Kaur and Arora [101] performed qualitative and quantitative phytochemical analyses on *F. vulgare* diachenes, demonstrating the presence of 2.80% alkaloids. Moreover, the presence of pyrrolizidine alkaloids was reported in *F. vulgare*, as depicted in Table 5 [102,103]. This large group of SMs was reported to be responsible for multiple cases of food and feed poisoning over the last 100 years [100,101].

5.4. Biological Activities

F. vulgare is officially noted in different national pharmacopoeias as an important part of polyherbal formulations in the treatment of many diseases and disorders like abdominal pains, arthritis, conjunctivitis, constipation, diarrhea, fever, gastritis, insomnia, irritable colon, mouth ulcer, stomach-ache, respiratory disorders, skin diseases, and so on [40]. Several pharmacological studies have reported that *F. vulgare* has an important variety of biological activities, comprising antioxidant, antimicrobial, antiviral anti-inflammatory, anti-cancer, hepato-protective, cardio-protective, gastro-protective, anti-cholesterol, anti-diabetic, estrogenic, anti-anxiety, and nootropic properties, as summarized in Table 6 [40,85].

Table 6. Pharmacological activities along with their phytochemical constituents in *F. vulgare*.

Pharmacological Activity	Phytochemicals	References
Antioxidant	Phenols	[87]
	Phenols	[88]
	Monoterpenes	[98]
	Monoterpenes	[93]
	Monoterpenes	[94]
Antimicrobial	Monoterpenes	[104]
	Monoterpenes	[105]
	Monoterpenes	[106]
Antiviral	Monoterpenes	[107]
Anti-inflammatory and anti-cancer	Monoterpenes	[108]
	-	[109]
	-	[110]
Hepato-protective	Monoterpenes	[111]
Cardio-protective	Monoterpenes	[95]
Gastro-protective	Monoterpenes	[95]
	-	[112]
	-	[113]
Anti-cholesterol and anti-atherogenic	-	[114]
	-	[115]

Table 6. *Cont.*

Pharmacological Activity	Phytochemicals	References
Anti-diabetic and hypoglycemic	-	[116]
	-	[117]
Estrogenic	Monoterpenes	[118]
	-	[119]
	-	[120]
Anti-anxiety	Monoterpenes	[121]
	Monoterpenes	[122]
	Monoterpenes	[123]
Nootropic	-	[124]

F. vulgare is known as an excellent source of natural antioxidants. Fennel extracts can inhibit free radicals due to their high content of phenolic acids and flavonoids, such as caffeoylquinic acid derivates, rosmarinic acid, eriodictyol-7-rutinoside, quercetin-3-*O*-galactoside, and kaempferol-3-*O*-glucoside [87,88]. Fennel essential oil was also reported to possess antioxidant activity associated with the monoterpene components [93,94,98], as itemized in Table 6.

Fennel is used to treat many bacterial, fungal, and viral infectious diseases. In particular, *F. vulgare* is characterized by antimicrobial effects on human pathogens and foodborne microorganisms. Among human pathogenic bacteria, Zellagui et al. [104] carried out the antimicrobial assay against Gram-positive (*Staphylococcus epidermidis, Staphylococcus saprophyticus, Staphylococcus blanc*) and Gram-negative bacteria (*E. coli, Proteus mirabilis, Proteus vulgaris*), and three fungal strains (*Aspergillus versicolor, Aspergillus fumigates* and *Penicillium camemberti*). Seven oxygenated monoterpenes, isolated and characterized from the aerial parts of *F. vulgare,* were tested and all microorganisms were inhibited [104]. The authors suggested that the antimicrobial activity of *F. vulgare* extracts can be attributed to the content of oxygenated monoterpenes by means of a mechanism that involves membrane disruption. Considering foodborne pathogens, Dadalioglu and Evrendilek [105] studied the chemical compositions and inhibitory effects of fennel essential oil on *E. coli, Listeria monocytogenes, Salmonella typhimurium,* and *S. aureus*. The results showed that the inhibitory effects of *F. vulgare* may be attributed to the main compound, trans-anethole (Table 6). These outcomes were also confirmed by Cetin et al. [106] who determined the chemical compositions of the essential oil from the inflorescence, leaf stems, and aerial parts of fennel, and their antimicrobial activities. The study revealed that trans-anethole, the main component, is responsible for the antimicrobial activity (Table 6).

Orhan et al. [107] studied the antiviral activity of the fennel essential oil against the DNA virus Herpes simplex type-1 (HSV-1) and the RNA virus parainfluenza type-3 (PI-3), recording a significant inhibition from *F. vulgare*. Moreover, trans-anethole was tested and was reported as the main compound for the antiviral activity of fennel (Table 6).

Monoterpenes present in *F. vulgare* are considered to be associated with the prevention of several disorders induced by oxidative stress, such as cardiovascular disease, cancer, and inflammation. In particular, Chainy et al. [108] showed that trans-anethole is responsible for the suppression of both inflammation and carcinogenesis (Table 6). This bioactive compound was reported to act at an early step in the cascade of TNF-dependent signal transduction, so inhibiting cytokine-induced cellular response was associated with both diseases. The in vitro cyto-protection activity of *F. vulgare* was also estimated against normal human blood lymphocytes and the B16F10 melanoma cell line [109]. These results suggest that fennel could be considered a natural source of antitumor agents as well as being cyto-protective to normal cells. Moreover, fennel was proven to have significant anticancer activity against breast cancer cells (MCF-7) and liver cancer (HepG), as reported by Mohamad et al. [110]; nevertheless, no information was provided about the phytochemicals of *F. vulgare,* presumably involved in the anti-cancer mechanism.

One of the most common uses of *F. vulgare* has been to lower blood pressure by causing diuresis and increasing the excretion of sodium and water from the body. Significant antithrombotic activity and inhibition of platelet aggregation were observed in mice after oral administration of fennel essential oil and its most abundant phytochemical trans-anethole [95].

In a study conducted by Ozbek et al. [111], the hepatotoxicity caused by CCl4 administration in rats was inhibited by *F. vulgare* essential oil. In this research, the decreased levels of serum aspartate aminotransferase (AST), alanine aminotransferase (ALT), alkaline phosphatase (ALP), and bilirubin were demonstrated, and D-limonene and β-myrcene were suggested to be the components responsible for the potent hepato-protective action (Table 6).

It has been shown that fennel has a positive effect on gastrointestinal disorders. In fact, *F. vulgare* plays a protective role against ethanol-induced gastric mucosal lesions, as a consequence of a reduction in lipid peroxidation and augmentation in the antioxidant activity, as reported by Birdane et al. [112]. Moreover, Al-Mofleh et al. [113] also demonstrated the protective effect of fennel on gastric ulcers. In both papers, it was proposed that this property was linked to the antioxidant capacity of fennel, but no investigation was carried out to exactly identify the phytochemicals involved in the mechanism. Instead, Tognolini et al. [95] tested trans-anethole in rats with ethanol-induced gastric lesions and demonstrated that this compound plays the role of a gastro-protecting molecule (Table 6).

The study of the anti-cholesterol and anti-atherogenic effect of methanolic extract from *F. vulgare* showed that the treatment significantly reduced plasma lipid levels, facilitating blood flow in the coronary and preventing fatty deposits in the arteries [114]. Further, fennel extracts were demonstrated to be useful for the control of blood glucose in diabetic patients. In fact, daily use of the extract could be effective in reducing chronic complications associated with diabetes [115]. *F. vulgare* was also reported to reduce blood glucose and triglycerides and, contemporarily, increase levels of liver and muscle glycogen [116]. Consequently, *F. vulgare* can be used in the pharmaceutical industry for the manufacture of anti-diabetic drugs [117], but further investigation is needed to understand the mechanism of action.

Fennel has been used for thousands of years as an estrogenic agent. As a consequence of this property, fennel increases milk secretion, reduces menstrual pain, facilitates birth, and increases sexual desire. Trans-anethole is the main estrogenic molecule in extract and essential oil from fennel, being the methyl ether of estrone [118]. Different quantities of fennel significantly decreased contraction intensity induced by oxytocin and prostaglandins, as showed by Ostad et al. [119]. On the other hand, Myrseyed et al. [120] demonstrated the effect of fennel extracts in reducing testosterone, FSH, and LH levels and sperm amount, thus suggesting a negative effect on male reproductive activity.

Fennel is also a drug used for the treatment of anxiety and stress. It relieves psychological and physical symptoms associated with these conditions. Mesfin et al. [121] evaluated the use of *F. vulgare* essential oil in stress and anxiety management in a mice model. They demonstrated that the group treated with fennel essential oil had much lower agitation and stress levels than the control group. The calming properties of fennel may be linked to phytoestrogens and to trans-anethole (methyl ether of estrone), which are involved in the phenomenon of anxiety mediated by the GABA-ergic system and estrogen receptors [122]. In another study [123], limonene, a minor component of the *F. vulgare* essential oil, has also been reported to have anxiety-relieving properties (Table 6).

Nootropic activity was also reported in *F. vulgare*. In fact, there is some evidence in favor of the use of *F. vulgare* for the treatment of cognitive disorders like dementia and Alzheimer's disease. Joshi and Parle [124] administered *F. vulgare* for eight successive days to mice. They registered an amelioration in the amnesic effect of scopolamine and in the aging-induced memory deficits, concluding that fennel may be employed in the treatment of cognitive disorders (dementia and Alzheimer's disease) as a nootropic and anticholinesterase agent. However, no information was provided on the fennel phytoconstituents involved.

5.5. Safety

Extracts and essential oils of fennel can be considered safe due to their long history of ethnomedicinal use with no reports of serious adverse effects. However, estragole (methyl-chavicol) has become a concern in recent years because of its structural similarity to methyl-eugenol present in *F. vulgare*. This has led the European Union (EU) to issue a new legal limit for estragole of 10 mg/kg in non-alcoholic beverages [125]. Further, the Scientific Committee on Food (SCF) of the European Union restricts the use of this substance.

The ability of estragole to cause genotoxicity and, thus, to be carcinogenic was first described by Drinkwater et al. [126] and then followed by numerous in vivo and in vitro studies [127–131]. To the present date, the potential of estragole to induce carcinogenesis in humans remains unclear. The critical factor for estragole's carcinogenicity is its metabolic activation, leading to the formation of unstable molecules that form adducts with nucleic acids, damaging DNA. Estragole metabolism is dose-dependent and elevated doses of estragole increase its biotransformation, leading to the formation of mutagenic metabolites [132].

6. Challenges and Future Perspectives

The Mediterranean is one of the most biologically diverse regions on the planet. It was recently declared an "Intangible Cultural Heritage of Humanity" by UNESCO for its rich cultures, customs, beliefs, environment, and diet. With 25,000 plant species, 13,000 of which are endemic, it is the third richest area in the world in terms of plant species and is considered one of the world's biodiversity hotspots. However, global climate change may pose a serious environmental threat to the region due to increased drought periods and heat waves. In order to survive in these worrying climates, plants have evolved various mechanisms, including the synthesis of an extraordinary array of secondary metabolites, which act mainly as plant defense compounds against environmental stress.

This paper reviewed the literature on the main SMs, phenols (phenolic acids, flavonoids), terpenes (monoterpenes, sesquiterpenes, saponins, carotenoids), and alkaloids, biosynthesized in Mediterranean *C. officinalis* and *F. vulgare*. Until the present, many papers have been published on bioactive compounds in calendula and fennel, but only a few of them also reported the biochemical/ecological aspect, and none of them in recent times. In fact, as SMs are strongly influenced by the genotype–environment interaction, it would be interesting to encourage the study of environmental features that maximize the production of these valuable biomolecules and to study how the climatic changes can modify the amounts and the type of SMs biosynthesized by these two Mediterranean species.

Nowadays, *C. officinalis* and *F. vulgare* are considered treasured sources of phenols, terpenes, and alkaloid compounds, with a wide array of therapeutic, pharmacological, and health-promoting properties. Particularly, antioxidant, antimicrobial, antiviral, anti-inflammatory, anti-cancer, anti-diabetic, cardio-protective, hepato-protective, nootropic, and skin-protective activities are the main biological properties reported for both species. Moreover, *C. officinalis* showed wound-healing and nephro-protective features, while *F. vulgare* exhibited estrogenic and anti-anxiety attributes.

One of the main outcomes of the study shows that among SMs with interesting biological activities, the sesquiterpene lactones, biomolecules that have been abundantly used as medicine, poison, flavoring, and fragrance for several millennia, are abundant in these two botanical families *Asteraceae* and *Apiaceae*. However, no studies have been carried out on *C. officinalis* and *F. vulgare* and further efforts are needed to identify and characterize these bioactive compounds in calendula and fennel.

It is interesting to notice that the use of only one plant part (flowers for *C. officinalis* and diachenes for *F. vulgare*) is the rule, while the rest of the biomass (leaves, stems, roots) is considered a waste that is typically unexploited and understudied. Hence, it is crucial that extensive research on all plant parts of calendula and fennel is conducted in the future. Plant wastes and by-products are high-added resources to obtain appreciated natural products, fully respecting the transition from a linear to a circular management present in the objectives of the European Union's Circular Economy Action Plan.

Author Contributions: Conceptualization, F.M.V. and B.L.; investigation, F.M.V.; original draft preparation, F.M.V.; review and editing, F.M.V., D.P. and B.L. All authors have read and agreed to the published version of the manuscript.

Funding: This research received no external funding.

Institutional Review Board Statement: Not applicable.

Informed Consent Statement: Not applicable.

Data Availability Statement: No new data were created or analyzed in this study.

Conflicts of Interest: The authors declare no conflicts of interest.

References

1. Chiocchio, I.; Mandrone, M.; Tomasi, P.; Marincich, L.; Poli, F. Plant secondary metabolites: An opportunity for circular economy. *Molecules* **2021**, *26*, 495. [CrossRef]
2. Giacometti, J.; Kovačević, D.B.; Putnik, P.; Gabrić, D.; Bilušić, T.; Krešić, G.; Stulić, V.; Barba, F.J.; Chemat, F.; Barbosa-Cánovas, G.; et al. Extraction of bioactive compounds and essential oils from Mediterranean herbs by conventional and green innovative techniques: A review. *Int. Food Res.* **2018**, *113*, 245–262. [CrossRef]
3. Kabera, J.N.; Semana, E.; Mussa, A.R.; He, X. Plant secondary metabolites: Biosynthesis, classification, function and pharmacological properties. *J. Pharm. Pharmacol.* **2014**, *2*, 377–392.
4. Bourgaud, F.; Gravot, A.; Milesi, S.; Gontier, E. Production of plant secondary metabolites: A historical perspective. *Plant Sci.* **2001**, *161*, 839–851. [CrossRef]
5. Demain, A.L.; Fang, A. The natural functions of secondary metabolites. In *History of Modern Biotechnology I, Advances in Biochemical Engineering/Biotechnology*; Fiechter, A., Ed.; Springer International Publishing: Cham, Switzerland, 2000; pp. 1–39. [CrossRef]
6. Abbas, M.; Saeed, F.; Anjum, F.M.; Afzaal, M.; Tufail, T.; Bashir, M.S.; Ishtiaq, A.; Hussain, S.; Suleria, H.A.R. Natural polyphenols: An overview. *Int. J. Food Prop.* **2017**, *20*, 1689–1699. [CrossRef]
7. Rasouli, H.; Farzaei, M.H.; Khodarahmi, R. Polyphenols and their benefits: A review. *Int. J. Food Prop.* **2017**, *20*, 1700–1741. [CrossRef]
8. Sharma, A.; Shahzad, B.; Rehman, A.; Bhardwaj, R.; Landi, M.; Zheng, B. Response of phenylpropanoid pathway and the role of polyphenols in plants under abiotic stress. *Molecules* **2019**, *24*, 2452. [CrossRef]
9. Radice, M.; Manfredini, S.; Ziosi, P.; Dissette, V.; Buso, P.; Fallacara, A.; Vertuani, S. Herbal extracts, lichens and biomolecules as natural photo-protection alternatives to synthetic UV filters. A systematic review. *Fitoterapia* **2016**, *114*, 144–162. [CrossRef]
10. Zillich, O.V.; Schweiggert-Weisz, U.; Eisner, P.; Kerscher, M. Polyphenols as active ingredients for cosmetic products. *Int. J. Cosmet. Sci.* **2015**, *37*, 455–464. [CrossRef]
11. da Silva, A.P.G.; Sganzerla, W.G.; John, O.D.; Marchiosi, R. A comprehensive review of the classification, sources, biosynthesis, and biological properties of hydroxybenzoic and hydroxycinnamic acids. *Phytochem. Rev.* **2023**, *2023*, 1–30. [CrossRef]
12. Kumar, N.; Goel, N. Phenolic acids: Natural versatile molecules with promising therapeutic applications. *Biotechnol. Rep.* **2019**, *24*, e00370. [CrossRef]
13. Falcone Ferreyra, M.L.; Rius, S.P.; Casati, P. Flavonoids: Biosynthesis, biological functions, and biotechnological applications. *Front. Plant Sci.* **2012**, *3*, 222. [CrossRef] [PubMed]
14. Das, A.K.; Islam, M.N.; Faruk, M.O.; Ashaduzzaman, M.; Dungani, R. Review on tannins: Extraction processes, applications and possibilities. *S. Afr. J. Bot.* **2020**, *135*, 58–70. [CrossRef]
15. Xavier, V.; Spréa, R.; Finimundy, T.C.; Heleno, S.A.; Amaral, J.S.; Barros, L.; Ferreira, I.C. Terpenes. In *Natural Secondary Metabolites: From Nature, through Science, to Industry*; Carocho, M., Heleno, S.A., Barros, L., Eds.; Springer International Publishing: Cham, Switzerland, 2023; pp. 107–156. [CrossRef]
16. Guimarães, A.G.; Serafini, M.R.; Quintans-Junior, L.J. Terpenes and derivatives as a new perspective for pain treatment: A patent review. *Expert Opin. Ther. Pat.* **2014**, *24*, 243–265. [CrossRef]
17. Li, C.; Zha, W.; Li, W.; Wang, J.; You, A. Advances in the biosynthesis of terpenoids and their ecological functions in plant resistance. *Int. J. Mol. Sci.* **2023**, *24*, 11561. [CrossRef] [PubMed]
18. Holopainen, J.K.; Himanen, S.J.; Yuan, J.S.; Chen, F.; Stewart, C.N. Ecological functions of terpenoids in changing climates. In *Natural Products Phytochemistry, Botany and Metabolism of Alkaloids, Phenolics and Terpenes*; Ramawat, K.G., Merillon, J.M., Eds.; Springer International Publishing: Cham, Switzerland, 2013; pp. 2913–2940. [CrossRef]
19. Yang, W.; Chen, X.; Li, Y.; Guo, S.; Wang, Z.; Yu, X. Advances in pharmacological activities of terpenoids. *Nat. Prod. Commun.* **2020**, *15*, 1934578X20903555. [CrossRef]
20. Barbulova, A.; Colucci, G.; Apone, F. New trends in cosmetics: By-products of plant origin and their potential use as cosmetic active ingredients. *Cosmetics* **2015**, *2*, 82–92. [CrossRef]

21. Aniszewski, T. *Alkaloids: Chemistry, Biology, Ecology, and Applications*, 2nd ed.; Elsevier: Amsterdam, The Netherlands, 2015; pp. 1–475. [CrossRef]
22. Debnath, B.; Singh, W.S.; Das, M.; Goswami, S.; Singh, M.K.; Maiti, D.; Manna, K. Role of plant alkaloids on human health: A review of biological activities. *Mater. Today Chem.* **2018**, *9*, 56–72. [CrossRef]
23. Ali, A.H.; Abdelrahman, M.; El-Sayed, M.A. Alkaloid role in plant defense response to growth and stress. In *Bioactive Molecules in Plant Defense*; Jogaiah, S., Abdelrahman, M., Eds.; Springer International Publishing: Cham, Switzerland, 2019; pp. 145–158. [CrossRef]
24. Garcia-Oliveira, P.; Barral, M.; Carpena, M.; Gullón, P.; Fraga-Corral, M.; Otero, P.; Prieto, M.A.; Simal-Gandara, J. Traditional plants from Asteraceae family as potential candidates for functional food industry. *Food Funct.* **2021**, *12*, 2850–2873. [CrossRef]
25. Thiviya, P.; Gamage, A.; Piumali, D.; Merah, O.; Madhujith, T. Apiaceae as an important source of antioxidants and their applications. *Cosmetics* **2021**, *8*, 111. [CrossRef]
26. Sayed-Ahmad, B.; Talou, T.; Saad, Z.; Hijazi, A.; Merah, O. The Apiaceae: Ethnomedicinal family as source for industrial uses. *Ind. Crops Prod.* **2017**, *109*, 661–671. [CrossRef]
27. Bessada, S.M.; Barreira, J.C.; Oliveira, M.B.P. Asteraceae species with most prominent bioactivity and their potential applications: A review. *Ind. Crops Prod.* **2015**, *76*, 604–615. [CrossRef]
28. Milosavljevic, S.; Bulatovic, V.; Stefanovic, M. Sesquiterpene lactones from the Yugoslavian wild growing plant families Asteraceae and Apiaceae. *J. Serb. Chem. Soc.* **1999**, *64*, 397–442. [CrossRef]
29. Holub, M.; Toman, J.; Herout, V. The phylogenetic relationships of the Asteraceae and Apiaceae based on phytochemical characters. *Biochem. Syst. Ecol.* **1987**, *15*, 321–326. [CrossRef]
30. Chadwick, M.; Trewin, H.; Gawthrop, F.; Wagstaff, C. Sesquiterpenoids lactones: Benefits to plants and people. *Int. J. Mol. Sci.* **2013**, *14*, 12780. [CrossRef] [PubMed]
31. Denisow-Pietrzyk, M.; Pietrzyk, L.; Denisow, B. Asteraceae species as potential environmental factors of allergy. *Environ. Sci. Pollut. Res.* **2019**, *26*, 6290–6300. [CrossRef] [PubMed]
32. Simonsen, H.T.; Weitzel, C.; Christensen, S.B. Guaianolide sesquiterpenoids: Pharmacology and biosynthesis. In *Natural Products*; Ramawat, K.G., Merillon, J.M., Eds.; Springer International Publishing: Cham, Switzerland, 2013; p. 3069. [CrossRef]
33. Chen, L.; Lu, X.; El-Seedi, H.; Teng, H. Recent advances in the development of sesquiterpenoids in the treatment of type 2 diabetes. *Trends Food Sci. Technol.* **2019**, *88*, 46–56. [CrossRef]
34. Faustino, M.V.; Seca, A.M.; Silveira, P.; Silva, A.M.; Pinto, D.C. Gas chromatography-mass spectrometry profile of four *Calendula* L. taxa: A comparative analysis. *Ind. Crops Prod.* **2017**, *104*, 91–98. [CrossRef]
35. Chitrakar, B.; Zhang, M.; Bhandari, B. Edible flowers with the common name "marigold": Their therapeutic values and processing. *Trends Food Sci. Technol.* **2019**, *89*, 76–87. [CrossRef]
36. Moghaddasi Mohammad, S.; Kashanisup, H.H. Pot marigold (*Calendula officinalis*) medicinal usage and cultivation. *Sci. Res. Essays* **2012**, *7*, 1468–1472. [CrossRef]
37. Arora, D.; Rani, A.; Sharma, A. A review on phytochemistry and ethnopharmacological aspects of genus *Calendula*. *Phcog. Rev.* **2013**, *7*, 179. [CrossRef] [PubMed]
38. Olennikov, D.N.; Kashchenko, N.I. Marigold metabolites: Diversity and separation methods of *Calendula* genus phytochemicals from 1891 to 2022. *Molecules* **2022**, *27*, 8626. [CrossRef] [PubMed]
39. Rather, M.A.; Dar, B.A.; Sofi, S.N.; Bhat, B.A.; Qurishi, M.A. *Foeniculum vulgare*: A comprehensive review of its traditional use, phytochemistry, pharmacology, and safety. *Arab. J. Chem.* **2016**, *9*, S1574–S1583. [CrossRef]
40. Badgujar, S.B.; Patel, V.V.; Bandivdekar, A.H. *Foeniculum vulgare* Mill: A review of its botany, phytochemistry, pharmacology, contemporary application, and toxicology. *Biomed Res. Int.* **2014**, *2014*, 842674. [CrossRef] [PubMed]
41. Malhotra, S.K. Fennel and fennel seed. In *Handbook of Herbs and Spices*, 2nd ed.; Peter, K.V., Ed.; Woodhead Publishing Series; Elsevier: Amsterdam, The Netherlands, 2012; pp. 275–302. [CrossRef]
42. Wilson, L. Spices and flavoring crops: Fruits and seeds. In *Encyclopedia of Food and Health*; Caballero, B., Finglas, P.M., Toldrá, F., Eds.; Woodhead Publishing Series; Elsevier: Amsterdam, The Netherlands, 2016; pp. 73–83. [CrossRef]
43. Siviero, P.; Esposito, C.; De Masi, L. Il finocchio (*Foeniculum vulgare* var. *dulce* Mill.). *Ess. Der. Agr.* **2005**, *75*, 15–20.
44. Ashwlayan, V.D.; Kumar, A.; Verma, M. Therapeutic potential of *Calendula officinalis*. *Pharm. Pharmacol. Int. J.* **2018**, *6*, 149–155. [CrossRef]
45. Shahane, K.; Kshirsagar, M.; Tambe, S.; Jain, D.; Rout, S.; Ferreira, M.K.M.; Mali, S.; Amin, P.; Srivastav, P.P.; Cruz, J.; et al. An updated review on the multifaceted therapeutic potential of *Calendula officinalis* L. *Pharmaceuticals* **2023**, *16*, 611. [CrossRef] [PubMed]
46. Meikle, R.D. *Calendula L. in Flora Europaea, 1*; Tutin, T.G., Heywood, V.H., Burges, N.A., Valentine, D.H., Walters, S.M., Webb, D.A., Eds.; Cambridge University Press: Cambridge, UK, 1976; pp. 206–207.
47. Safdar, W.; Majeed, H.; Naveed, I.; Kayani, W.K.; Ahmed, H.; Hussain, S.; Kamal, A. Pharmacognostical study of the medicinal plant *Calendula officinalis* L. (family Compositae). *Int. J. Cell Mol. Biol.* **2010**, *1*, 108–116.
48. de Lima Franzen, F.; Rodríguez de Oliveira, M.S.; Lidório, H.F.; Farias Menegaes, J.; Martins Fries, L.L. Chemical composition of rose, sunflower and calendula flower petals for human food use. *Cienc. Tecnol. Agropecu.* **2019**, *20*, 149–168.
49. Sapkota, B.; Kunwar, P. A Review on Traditional Uses, Phytochemistry and Pharmacological Activities of Calendula officinalis Linn. *Nat. Prod. Commun.* **2024**, *19*, 1934578X241259021. [CrossRef]

50. Swiatek, L.; Gora, J. Phenolic acids in the inflorescences of *Arnica montana* L. and *Calendula officinalis* L. *Herba Pol.* **1978**, *24*, 187–192.
51. Ak, G.; Zengin, G.; Sinan, K.I.; Mahomoodally, M.F.; Picot-Allain, M.C.N.; Cakır, O.; Bensari, S.; Yılmaz, M.A.; Gallo, M.; Montesano, D. A comparative bio-evaluation and chemical profiles of *Calendula officinalis* L. extracts prepared via different extraction techniques. *Appl. Sci.* **2020**, *10*, 5920. [CrossRef]
52. Ukiya, M.; Akihisa, T.; Yasukawa, K.; Tokuda, H.; Suzuki, T.; Kimura, Y. Anti-inflammatory, anti-tumour-promoting, and cytotoxic activities of constituents of marigold (*Calendula officinalis*) flowers. *J. Nat. Prod.* **2006**, *69*, 1692–1696. [CrossRef] [PubMed]
53. Tavallali, V.; Rahmati, S.; Bahmanzadegan, A.; Lasibi, M.J.M. Phenolic profile and evaluation of antimicrobial and anticancer activities of *Calendula officinalis* L. using exogenous polyamines application. *Ind. Crops Prod.* **2024**, *214*, 118571. [CrossRef]
54. Yoshikawa, M.; Murakami, T.; Kishi, A.; Kageura, T.; Matsuda, H. Medicinal flowers. III. Marigold. (1): Hypoglycemic, gastric emptying inhibitory, and gastroprotective principles and new oleanane-type triterpene oligoglycosides, calendasaponins A, B, C, and D, from Egyptian *Calendula officinalis*. *Chem. Pharm. Bull.* **2001**, *49*, 863–870. [CrossRef] [PubMed]
55. Olennikov, D.N.; Kashchenko, N.I. New isorhamnetin glycosides and other phenolic compounds from *Calendula officinalis*. *Chem. Nat. Compd.* **2013**, *49*, 833–840. [CrossRef]
56. Okoh, O.O.; Sadimenko, A.A.; Afolayan, A.J. The effects of age on the yield and composition of the essential oils of *Calendula officinalis*. *J. Appl. Sci.* **2007**, *7*, 3806–3810. [CrossRef]
57. Ak, G.; Zengin, G.; Ceylan, R.; Fawzi Mahomoodally, M.; Jugreet, S.; Mollica, A.; Stefanucci, A. Chemical composition and biological activities of essential oils from *Calendula officinalis* L. flowers and leaves. *Flavour Fragr. J.* **2021**, *36*, 554–563. [CrossRef]
58. Adler, G.; Kasprzyk, Z. Free sterols, steryl esters, glycosides, acelyted glycosides and water soluble complexes in *Calendula officinalis*. *Phytochemistry* **1975**, *14*, 627–631. [CrossRef]
59. Wilkomirski, B. Pentacyclic triterpene triols from *Calendula officinalis* flowers. *Phytochemistry* **1985**, *24*, 3066–3067. [CrossRef]
60. Zitterl-Eglseer, K.; Reznicek, G.; Jurenitsch, J.; Novak, J.; Zitterl, W.; Franz, C. Morphogenetic variability of faradiol monoesters in marigold *Calendula officinalis* L. *Phytochem. Anal.* **2001**, *12*, 199–201. [CrossRef]
61. Sliwowski, J.; Dziewanowska, K.; Kasprzyk, E. Ursadiol: A new triterpene diol from *Calendula officinalis* flowers. *Khimija Prir. Soyedineniy* **1973**, *12*, 157–160. [CrossRef]
62. Wojciechowski, Z.; Bochenska-Hryniewicz, M.; Kurcharezak, B.; Kasprzyk, Z. Sterol and triterpene alcohol esters from *Calendula officinalis*. *Phytochemistry* **1972**, *11*, 1165–1168. [CrossRef]
63. Neukirch, H.; D'Ambrosio, M.; Via, J.D.; Guerriero, A. Simultaneous quantitative determination of eight triterpenoid monoesters from flowers of 10 varieties of *Calendula officinalis* L. and characterisation of a new triterpenoid monoester. *Phytochem. Anal.* **2004**, *15*, 30–35. [CrossRef]
64. Bakò, E.; Deli, J.; Toth, G. HPLC study on the carotenoid composition of calendula products. *J. Biochem. Biophys. Methods* **2002**, *53*, 241–250. [CrossRef]
65. Kishimoto, S.; Maoka, T.; Sumitomo, K.; Ohmiya, A. Analysis of carotenoid composition in petals of calendula (*Calendula officinalis* L.). *Biosci. Biotechnol. Biochem.* **2005**, *69*, 2122–2128. [CrossRef]
66. Hernández-Saavedra, D.; Pérez-Ramírez, I.F.; Ramos-Gómez, M.; Mendoza-Díaz, S.; Loarca-Pina, G.; Reynoso-Camacho, R. Phytochemical characterization and effect of *Calendula officinalis*, *Hypericum perforatum*, and *Salvia officinalis* infusions on obesity-associated cardiovascular risk. *Med. Chem. Res.* **2016**, *25*, 163–172. [CrossRef]
67. Martinez, L.G. Preclinical vascular activity of an aqueous extract from flowers of *Calendula officinalis*. *J. Pharm. Pharmacol.* **2020**, *8*, 339–344. [CrossRef]
68. Ray, D.; Mukherjee, S.; Falchi, M.; Bertelli, A.; Braga, P.C.; Das, K.D. Amelioration of myocardial ischemic reperfusion injury with *Calendula officinalis*. *Curr. Pharm. Biotechnol.* **2010**, *11*, 849–854. [CrossRef]
69. Rigane, G.; Younes, S.B.; Ghazghazi, H.; Salem, R.B. Investigation into the biological activities and chemical composition of *Calendula officinalis* L. growing in Tunisia. *Int. Food Res. J.* **2013**, *20*, 3001.
70. Frankic, T.; Salobir, K.; Salobir, J. The comparison of in vivo antigenotoxic and antioxidative capacity of two propylene glycol extracts of *Calendula officinalis* (marigold) and vitamin E in young growing pigs. *J. Anim. Physiol. Anim. Nutr.* **2008**, *41*, 688–694. [CrossRef]
71. Braga, P.C.; Dal Sasso, M.; Culici, M.; Spallino, A.; Falchi, M.; Bertelli, A.; Morelli, R.; Lo Scalzo, R. Antioxidant activity of *Calendula officinalis* extract: Inhibitory effects on chemiluminescence of human neutrophil bursts and electron paramagnetic resonance spectroscopy. *Pharmacology* **2009**, *83*, 348–355. [CrossRef]
72. Efstratiou, E.; Hussain, A.I.; Nigam, P.S.; Moore, J.E.; Ayub, M.A.; Rao, J.R. Antimicrobial activity of *Calendula officinalis* petal extracts against fungi, as well as Gram-negative and Gram-positive clinical pathogens. *Complement. Ther. Clin. Pract.* **2012**, *18*, 173–176. [CrossRef]
73. Darekar, D.; Hate, M. Phytochemical screening of *Calendula officinalis* (Linn.) using gas-chromatography-mass spectroscopy with potential antibacterial activity. *J. Sci. Res.* **2021**, *65*, 131–134. [CrossRef]
74. Elias, R.; De Meo, M.; Vidal-Ollivier, E.; Laget, M.; Balansard, G.; Dumenil, G. Antimutagenic activity of some saponins isolated from *Calendula officinalis* L., *C. arvensis* L. and *Hedera helix* L. *Mutagenesis* **1990**, *5*, 327–331. [CrossRef]
75. Jimenez-Medina, E.; Garcia-Lora, A.; Paco, L.; Algarra, I.; Collado, A.; Garrido, F. A new extract of the plant *Calendula officinalis* produces a dual in vitro effect: Cytotoxic antitumor activity and lymphocyte activation. *BMC Cancer* **2006**, *6*, 119–132. [CrossRef]
76. Cruceriu, D.; Balacescu, O.; Rakosy, E. *Calendula officinalis*: Potential roles in cancer treatment and palliative care. *Integr. Cancer Ther.* **2018**, *17*, 1068–1078. [CrossRef]

77. Ercetin, T.; Senol, F.S.; Orhan, I.E.; Toker, G. Comparative assessment of antioxidant and cholinesterase inhibitory properties of the marigold extracts from *Calendula arvensis* L. and *Calendula officinalis* L. *Ind. Crops Prod.* **2012**, *36*, 203–208. [CrossRef]
78. Silva, D.; Ferreira, M.S.; Sousa-Lobo, J.M.; Cruz, M.T.; Almeida, I.F. Anti-inflammatory activity of *Calendula officinalis* L. flower extract. *Cosmetics* **2021**, *8*, 31. [CrossRef]
79. Nicolaus, C.; Junghanns, S.; Hartmann, A.; Murillo, R.; Ganzera, M.; Merfort, I. In vitro studies to evaluate the wound healing properties of *Calendula officinalis* extracts. *J. Ethnopharmacol.* **2017**, *196*, 94–103. [CrossRef]
80. Preethi, K.C.; Kuttan, R. Hepato and reno protective action of *Calendula officinalis* L. flower extract. *Indian J. Exp. Biol.* **2009**, *47*, 163–168.
81. Bogdanova, N.S.; Nikolaeva, I.S.; Shcherbakova, L.I.; Tolstova, T.I.; Moskalenko, N.I.; Pershin, G.N. Study of antiviral properties of *Calendula officinalis*. *Farmakol. Toksikol.* **1970**, *33*, 349.
82. Kalvatchev, Z.; Walder, R.; Garzaro, D. Anti-HIV activity of extracts from *Calendula officinalis* flowers. *Biomed. Pharmacother.* **1997**, *51*, 176–180. [CrossRef]
83. Reider, N.; Comericki, P.; Hausen, B.M.; Fritsch, P.; Aberer, W. The seamy side of natural medicines: Contact sensitization to arnica (*Arnica montana* L.) and marigold (*Calendula officinalis* L.). *Contact Dermat.* **2001**, *45*, 269–272. [CrossRef]
84. Salapovic, H.; Geier, J.; Reznicek, G. Quantification of sesquiterpene lactones in Asteraceae plant extracts: Evaluation of their allergenic potential. *Sci. Pharm.* **2013**, *81*, 807–818. [CrossRef]
85. Kooti, W.; Moradi, M.; Ali-Akbari, S.; Sharafi-Ahvazi, N.; Asadi-Samani, M.; Ashtary-Larky, D. Therapeutic and pharmacological potential of *Foeniculum vulgare* Mill: A review. *J. Herbmed Pharmacol.* **2015**, *4*, 1–9.
86. Diaz-Maroto, M.C.; Hidalgo, I.J.; Sanchez-Palomo, E.; Perez-Coello, M.S. Volatile components and key odorants of fennel (*Foeniculum vulgare* Mill.) and thyme (*Thymus vulgaris* L.) oil extracts obtained by simultaneous distillation—Extraction and supercritical fluid extraction. *J. Agric. Food Chem.* **2005**, *53*, 5385–5389. [CrossRef]
87. Faudale, M.; Viladomat, F.; Bastida, J.; Poli, F.; Codina, C. Antioxidant activity and phenolic composition of wild, edible, and medicinal fennel from different Mediterranean countries. *J. Agric. Food Chem.* **2008**, *56*, 1912–1920. [CrossRef]
88. Roby, M.H.H.; Sarhan, M.A.; Selim, K.A.; Khalel, K.I. Antioxidant and antimicrobial activities of essential oil and extracts of fennel (*Foeniculum vulgare* L.) and chamomile (*Matricaria chamomilla* L.). *Ind. Crops Prod.* **2013**, *44*, 437–445. [CrossRef]
89. Parejo, I.; Jauregui, O.; Sánchez-Rabaneda, F.; Viladomat, F.; Bastida, J.; Codina, C. Separation and characterization of phenolic compounds in fennel (*Foeniculum vulgare*) using liquid chromatography-negative electrospray ionization tandem mass spectrometry. *J. Agric. Food Chem.* **2004**, *52*, 3679–3687. [CrossRef]
90. Marino, S.D.; Gala, F.; Borbone, N.; Zollo, F.; Vitalini, S.; Visioli, F.; Iorizzi, M. Phenolic glycosides from *Foeniculum vulgare* fruit and evaluation of antioxidative activity. *Phytochemistry* **2007**, *68*, 1805–1812. [CrossRef]
91. Diaz-Maroto, M.C.; Perez-Coello, M.S.; Esteban, J.; Sanz, J. Comparison of the volatile composition of wild fennel samples (*Foeniculum vulgare* Mill.) from Central Spain. *J. Agric. Food Chem.* **2006**, *54*, 6814–6818. [CrossRef]
92. Damjanovic, B.; Lepojevic, Z.; Zivkovic, V.; Tolic, A. Extraction of fennel (*Foeniculum vulgare* Mill.) seeds with supercritical CO2: Comparison with hydrodistillation. *Food Chem.* **2005**, *92*, 143–149. [CrossRef]
93. Vella, F.M.; Calandrelli, R.; Cautela, D.; Fiume, I.; Pocsfalvi, G.; Laratta, B. Chemometric screening of fourteen essential oils for their composition and biological properties. *Molecules* **2020**, *25*, 5126. [CrossRef]
94. Cautela, D.; Vella, F.M.; Castaldo, D.; Laratta, B. Characterization of essential oil recovered from fennel horticultural wastes. *Nat. Prod. Res.* **2019**, *33*, 1964–1968. [CrossRef]
95. Tognolini, M.; Ballabeni, V.; Bertoni, S.; Bruni, R.; Impicciatore, M.; Barocelli, E. Protective effect of Foeniculum vulgare essential oil and anethole in an experimental model of thrombosis. *Pharmacol. Res.* **2007**, *56*, 254–260. [CrossRef]
96. Senatore, F.; Oliviero, F.; Scandolera, E.; Taglialatela-Scafati, O.; Roscigno, G.; Zaccardelli, M.; De Falco, E. Chemical composition, antimicrobial and antioxidant activities of anethole-rich oil from leaves of selected varieties of fennel [*Foeniculum vulgare* Mill. ssp. *vulgare* var. *azoricum* (Mill.) Thell]. *Fitoterapia* **2013**, *90*, 214–219. [CrossRef]
97. Diao, W.; Hu, Q.; Zhang, H.; Xu, J. Chemical composition, antibacterial activity and mechanism of action of essential oil from seeds of fennel (*Foeniculum vulgare* Mill.). *Food Control* **2014**, *35*, 109–116. [CrossRef]
98. Shahat, A.A.; Ibrahim, A.Y.; Hendawy, S.F.; Omer, E.A.; Hammouda, F.M.; Rahman, F.H.A.; Saleh, M.A. Chemical composition, antimicrobial and antioxidant activities of essential oils from organically cultivated fennel cultivars. *Molecules* **2011**, *16*, 1366–1377. [CrossRef]
99. Servi, H.; Şen, A.; Yildirim, S.; Doğan, A. Chemical composition and biological activities of essential oils of *Foeniculum vulgare* Mill. and *Daucus carota* L. growing wild in Turkey. *J. Res. Pharm.* **2021**, *25*, 142–152.
100. Afifi, S.M.; El-Mahis, A.; Heiss, A.G.; Farag, M.A. Gas chromatography-mass spectrometry-based classification of 12 fennel (*Foeniculum vulgare* Miller) varieties based on their aroma profiles and estragole levels as analyzed using chemometric tools. *ACS Omega* **2021**, *6*, 5775–5785. [CrossRef]
101. Kaur, G.J.; Arora, D.S. Antibacterial and phytochemical screening of *Anethum graveolens*, *Foeniculum vulgare* and *Trachyspermum ammi*. *BMC Complement. Med. Ther.* **2009**, *9*, 30. [CrossRef]
102. Schulz, M.; Meins, J.; Diemert, S.; Zagermann-Muncke, P.; Goebel, R.; Schrenk, D.; Schubert-Zsilavecz, M.; Abdel-Tawab, M. Detection of pyrrolizidine alkaloids in German licensed herbal medicinal teas. *Phytomedicine* **2015**, *22*, 648–656. [CrossRef]
103. Kwon, Y.; Koo, Y.; Jeong, Y. Determination of pyrrolizidine alkaloids in teas using liquid chromatography-tandem mass spectrometry combined with rapid-easy extraction. *Foods* **2021**, *10*, 2250. [CrossRef]

104. Zellagui, A.; Gherraf, N.; Elkhateeb, A.; Hegazy, M.E.F.; Mohamed, T.A.; Touil, A.; Shahat, A.A.; Rhouati, S. Chemical constituents from Algerian *Foeniculum vulgare* aerial parts and evaluation of antimicrobial activity. *J. Chil. Chem. Soc.* **2011**, *56*, 759–763. [CrossRef]
105. Dadalioglu, I.; Evrendilek, G.A. Chemical compositions and antibacterial effects of essential oils of turkish oregano (*Origanum minutiflorum*), bay laurel (*Laurus nobilis*), spanish lavender (*Lavandula stoechas* L.), and fennel (*Foeniculum vulgare*) on common foodborne pathogens. *J. Agric. Food Chem.* **2004**, *52*, 8255–8260. [CrossRef]
106. Cetin, B.; Ozer, H.; Cakir, A.; Polat, T.; Dursun, A.; Mete, E.; Oztürk, E.; Ekinci, M. Antimicrobial activities of essential oil and hexane extract of Florence fennel [*Foeniculum vulgare* var. *azoricum* (Mill.) Thell.] against foodborne microorganisms. *J. Med. Food* **2010**, *13*, 196–204. [CrossRef]
107. Orhan, I.E.; Ozcelik, B.; Kartal, M.; Kan, Y. Antimicrobial and antiviral effects of essential oils from selected Umbelliferae and Labiatae plants and individual essential oil components. *Turk. J. Biol.* **2012**, *36*, 239–246. [CrossRef]
108. Chainy, G.B.; Manna, S.K.; Chaturvedi, M.M.; Aggarwal, B.B. Anethole blocks both early and late cellular responses transduced by tumor necrosis factor: Effect on NF-κB, AP-1, JNK, MAPKK and apoptosis. *Oncogene* **2000**, *19*, 2943–2950. [CrossRef]
109. Pradhan, M.; Sribhuwaneswari, S.; Karthikeyan, D.; Minz, S.; Sure, P.; Chandu, A.N.; Mishra, U.; Kamalakannan, K.; Saravanankumar, A.; Sivakumar, T. In vitro cytoprotection activity of *Foeniculum vulgare* and *Helicteres isora* in cultured human blood lymphocytes and antitumor activity against B16F10 melanoma cell line. *Res. J. Pharm. Technol.* **2008**, *1*, 450–452.
110. Mohamad, R.H.; El-Bastawesy, A.M.; Abdel-Monem, M.G.; Noor, A.M.; Al-Mehdar, H.A.R.; Sharawy, S.M.; El-Merzabani, M.M. Antioxidant and anticarcinogenic effects of methanolic extract and volatile oil of fennel seeds (*Foeniculum vulgare*). *J. Med. Food* **2011**, *14*, 986–1001. [CrossRef]
111. Ozbek, H.; Ugraş, S.; Dulger, H.; Bayram, I.; Tuncer, I.; Ozturk, G.; Ozturk, A. Hepatoprotective effect of *Foeniculum vulgare* essential oil. *Fitoterapia* **2003**, *74*, 317–319. [CrossRef]
112. Birdane, F.M.; Cemek, M.; Birdane, Y.O.; Gülçin, I.; Büyükokuroğlu, M.E. Beneficial effects of Foeniculum vulgare on ethanol-induced acute gastric mucosal injury in rats. *World J. Gastroenterol.* **2007**, *13*, 607–611. [CrossRef]
113. Al-Mofleh, I.; Al-Sobaihani, M.; Alqasoumi, S.; Al-Said, M.; Al-Dosari, M.; Al-Yahya, M.; Rafatullah, S. Fennel *Foeniculum vulgare* treatment protects the gastric mucosa of rats against chemically-induced histological lesions. *Int. J. Pharm.* **2013**, *9*, 182. [CrossRef]
114. Oulmouden, F.; Ghalim, N.; El Morhit, M.; Benomar, H.; Daoudi, E.M.; Amrani, S. Hypolipidemic and anti-atherogenic effect of methanol extract of fennel (*Foeniculum vulgare*) in hypercholesterolemic mice. *Int. J. Sci. Knowl.* **2014**, *3*, 42–52.
115. Sushruta, K.; Satyanarayana, S.; Srinivas, S.; Sekhar, J.R. Evaluation of the blood-glucose reducing effects of aqueous extracts of the selected umbelliferous fruits used in culinary practices. *Trop. J. Pharm. Res.* **2007**, *5*, 613–617. [CrossRef]
116. Dongare, V.; Arvindekar, A.; Magadum, C. Hypoglycemic effect of *Foeniculum vulgare* Mill. fruit on dexamethasone induced insulin resistance rats. *Res. J. Pharmacogn. Phytochem.* **2010**, *2*, 163–165.
117. El-Soud, N.; El-Laithy, N.; El-Saeed, G.; Wahby, M.; Khalil, M.; Morsy, F.; Shaffie, N. Antidiabetic activities of *Foeniculum vulgare* Mill. essential oil in streptozotocin-induced diabetic rats. *Maced. J. Med. Sci.* **2011**, *4*, 139–146.
118. Albert-Puleo, M. Fennel and anise as estrogenic agents. *J. Ethnopharmacol.* **1980**, *2*, 337–344. [CrossRef]
119. Ostad, S.; Soodi, M.; Shariffzadeh, M.; Khorshidi, N.; Marzban, H. The effect of fennel essential oil on uterine contraction as a model for dysmenorrhea, pharmacology and toxicology study. *J. Ethnopharmacol.* **2011**, *76*, 299–304. [CrossRef]
120. Myrseyed, F.; Shiravi, A.; Nasr-Abadi, M. The effect of intraperitoneal injection of alcoholic extract *Foeniculum vulgare* seed on gonadotropic and testosterone hormones in male wistar rats. *J. Anim. Sci.* **2008**, *1*, 49–56.
121. Mesfin, M.; Asres, K.; Shibeshi, W. Evaluation of anxiolytic activity of the essential oil of the aerial part of *Foeniculum vulgare* Miller in mice. *BMC Complement. Altern. Med.* **2014**, *14*, 310. [CrossRef]
122. Pourabbas, S.; Kesmati, M.; Rasekh, A. Study of the anxiolytic effects of fennel and possible roles of both gabaergic system and estrogen receptors in these effects in adult female rat. *Physiol. Pharmacol.* **2011**, *15*, 134–143.
123. Lima, N.G.; De Sousa, D.P.; Pimenta, F.C.F.; Alves, M.F.; De Souza, F.S.; Macedo, R.O.; Cardoso, R.B.; de Morais, L.C.; De Almeida, R.N. Anxiolytic-like activity and GC-MS analysis of (R)-(+)-limonene fragrance, a natural compound found in foods and plants. *Pharmacol. Biochem. Behav.* **2013**, *103*, 450–454. [CrossRef]
124. Joshi, H.; Parle, M. Cholinergic basis of memory-strengthening effect of *Foeniculum vulgare* Linn. *J. Med. Food* **2006**, *9*, 413–417. [CrossRef]
125. Zeller, A.; Rychlik, M. Character impact odorants of fennel fruits and fennel tea. *J. Agric. Food Chem.* **2006**, *54*, 3686–3692. [CrossRef]
126. Drinkwater, N.R.; Miller, E.C.; Miller, J.A.; Pitot, H.C. Hepato-carcinogenicity of estragole (1-allyl-4-methoxybenzene) and 1-hydroxyestragole in the mouse and mutagenicity of 1-acetoxyestragole in bacteria. *J. Natl. Cancer Inst.* **1976**, *57*, 1323–1331. [CrossRef]
127. Riboli, E.; Beland, F.A.; Lachenmeier, D.W.; Marques, M.M.; Phillips, D.H.; Schernhammer, E.; Afghan, A.; Assunção, R.; Caderni, G.; Corton, J.C.; et al. Carcinogenicity of aspartame, methyleugenol, and isoeugenol. *Lancet Oncol.* **2023**, *24*, 848–850. [CrossRef]
128. Paini, A.; Punt, A.; Viton, F.; Scholz, G.; Delatour, T.; Marin-Kuan, M.; Schilter, B.; van Bladeren, P.J.; Rietjens, I.M. A physiologically based biodynamic (PBBD) model for estragole DNA binding in rat liver based on in vitro kinetic data and estragole DNA adduct formation in primary hepatocytes. *Toxicol. Appl. Pharmacol.* **2010**, *245*, 57–66. [CrossRef]

129. Miller, E.C.; Swanson, A.B.; Phillips, D.H.; Fletcher, T.L.; Liem, A.; Miller, J.A. Structure-activity studies of the carcinogenicities in the mouse and rat of some naturally occurring and synthetic alkenyl-benzene derivatives related to safrole and estragole. *Cancer Res.* **1983**, *43*, 1124–1134.
130. Swanson, A.B.; Miller, E.C.; Miller, J.A. The side-chain epoxidation and hydroxylation of the hepatocarcinogens safrole and estragole and some related compounds by rat and mouse liver microsomes. *Biochim. Biophys. Acta* **1981**, *673*, 504–516. [CrossRef] [PubMed]
131. Swanson, A.B.; Chambliss, D.D.; Blomquist, J.C.; Miller, E.C.; Miller, J.A. The mutagenicities of safrole, estragole, eugenol, trans-anethole, and some of their known or possible metabolites for *Salmonella typhimurium* mutants. *Mutat. Res.* **1979**, *60*, 143–153. [CrossRef]
132. Punt, A.; Freidig, A.P.; Delatour, T.; Scholz, G.; Boersma, M.G.; Schilter, B.; van Bladeren, P.J.; Rietjens, I.M. A physiologically based biokinetic (PBBK) model for estragole bioactivation and detoxification in rat. *Toxicol. Appl. Pharmacol.* **2008**, *231*, 248–259. [CrossRef]

molecules

MDPI

Article

Dentifragilones A–B and Other Benzoic Acid Derivatives from the European Basidiomycete *Dentipellis fragilis*

Winnie Chemutai Sum [1,2], Sherif S. Ebada [1,3], Mahmoud A. A. Ibrahim [4,5], Harald Kellner [6] and Marc Stadler [1,2,*]

1 Department of Microbial Drugs, Helmholtz Centre for Infection Research (HZI) and German Centre for Infection Research (DZIF), DZIF Partner Site Hannover-Braunschweig, Inhoffenstrasse 7, 38124 Braunschweig, Germany; winnie.sumchemutai@helmholtz-hzi.de (W.C.S.); sherif.elsayed@helmholtz-hzi.de or sherif_elsayed@pharma.asu.edu.eg (S.S.E.)
2 Institute of Microbiology, Technische Universität Braunschweig, Spielmannstraße 7, 38106 Braunschweig, Germany
3 Department of Pharmacognosy, Faculty of Pharmacy, Ain Shams University, Cairo 11566, Egypt
4 Computational Chemistry Laboratory, Chemistry Department, Faculty of Science, Minia University, Minia 61519, Egypt; m.ibrahim@compchem.net
5 School of Health Sciences, University of KwaZulu-Natal, Westville, Durban 4000, South Africa
6 Department of Bio- and Environmental Sciences, Technische Universität Dresden-International Institute Zittau, Markt 23, 02763 Zittau, Germany; harald.kellner@tu-dresden.de
* Correspondence: marc.stadler@helmholtz-hzi.de; Tel.: +49-531-6181-4240; Fax: +49-531-6181-9499

Abstract: A chemical and biological exploration of the European polypore *Dentipellis fragilis* afforded two previously undescribed natural products (**1** and **2**), together with three known derivatives (**3**–**5**). Chemical structures of the isolated compounds were confirmed through 1D/2D NMR spectroscopic analyses, mass spectrometry, and by comparison with the reported literature. The relative and absolute configurations of **1** were determined according to the ROESY spectrum and time-dependent density functional theory electronic circular dichroism (TDDFT-ECD), respectively. Furthermore, the absolute configuration of dentipellinol (**3**) was revisited and revealed to be of (*R*) configuration. All the isolated compounds were assessed for their cytotoxic and antimicrobial activities, with some being revealed to have weak to moderate antimicrobial activity, particularly against Gram-positive bacteria.

Keywords: Basidiomycota; Hericiaceae; crassinervic acid

1. Introduction

Fungal-based natural products have made an immense contribution to the modern-day medical and agrochemistry sectors, with their vast chemical novelties unmatched by many sources. Over the last two centuries, regions rich in biodiversity have undoubtedly proved to be invaluable sources of therapeutic targets useful for the global pharmaceutical industries [1]. Nonetheless, Basidiomycota of the temperate zones, assumed to be well sampled, still offer great opportunities for novelty [2]. The rare Basidiomycota of Europe in particular have been neglected, but constitute potential reservoirs of new pharmacotherapeutic agents [3].

During our investigations of the seldom-found Basidiomycetes of Germany, we encountered *Dentipellis fragilis*, a red-listed member of the wood-rot fungi of the Hericiaceae family. The fungal genus *Dentipellis* was coined by Anton Donk in 1962 and, thus far, *D. fragilis* is the only studied species of the genus with regards to its secondary metabolites. It is worth mentioning that *D. fragilis* has been demonstrated to be quite 'talented' in relation to the diversity of its produced natural compounds, with bioactive phthalide, cyathane, benzofuranone, and drimane derivatives having recently been reported from the fungus [4–9]. In particular, the discovery of cyathanes is in strong accordance with the placement of *Dentipellis* in the Hericiaceae family, since the same compound class is also characteristic of cultures of other genera in this family like *Hericium* and *Laxitextum* [2,6]. The current paper is dedicated to describing additional secondary metabolites (SMs) of the strain studied by Sum et al. [6] that were obtained upon modifying the culture conditions.

2. Results and Discussion

2.1. Chemical Characterization of 1–5 (Figure 1)

Compound **1** was isolated as a white amorphous solid. The HR-ESI-MS of **1** revealed a protonated molecular ion and sodium adduct peaks at *m/z* 227.1277 $[M+H]^+$ (calculated 227.1278) and 249.1098 $[M+Na]^+$ (calculated 249.1097), respectively, determining its molecular formula as $C_{12}H_{18}O_4$ and hence indicating four degrees of unsaturation. The ^{13}C NMR and HSQC spectral data of **1** (Table 1, Figures S4 and S7) unveiled the presence of twelve carbon resonances that can be classified into five unprotonated carbon atoms separated into one carbonyl carbon at δ_C 195.8 (C-1), two olefinic carbon atoms at δ_C 144.0 (C-2), 130.0 (C-3), and two sp^3 carbon atoms at δ_C 44.8 (C-7) and 80.4 (C-6). In addition, the ^{13}C NMR spectral data revealed one methylene sp^3 carbon atom at δ_C 36.7 (C-4) and one methylidene sp^2 carbon atom at δ_C 113.2 (C-9). The 1D (^1H and ^{13}C) NMR data and HSQC spectrum of **1** (Table 1, Figure S7) revealed the presence of three olefinic protons at δ_H 6.04 (dd, J = 17.4, 10.7 Hz, H-8; δ_C 143.6), δ_H 5.02 (d, J = 10.7 Hz, H-9α), and δ_H 5.09 (d, J = 17.4 Hz, H-9β) that were both correlated to sp^2 methylene carbon at δ_C 113.2 (C-9). The ^1H–^1H COSY spectrum of **1** (Figures 2 and S5) revealed the presence of one spin system between three olefinic protons ascribed to H-8 and H_2-9, indicating their presence as an allyl moiety. The ^1H–^1H COSY spectrum of **1** (Figures 2 and S5) also revealed an additional spin system from H-5 (δ_H 4.18, ddd, J = 9.4, 5.8, 2.8 Hz) to two diastereotopic methylene protons at δ_H 2.56/2.74 (H_2-4). The HMBC spectrum of **1** (Figures 2 and S6) revealed the presence of two diastereotopic singlet methyl groups at δ_H 1.07 (H_3-11; δ_C 24.4) and 1.26 (H_3-12; δ_C 23.9) that revealed correlations with three different carbon atoms assigned as C-6, and C-7 and two allylic carbon atoms, C-8/C-9. Furthermore, the HMBC spectrum of **1** (Figure 2) revealed key correlations between H_3-10 and two unprotonated sp^2 carbons C-2/C-3, together with a long-range "ω" correlation with the carbonyl carbon (C-1), indicating their existence as an α,β-unsaturated carbonyl moiety in **1**.

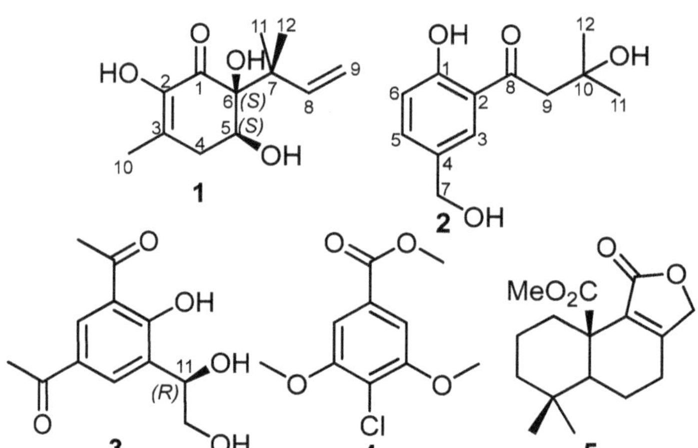

Figure 1. Chemical structures of **1**–**5**.

Table 1. (^{1}H and ^{13}C) 1D NMR data of **1**.

pos.	δ_C, ^{a,c} Type	δ_H ^{b} Multi (J[Hz])
1	195.8, CO	
2	144.0, C	
3	130.0, C	
4	36.7, CH_2	α 2.56 dd (18.4, 5.8) β 2.74 dd (18.4, 9.8)
5	75.7, CH	4.18 ddd (9.4, 5.8, 2.8)
6	80.4, C	
7	44.8, C	
8	143.6, CH	6.04 dd (17.4, 10.7)
9	113.2, CH_2	α 5.02 d (10.7) β 5.09 d (17.4)
10	17.0, CH_3	1.92 s
11	24.4, CH_3	1.07 s
12	23.9, CH_3	1.26 s
1-OH	-	-
2-OH	-	5.76 s
5-OH	-	3.44 br s
6-OH	-	2.37 br s

Measured in chloroform-d at ^{a} 125/^{b} 500 MHz. ^{c} Assignment confirmed by HMBC and HSQC spectra.

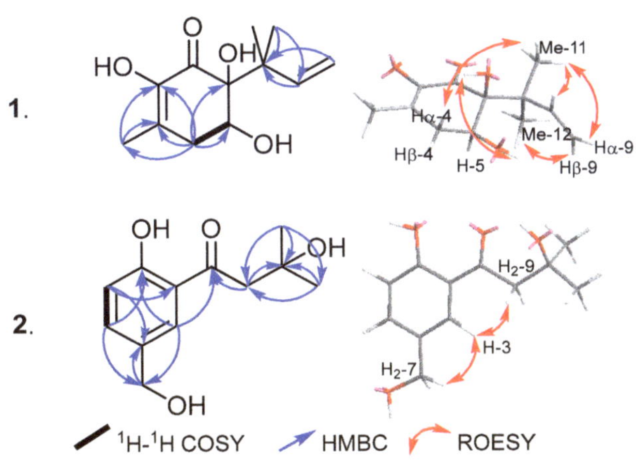

Figure 2. Key ^{1}H–^{1}H COSY, HMBC, and ROESY correlations of **1** and **2**.

A literature search of **1** revealed its common structural features of strobiloscyphones [10], pestallic acids [11], dentipellin [4], and the recently reported lachnoic acids [12], where they shared the presence of a 2-cyclohexenone moiety in their structures. A careful interpretation of the obtained 1D and 2D NMR spectral data of **1** (Table 1, Figure 2) suggested the structure depicted in Figure 1, with a 1,1-dimethyl-2-propenyl functionality being attached at C-6. The relative configuration of **1** at C-5 and C-6 was determined by its ROESY spectrum, which revealed an ROE correlation between two exchangeable broad singlet proton signals at δ_H 3.44 and 2.37 assigned to 5-OH and 6-OH, respectively. This thus indicated their cofacial orientation, while the 1,1-dimethyl-2-propenyl moiety is projected toward the opposite face of the molecule. The ROESY spectrum of **1** (Figures 2 and S8) also revealed key ROE correlations between the two diastereotopic methyl groups, H_3-11 and H_3-12, and H-4β, H-5, H-8, and H_2-9, confirming the depicted structure of **1**. Accordingly, the ROESY spectrum suggested the relative configurations at C-5 and C-6 were either (5*S**,6*S**) or (5*R**,6*R**). To determine the absolute configuration of **1**, its ECD spectrum was acquired and, hence, compared to the calculated TDDFT-ECD spectra of (5*S*,6*S*) and (5*R*,6*R*) enantiomers. The obtained results (Figure 3) revealed a coherence between both the experimental and calculated ECD spectra of (5*S*,6*S*) configuration. Based on the obtained results, compound **1** was identified as a previously undescribed natural product named dentifragilone A.

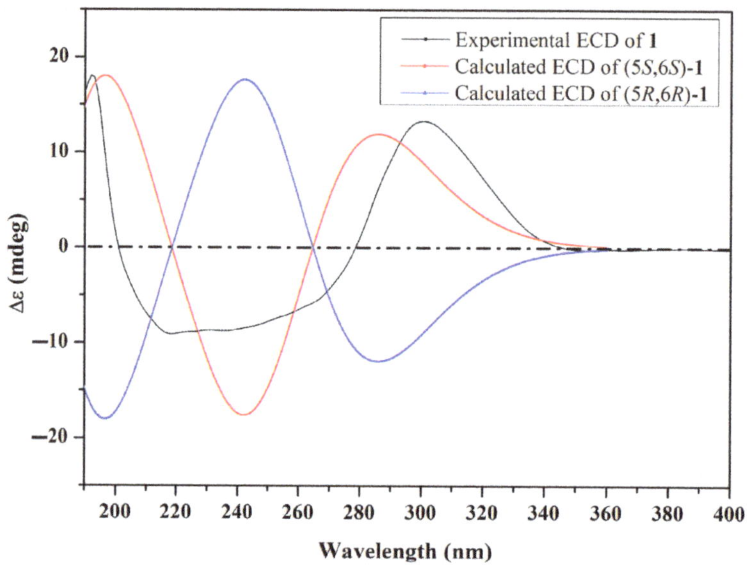

Figure 3. Experimental and calculated ECD spectra of **1**.

Compound **2** was purified as an off-white amorphous solid. The HR-ESI-MS revealed a protonated molecular ion peak and a sodium adduct peak at *m/z* 225.1117 $[M+H]^+$ (calculated 225.1121) and 247.0937 $[M+Na]^+$ (calculated 247.0941), respectively, determining its molecular formula as $C_{12}H_{16}O_4$ and indicating five degrees of unsaturation. The ^1H NMR spectral data and the ^1H–^1H COSY spectrum of **2** (Table 2, Figure S12) revealed three aromatic proton signals at δ_H 7.75 (d, *J* = 2.1 Hz), 7.49 (dd, *J* = 8.5, 2.1 Hz), and 7.00 (d, *J* = 8.5 Hz) that were correlated as one spin system, suggesting their presence on a 1,2,4-trisubstituted aromatic ring. In addition, the ^1H NMR spectral data of **2** (Table 2) revealed the presence of three singlet proton resonances that were categorized, according to their integration indices, into two methylene groups at δ_H 4.66 and δ_H 3.20 (each of an integration index of two), together with two magnetically equivalent methyl groups at δ_H 1.37 integrated for six protons. The ^{13}C NMR spectral data and DEPTQ spectrum (Table 2, Figure S16) revealed the presence of five unprotonated carbon signals resolved into one carbonyl carbon at δ_C 207.2, three sp^2 carbon atoms (δ_C 162.3, 131.4, and 119.2), and one oxygenated sp^3 carbon (δ_C 70.0). A literature search for **2** revealed its close resemblance to methyl 4-hydroxy-3-(3-methylbutanoyl) benzoate, a fungal metabolite that has been previously reported from scrap cultivation beds of *Hericium erinaceus* (published by Ueda et al. [13] under the grammatically incorrect name "*H. erinaceum*").

Table 2. (^{1}H and ^{13}C) 1D NMR data of **2**.

pos.	δ_C, ^{a,e} Type	δ_H ^{b} multi (J[Hz])	δ_C, ^{c,e} Type	δ_H ^{d} multi (J[Hz])
1	162.3, C		163.0, C	
2	119.2, C		121.6, C	
3	128.5, CH	7.75 d (2.1)	131.1, CH	7.91 d (2.1)
4	131.4, C		133.4, C	
5	135.8, CH	7.49 dd (8.5, 2.1)	136.8, CH	7.51 dd (8.5, 2.1)
6	118.7, CH	7.00 d (8.5)	119.1, CH	6.93 d (8.5)
7	64.2, CH_2	4.66 s	64.5, CH_2	4.57 s
8	207.2, CO		207.5, CO	
9	48.1, CH_2	3.20 s	50.8, CH_2	3.22 s
10	70.0, C		71.3, C	
11,12	29.4, CH_3	1.37 s	29.9, CH_3	1.36 s
1-OH	-	12.13 br s	-	-

Measured in chloroform-*d* ^{a} at 125 MHz/^{b} at 500 MHz. Measured in methanol-d_4 ^{c} at 125 MHz/^{d} at 500 MHz. ^{e} Assignment confirmed by HMBC and HSQC spectra.

Further characterization for the suggested structure of **2** was obtained via the HMBC spectrum (Figures 2 and S13), which revealed key correlations between two aromatic protons assigned to H-3 (δ_H 7.75) and H-5 (δ_H 7.49), the hydoxymethylene carbon at δ_C 64.2 (C-7), and an oxygenated aromatic carbon at δ_C 162.3 (C-1), indicating that the hydroxymethylene group is bound at C-4. The HMBC spectrum of **2** also revealed key correlations from two magnetically equivalent methyl groups (H_3-11/H_3-12) to C-10 (δ_C 70.0) and C-9 (δ_C 48.1), whereas the methylene group at δ_H 3.20 (H_2-9) and H-3 revealed correlations to a carbonyl carbon (C-8), indicating the presence of a 3-hydroxy-3-methylbutanoyl moiety at C-2 on the aromatic ring. The ROESY spectrum of **2** (Figure 2) revealed key correlations from H-3 to both H_2-7 and H_2-9, confirming the depicted arrangement of substituents on the aromatic ring at C-1, C-2, and C-4. A literature search revealed that compound **2** revealed a reduced primary alcohol derivative related to crassinervic acid, an antifungal metabolite from *Piper crassinervium* [14]. Based on the obtained results, compound **2** was identified as a previously undescribed natural product and it was given a trivial name, dentifragilone B.

Compound **3** was obtained as an off-white amorphous solid. Its molecular formula was established to be $C_{12}H_{14}O_5$, indicating six degrees of unsaturation. A literature search of **3**, based on its molecular formula and the 1D (1H and ^{13}C) NMR spectral data (Table S1), revealed that the measured values were in close accordance with those recently reported for dentipellinol [5]. Although the absolute configuration of dentipellinol (**3**) was reported to be (*S*) configuration [5], herein its structure was revisited and determined by comparing its experimental and calculated ECD spectra (Figure S4). The results in Figure S4 indicate a closer coherence to the calculated ECD spectrum of (*R*) configuration than that of (*S*). Moreover, the 3D coordinates provided in the Supplementary Material of Ki et al.'s study confirmed the (*R*) configuration of dentipellinol (**3**), which was misinterpreted as (*S*) [5].

In addition, compounds **4** and **5** were identified as methyl 4-chloro-3,5-dimethoxybenzoate and 10-methoxycarbonyl-10-norisodrimenin, respectively, based on their HR-ESI-MS data and 1D/2D NMR spectral analyses compared to the reported literature [14,15].

2.2. Biological Assays

To assess the antimicrobial activity of compounds **1–4**, a serial dilution assay was conducted against several Gram-positive and Gram-negative bacteria as well as fungal strains. Compound **5** was not tested, since similar activities had been reported by our group in a recent study [15]. Notably, compound **4** demonstrated moderate or weak antibiotic effects against *Staphylococcus aureus* with a MIC value of 66.6 μg/mL. Compounds **1–3** were inactive in the antimicrobial tests.

An evaluation of the cytotoxic activities of the isolated compounds was first conducted against the two most sensitive cell lines, namely mouse fibroblast (L929) and human endocervical adenocarcinoma (KB3.1). The compounds had no apparent cytotoxic effects; hence, further tests were not conducted.

3. Materials and Methods

3.1. General Experimental Procedures

Optical rotations (OR) were recorded on a MCP-150 polarimeter (Anton Paar; Seelze, Germany) at 20 °C using methanol (Uvasol, Merck; Darmstadt, Germany). UV-Vis spectra measurements were acquired using a UV-Vis spectrophotometer UV-2450 (Shimadzu; Kyoto, Japan), while electronic circular dichroism (ECD) spectra were measured using a J-815 spectropolarimeter (Jasco, Pfungstadt, Germany).

Nuclear magnetic resonance (NMR) spectra were recorded using an Avance III 500 MHz spectrometer equipped with a BBFO (plus) SmartProbe (^{1}H: 500 MHz, ^{13}C: 125 MHz; Bruker, Billerica, MA, USA) and an Avance III 700 MHz spectrometer equipped with a 5 mm TCI cryoprobe (^{1}H: 700 MHz, ^{13}C: 175 MHz; Bruker, Billerica, MA, USA) (sample temperature: 298 K). The NMR data were referenced to selected chemical shifts δ of chloroform-*d* (^{1}H, δ = 7.27 ppm; ^{13}C, δ = 77.2 ppm) and methanol-d_4 (^{1}H, δ = 3.31 ppm; ^{13}C, δ = 49.0 ppm). Electrospray ionization mass (ESI-MS) spectra were acquired with an UltiMate® 3000 Series uHPLC (Thermo Fisher Scientific; Waltman, MA, USA) employing a C_{18} Acquity® UPLC BEH column (50 × 2.1 mm, 1.7 μm; Waters, Milford, MA, USA) (temperature of the column: 40 °C) and connected to an amaZon® speed ESI-Iontrap-MS (Bruker; Billerica, MA, USA). The following parameters were used to set up the HPLC system: solvent A: Deionized H_2O + 0.1% formic acid (FA) (*v*/*v*); solvent B: acetonitrile (MeCN) + 0.1% FA (*v*/*v*) as the mobile phase; gradient: 5% B for 0.5 min increasing to 100% B in 19.5 min and maintaining isocratic conditions at 100% B for 5 min; flow rate: 0.6 mL/min; and Diode-Array Detection (DAD) at 190–600 nm. The crude extracts and pure compounds were dissolved in a solution of acetone and methanol (1:1) to achieve concentrations of 4.5 mg/mL and 1.0 mg/mL, respectively. High-resolution electrospray ionization mass spectrometry (HR-ESI-MS) spectra were measured through an Agilent 1200 Infinity Series HPLC–UV system (Agilent Technologies, Böblingen, Germany) with the same conditions as for ESI-MS spectra, connected to a maXis® ESI-TOF mass spectrometer (Bruker; Daltonics, Bremen, Germany)) (scan range 100–2500 *m/z*, capillary voltage 4500 V, dry temperature 200 °C).

3.2. Fungal Material

The fungus was collected on a decaying beech (*Fagus sylvatica*) in the Bavarian Forest National Park (49.098387 N, 13.246003 E) in August 2015 [16] and cultured by one of the authors (HK). The mycelial culture was deposited at the Deutsche Sammlung von Mikroorganismen und Zellkulturen (DSMZ), Braunschweig, designated as DSM 105465. Its identification was reported in our previous study [15] and an ITS-nRDNA sequence of the strain is deposited at the GenBank under the accession number MK463979. We would, however, like to point out that ITS sequences are unreliable for fungal identification and the morphological characters of the specimen already allowed for an unambiguous assignment to the taxon *Dentipellis fragilis*.

3.3. Fermentation and Extraction

The fungal strain was cultured in Erlenmeyer flasks containing either YMG or rice media. For submerged YMG cultivation, fermentation was carried out in 18 × 1 L shaker flasks containing 400 mL of medium (10 g/L malt extract, 4 g/L D-glucose, 4 g/L yeast extract, pH 6.3 before autoclaving), as previously described [6], with each inoculum consisting of 10 well-grown mycelial plugs. The cultures were incubated under shaking conditions in the dark at 140 rpm and 23 °C, and the fermentation process was monitored by checking the concentration of free glucose with Medi-Test glucose (Macherey-Nagel, Düren, Germany). The free glucose was fully consumed after 45 days, and extraction was performed after 3 days of glucose depletion. Alternatively, the rice substrate cultures were cultivated as previously reported [15]. Basically, 10 × 500 mL Erlenmeyer flasks consisting of 90 mg of rice in 90 mL distilled water were prepared and autoclaved. These were used to inoculate fungal plugs, as similarly carried out for the YMG cultures. However, the rice medium cultures were cultivated under static conditions at 23 °C, and the cultures were extracted after 30 days.

To extract the secondary metabolites from the liquid cultures, the supernatant and mycelia were first separated by vacuum filtration. The supernatant was decanted with an equal amount of EtOAc in a separatory funnel. The obtained organic phase was filtered through anhydrous sodium sulfate and the filtrate was evaporated to dryness under a vacuum at 40 °C with a rotary evaporator (Heidolph Instruments GmbH & Co. KG, Schwabach, Germany; pump: Vacuubrand GmbH & Co. KG, Wertheim am Main, Germany) in order to produce a solid residue of the total extract. The secondary metabolites from the mycelia (from either submerged or rice medium cultures) were extracted by initial soaking the mycelia in acetone, followed by immediate sonication for 30 min at 40 °C using an ultrasonic bath (Sonorex Digital 10 P, Bandelin Electronic GmbH & Co. KG, Berlin, Germany). The acetone was evaporated under reduced pressure at 40 °C, the resulting aqueous phase was decanted with an equal amount of ethyl acetate, and the total extract was obtained, as previously described for the supernatant phase. The overall process yielded 881 mg, 367 mg, and 1.6 g of supernatant, mycelia, and rice crude extracts, respectively.

3.4. Isolation of Compounds **1–5**

To further separate the compounds, the mycelial and supernatant crude extracts were first combined due to their similar chemical profiles. The total extract (1.25 g) was dissolved in methanol (MeOH) and pre-fractionated using a Reveleris X2 flash chromatography system (W.R. Grace and Co., Columbia, MD, USA) equipped with a 40 g silica pre-packed column (Reveleris®). Dichloromethane (CH_2Cl_2) (solvent A) and CH_2Cl_2:MeOH (ratio 8:2) (solvent B) were used as eluents, with a flow rate of 60 mL/min. The gradient of separation started with 0% to 30% B in 30 min, isocratic holding at 30% B for 2 min, 30% to 60% B in 15 min, isocratic holding at 60% B for 2 min, and 60% to 100% B in 10 min. UV detections were obtained at 190, 210, and 280 nm, and several fractions were obtained from this separation and further purified on the Gilson preparative reversed-phase HPLC (PLC 2020, Gilson, Middleton, WI, USA). A Synergi™ 10 μm Polar-RP 80 Å (250 × 50 mm) AXIA™ packed column (Phenomenex Inc., Aschaffenburg, Germany) was used as the stationary phase. Deionized H_2O + 0.1% formic FA (*v*/*v*) (solvent A) and acetonitrile (MeCN) + 0.1% FA (*v*/*v*) (solvent B) were used as the mobile phase with a flow rate of 40 mL/min. Separation was carried out with an elution gradient beginning isocratically at 5% solvent B for 10 min, followed by a gradient increase to 65% B in 30 min, then an increase from 65% B to 100% B in 10 min, and ending with an isocratic hold at 100% B for 15 min. UV detection was performed at 190, 210, and 280 nm to yield compounds **1** (t_R = 14.0 min), **3** (t_R = 12.0 min), and **2** (t_R = 16.0 min). The purity of the fractions was checked using HPLC-DAD-ESI-MS.

The rice crude extract was divided into two portions. One portion was fractionated directly on the Gilson system after initially passing it through a RP solid-phase cartridge (Strata-X 33 μm Polymeric Reversed Phase; Phenomenex, Aschaffenburg, Germany) to remove fatty acids. A similar solvent system and UV detections were used on the instrument as mentioned for the YMG extract. However, the stationary phase in this case was a C_{18} VP-Nucleodur column 100-5 (250 × 40 mm, 7 μm: Machery-Nagel, Düren, Germany). The gradient was operated with isocratic conditions at 5% B for 10 min, followed by an increase from 5% B to 10% B in 10 min, from 10% B to 80% B in 40 min, from 80% to 100% B in 5 min, and a final isocratic step at 100% B for 10 min. This yielded compounds **4** (t_R = 41.0 min) and **5** (t_R = 55.0 min).

Dentifragilone A (**1**): White amorphous solid; 1.14 mg; $[\alpha]_D^{20}$ = +69° (*c* 0.1, methanol); UV/Vis (MeOH): λ_{max} (log ε) = 196.6 (1.7), 279.6 (0.6) nm; NMR data (^{1}H NMR: 500 MHz, ^{13}C NMR: 125 MHz in chloroform-*d*) see Table 1; HR-(+)ESIMS: *m/z* 209.1167 $[M-H_2O+H]^+$ (calcd. 209.1172 for $C_{12}H_{17}O_3^+$), 227.1277 $[M+H]^+$ (calcd. 227.1278 for $C_{12}H_{19}O_4^+$), 249.1098 $[M+Na]^+$ (calcd. 249.1097 for $C_{12}H_{18}NaO_4^+$); t_R = 3.90 min (HR-LC-ESIMS). $C_{12}H_{18}O_4$ (226.11 g/mol).

Dentifragilone B (**2**): Off-white amorphous solid; 0.84 mg; UV/Vis (MeOH): λ_{max} (log ε) = 196.6 (1.7); NMR data (^{1}H NMR: 500 MHz, ^{13}C NMR: 125 MHz in chloroform-*d* and methanol-d_4) see Table 2; HR-(+)ESIMS: *m/z* 225.1117 $[M+H]^+$ (calcd. 225.1121 for $C_{12}H_{17}O_4^+$), 247.0937 $[M+Na]^+$ (calcd. 247.0941 for $C_{12}H_{16}NaO_4^+$); t_R = 4.04 min (HR-LC-ESIMS). $C_{12}H_{16}O_4$ (224.09 g/mol).

Dentipellinol (**3**): Off-white amorphous solid; 0.62 mg; $[\alpha]_D^{20}$ = −42° (*c* 0.1, methanol); UV/Vis (MeOH): λ_{max} (log ε) = 196.6 (1.7); NMR data (^{1}H NMR: 700 MHz, ^{13}C NMR: 125 MHz in chloroform-*d*) see Table S1 comparable to the reported literature [5]; *m/z* 221.0805 $[M-H_2O+H]^+$ (calcd. 221.0808 for $C_{12}H_{13}O_4^+$), 239.0911 $[M+H]^+$ (calcd. 239.0914 for $C_{12}H_{15}O_5^+$), 261.0734 $[M+Na]^+$ (calcd. 261.0733 for $C_{12}H_{14}NaO_5^+$); t_R = 2.86 min (HR-LC-ESIMS). $C_{12}H_{14}O_5$ (238.10 g/mol).

Methyl 4-chloro-3,5-dimethoxybenzoate (**4**): Off-white amorphous solid; 5.43 mg; UV/Vis (MeOH): λ_{max} = 217, 263, 301 nm; NMR data (^{1}H NMR: 500 MHz, ^{13}C NMR: 125 MHz in chloroform-*d*) comparable to the reported literature [14]; HR-(+)ESIMS: *m/z* 231.4012 $[M+H]^+$ (calcd. 231.4019 for $C_{10}H_{12}ClO_4^+$), 253.0236 $[M+Na]^+$ (calcd. 253.0238 for $C_{10}H_{11}ClNaO_4^+$); t_R = 8.56 min (HR-LC-ESIMS). $C_{10}H_{11}ClO_4$ (230.37 g/mol).

10-Methoxycarbonyl-10-norisodrimenin (**5**): Off-white amorphous solid; 1.43 mg; UV/Vis (MeOH): λ_{max} = 219 nm; NMR data (^{1}H NMR: 500 MHz, ^{13}C NMR: 125 MHz in chloroform-*d*) comparable to the reported literature [16]; HR-(+)ESIMS: *m/z* 279.1588 $[M+H]^+$ (calcd. 279.1591 for $C_{16}H_{23}O_5^+$), 301.1411 $[M+Na]^+$ (calcd. 301.1410 for $C_{16}H_{22}NaO_5^+$); t_R = 9.75 min (HR-LC-ESIMS). $C_{16}H_{22}O_5$ (278.14 g/mol).

3.5. Antimicrobial Assay

The Minimum Inhibitory Concentration (MIC) of the isolated compounds was determined following the method previously described [16]. Accordingly, the compounds were tested against bacteria and fungi on a serial dilution assay performed on 96-well microtiter plates. YMG medium was used to culture the yeasts and filamentous fungi, whereas MHB media (Müller–Hinton Broth: SNX927.1, Carl Roth GmbH, Karlsruhe, Germany) was used for bacteria.

3.6. Cytotoxicity Assay

The in vitro cytotoxicity (IC_{50}) of isolated compounds was evaluated against an array of mammalian cell lines using a colorimetric tetrazolium dye MTT assay with epothilone B as a positive control. The cell lines, L929 (mouse fibroblasts) and KB3.1 (human endocervical adenocarcinoma), were employed, following established methodologies [16].

3.7. Density Functional Theory Calculations

In order to elucidate the electronic circular dichroism (ECD) spectra, a conformational analysis was principally executed to extract all possible conformations of compounds **1** and **2**, employing Omega2 software 2.5.1.4 [17] within an energy window value of 10 kcal/mol [18]. The resulting configurations were set to the geometry optimization process and then frequency computations at the B3LYP/6-31G* level of theory. The time-dependent density functional theory (TDDFT) calculations were then performed in methanol to determine the first fifty excitation states. The solvent effect was incorporated using the polarizable continuum model (PCM). The calculated ECD spectra were graphed using the SpecDis 1.71 [18,19]. The extracted ECD spectra were finally Boltzmann-averaged. All quantum calculations were performed using Gaussian09 software [20].

4. Conclusions

Two previously undescribed and two known benzoic acid derivatives (**1**–**4**), in addition to a previously isolated drimane sesquiterpenoid (**5**), were derived from submerged and solid-state cultures of *D. fragilis*. The fungus is rarely encountered in temperate zones and has proven to be a significant source of novel chemical components. However, compounds **1** and **2**, herein isolated, were unprecedented. Thus, it would not be surprising to discover new useful compounds from *D. fragilis* in the future. Although only moderate to weak antibiotics effects of the compounds have been realized, the compounds are potential candidates for alternative bioactivity studies that were not attained within the realm of the current study. Our findings also provide further insights into the chemodiversity of *D. fragilis*, highlighting the potential role of alternative media components in the versatile production of secondary metabolites.

Supplementary Materials: The following supporting information can be downloaded at https://www.mdpi.com/article/10.3390/molecules29122859/s1: Figures S1–S8: HPLC, LR-/HR-ESI-MS, 1D/2D NMR spectra of **1**; Figure S9–S20: HPLC, LR-/HR-ESI-MS, 1D/2D NMR spectra of **2**; Figure S21–S29: HPLC, LR-/HR-ESI-MS, 1D/2D NMR, ECD spectra of **3**; Table S1: 1D (^{1}H and ^{13}C) NMR data of compound **3** and dentipellinol; Figures S30–S37: HPLC, LR-/HR-ESI-MS, 1D/2D NMR spectra of **4**; Figures S38–S45: HPLC, LR-/HR-ESI-MS, 1D/2D NMR spectra of **5**.

Author Contributions: W.C.S.: conceptualization, large-scale fermentation, isolation and structure elucidation of compounds, and preparation of the original draft; S.S.E.: structure elucidation of compounds, preparation, and polishing the original draft; M.A.A.I.: TDDFT-ECD calculations and reporting findings; H.K.: isolation and identification of the producer strain; M.S.: supervision, funding acquisition, correcting, editing, and polishing the draft. All authors have read and agreed to the published version of the manuscript.

Funding: W.C.S. was supported by doctoral scholarship funding from the German Academic Exchange Service (DAAD), program number 57507871. S.S.E. thanks the Alexander von Humboldt (AvH) Foundation for the Georg Forster Fellowship for Experienced Researchers grant, reference number 3.4-1222288-EGY-GF-E.

Institutional Review Board Statement: N/A.

Informed Consent Statement: N/A.

Data Availability Statement: Data are contained within the article and supplementary materials.

Acknowledgments: The authors gratefully acknowledge Wera Collisi for conducting the bioassays, as well as Esther Surges and Aileen Gollasch for the NMR and HR-ESI-MS measurements. We also extend our acknowledgements to the Bavarian Forest National Park and the Mycology group of the Department of Conservation and Research for their support and sample permissions.

Conflicts of Interest: The authors declare no conflicts of interest.

References

1. Newman, D.J.; Cragg, G.M. Natural products as sources of new drugs from 1981 to 2014. *J. Nat. Prod.* **2016**, *79*, 629–661. [CrossRef] [PubMed]
2. Sum, W.C.; Ebada, S.S.; Matasyoh, J.C.; Stadler, M. Recent progress in the evaluation of secondary metabolites from Basidiomycota. *Curr. Res. Biotechnol.* **2023**, *6*, 100155. [CrossRef]
3. Sandargo, B.; Chepkirui, C.; Cheng, T.; Chaverra-Muñoz, L.; Thongbai, B.; Stadler, M.; Hüttel, S. Biological and chemical diversity go hand in hand: Basidiomycota as source of new pharmaceuticals and agrochemicals. *Biotechnol. Adv.* **2019**, *37*, 107344. [CrossRef] [PubMed]
4. Ha, L.S.; Ki, D.W.; Kim, J.Y.; Choi, D.C.; Lee, I.K.; Yun, B.S. Dentipellin, a new antibiotic from culture broth of *Dentipellis fragilis*. *J. Antibiot.* **2021**, *74*, 538–541. [CrossRef] [PubMed]
5. Ki, D.-W.; Kim, C.-W.; Choi, D.-C.; Oh, G.W.; Doan, T.-P.; Kim, J.-Y.; Oh, W.-K.; Lee, I.-K.; Yun, B.-S. Chemical constituents of the culture broth of *Dentipellis fragilis* and their anti-inflammatory activities. *Phytochemistry* **2023**, *214*, 113828. [CrossRef] [PubMed]

6. Sum, W.C.; Mitschke, N.; Schrey, H.; Wittstein, K.; Kellner, H.; Stadler, M.; Matasyoh, J.C. Antimicrobial and cytotoxic cyathane-xylosides from cultures of the basidiomycete *Dentipellis fragilis*. *Antibiotics* **2022**, *11*, 1072. [CrossRef] [PubMed]
7. Bird, C.W.; Chauhan, Y.-P.S. A convenient synthesis of *p*-hydroxybenzaldehydes. *Org. Prep. Proced. Int.* **1980**, *12*, 201–202. [CrossRef]
8. Ki, D.-W.; Choi, D.-C.; Won, Y.-S.; Lee, S.-J.; Kim, Y.-H.; Lee, I.-K.; Yun, B.-S. Three new phthalide derivatives from culture broth of *Dentipellis fragilis* and their cytotoxic activities. *J. Antibiot.* **2024**, *77*, 338–344. [CrossRef] [PubMed]
9. Ki, D.-W.; Yun, B.-S. A new antibiotic from the culture broth of *Dentipellis fragilis*. *J. Antibiot.* **2023**, *76*, 351–354. [CrossRef] [PubMed]
10. Qu, W.; Wijeratne, E.M.K.; Bashyal, B.P.; Xu, J.; Xu, Y.-M.; Liu, M.X.; Inácio, M.C.; Arnold, A.E.; U'Ren, J.M.; Gunatikala, A.A.L. Strobiloscyphones A-F, 6-isopentylsphaeropsidones and other metabolites from *Strobiloscypha* sp. AZ0266, a leaf-associated fungus of Douglas Fir. *J. Nat. Prod.* **2021**, *84*, 2575–2586. [CrossRef] [PubMed]
11. Li, C.-S.; Yang, B.-J.; Turkson, J.; Cao, S. Antiproliferative ambuic acid derivatives from Hawaiian endophytic fungus *Pestalotiopsis* sp. FT172. *Phytochemistry* **2017**, *140*, 77–82. [CrossRef] [PubMed]
12. Phutthacharoen, K.; Toshe, R.; Khalid, S.J.; Llanos-López, N.A.; Wennrich, J.-P.; Schrey, H.; Ebada, S.S.; Hyde, K.D.; Stadler, M. Lachnuoic acids A–F: Ambuic acid congeners from a saprotrophic *Lachnum* Species. *Chem. Biodivers.* **2024**, *2024*, e202400385. [CrossRef] [PubMed]
13. Ueda, K.; Kodani, S.; Kubo, M.; Masuno, K.; Sekiya, A.; Nagai, K.; Kawagishi, H. Endoplasmic reticulum (ER) stress-suppressive compounds from scrap cultivation beds of the mushroom *Hericium erinaceum*. *Biosci. Biotechnol. Biochem.* **2009**, *73*, 1908–1910. [CrossRef] [PubMed]
14. Lao, J.H.G.; Ramos, C.S.; Casanova, D.C.C.; de, A. Morandim, A.; Bergamo, D.C.B.; Cavalheiro, A.J.; da S. Bolzani, V.; Furlan, M.; Guimarães, E.F.; Young, M.C.M.; et al. Benzoic acid derivatives from *Piper* species and their antifungal activity against *Cladosporium cladosporioides* and *C. sphaerospermum*. *J. Nat. Prod.* **2004**, *67*, 1783–1788.
15. Mitschke, N.; Sum, W.C.; Hassan, K.; Kirchenwitz, M.; Schrey, H.; Gerhards, L.; Kellner, H.; Stradal, T.E.B.; Matasyoh, J.C.; Stadler, M. Biologically active drimane derivatives isolated from submerged cultures of the wood-inhabiting basidiomycete *Dentipellis fragilis*. *RSC Adv.* **2023**, *13*, 25752–25761. [CrossRef] [PubMed]
16. Wennrich, J.-P.; Ebada, S.S.; Sepanian, E.; Holzenkamp, C.; Khalid, S.J.; Schrey, H.; Maier, W.; Mándi, A.; Kurtán, T.; Ashrafi, S.; et al. Omnipolyphilins A and B: Chlorinated cyclotetrapeptides and naphtho-α-pyranones from the plant nematode-derived fungus *Polyphilus sieberi*. *J. Agric. Food Chem.* **2024**, *72*, 6998–7009. [CrossRef]
17. *OMEGA*; Version 2.5.1.4; OpenEye Scientific Software: Santa Fe, NM, USA, 2017.
18. Bruhn, T. SpecDis, Version 1.71; Berlin, Germany, 2017.
19. Bruhn, T.; Schaumlöffel, A.; Hemberger, Y.; Bringmann, G. SpecDis 1.53, Würzburg, Germany: University of Würzburg. *Chirality* **2013**, *25*, 243–249. [CrossRef]
20. Frisch, G.W.T.J.; Schlegel, H.B.; Scuseria, G.E.; Robb, M.A.; Cheeseman, J.R.; Scalmani, G.; Barone, V.; Mennucci, B.; Petersson, G.A.; Nakatsuji, H.; et al. *Gaussian*; Gaussian, Inc.: Wallingford, CT, USA, 2009.

molecules

MDPI

Article

New Pyranone Derivatives and Sesquiterpenoid Isolated from the Endophytic Fungus *Xylaria* sp. Z184

Yan Zhang [1,†], Yang Jin [1,†], Wensi Yan [1], Peishan Gu [1], Ziqian Zeng [1], Ziying Li [2], Guangtao Zhang [2], Mi Wei [3] and Yongbo Xue [1,*]

[1] School of Pharmaceutical Sciences (Shenzhen), Shenzhen Campus of Sun Yat-sen University, Shenzhen 518107, China; zhangy2328@mail2.sysu.edu.cn (Y.Z.); jiny67@mail2.sysu.edu.cn (Y.J.); yanws3@mail2.sysu.edu.cn (W.Y.); gupsh@mail2.sysu.edu.cn (P.G.); zengzq8@mail2.sysu.edu.cn (Z.Z.)
[2] School of Pharmacy, Binzhou Medical University, Yantai 264003, China; 15853386606@163.com (Z.L.); greatzhangtao@hotmail.com (G.Z.)
[3] School of Agriculture, Shenzhen Campus of Sun Yat-sen University, Shenzhen 518107, China; weim29@mail.sysu.edu.cn
* Correspondence: xueyb@mail.sysu.edu.cn
† These authors contributed equally to this work.

Abstract: The fungus *Xylaria* sp. Z184, harvested from the leaves of *Fallopia convolvulus* (L.) Á. Löve, has been isolated for the first time. Chemical investigation on the methanol extract of the culture broth of the titles strain led to the discovery of three new pyranone derivatives, called fallopiaxylaresters A–C (**1–3**), and a new bisabolane-type sesquiterpenoid, named fallopiaxylarol A (**4**), along with the first complete set of spectroscopic data for the previously reported pestalotiopyrone M (**5**). Known pyranone derivatives (**6–11**), sesquiterpenoids (**12–14**), isocoumarin derivatives (**15–17**), and an aromatic allenic ether (**18**) were also co-isolated in this study. All new structures were elucidated by the interpretation of HRESIMS, 1D, 2D NMR spectroscopy, and quantum chemical computation approach. The in vitro antimicrobial, anti-inflammatory, and α-glucosidase-inhibitory activities of the selected compounds and the crude extract were evaluated. The extract was shown to inhibit nitric oxide (NO) production induced by lipopolysaccharide (LPS) in murine RAW264.7 macrophage cells, with an inhibition rate of 77.28 ± 0.82% at a concentration of 50 μg/mL. The compounds **5**, **7**, and **8** displayed weak antibacterial activity against *Staphylococcus areus* subsp. *aureus* at a concentration of 100 μM.

Keywords: *Xylaria* sp.; pyranone derivative; sesquiterpenoid; antimicrobial activity

1. Introduction

Natural products (NPs) have always been an indispensable source of new drugs [1]. As an important source of NPs with novel structures and high-value biological activities, plant endophytic fungi have always been attracting broad attention from natural product chemists and pharmacologists [2,3]. Xylariaceae is one of the largest, most commonly encountered, and highly diverse fungal families of the Ascomycota [4]. The genus *Xylaria*, belonging to the family Xylariaceae, is medicinal fungi commonly found in decaying plant tissues and is widely distributed in temperate, tropical, and subtropical regions [5,6]. So far, more than 200 bioactive compounds (>100 new ones) were isolated from *Xylaria*, including cytochalasins, α-pyrones, cyclopeptides, terpenoids, lactones, and succinic acid derivatives [7]. Furthermore, the secondary metabolites produced by species of *Xylaria* have been found to exert a wide range of biological activities, such as anti-inflammatory, antifungal, antibacterial, anti-tumor, and α-glucosidase-inhibitory activities [7,8].

As part of our group's ongoing effort to identify bioactive natural products from medicinal plants and endophytic fungi [9–11], the fungus *Xylaria* sp. Z184, isolated from the leaves of *Fallopia convolvulus* (L.) Á. Löve for the first time, has attracted our attention for its impressive compound abundance in TLC and HPLC analyses (Figure S106). Our current investigation on this strain led to the isolation of three new pyranone derivatives, called fallopiaxylaresters A–C (**1–3**), a new bisabolane-type sesquiterpenoid fallopiaxylarol A (**4**) (Figure 1), and the first complete set of the spectroscopic data for the previously disclosed pestalotiopyrone M (**5**), as well as a suite of known compounds consisting of six pyranone derivatives (**6–11**), three sesquiterpenoids (**12–14**), three isocoumarin derivatives (**15–17**), and one aromatic allenic ether (**18**). Herein, the details of the isolation, structure elucidation of all new compounds and their anti-inflammatory, antimicrobial, and α-glucosidase-inhibitory activities were described.

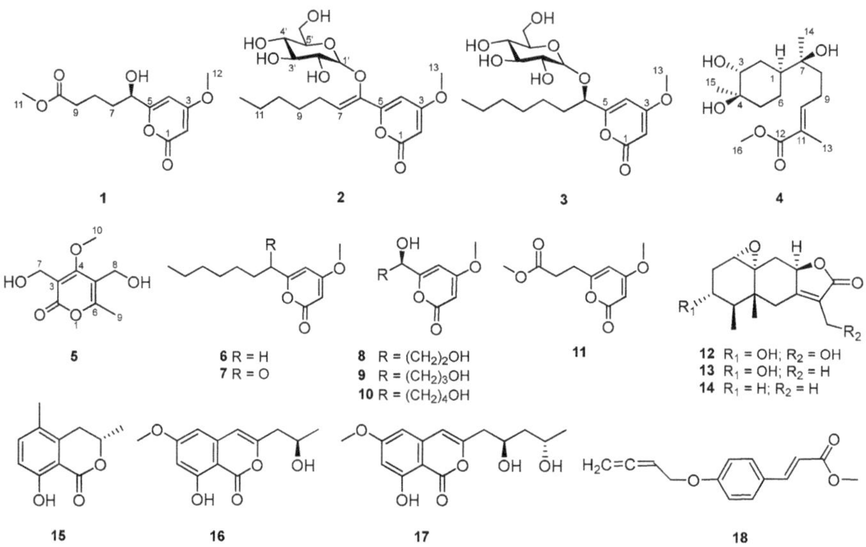

Figure 1. Chemical structures of compounds **1–18**.

2. Results and Discussion

Compound **1**, named fallopiaxylarester A, was obtained as a white solid. Its molecular formula $C_{12}H_{16}O_6$ was determined by the HRESIMS molecular ion peak at m/z 279.0837 $[M + Na]^+$ (calcd for $C_{12}H_{16}O_6Na$, 279.0839). The IR absorption bands showed the presence of hydroxyl (3426 cm^{-1}) and carbonyl (1733 cm^{-1}) functionalities. Detailed comparison of ^1H and ^{13}C NMR spectra of **1** and **10** revealed that the structure of **1** is almost identical with that of **10**, which was supported by the further analysis of 2D NMR spectra (Table 1). The HMBC correlations of **1** from H-2 (δ_H 5.56) to C-1/C-3/C-4, from H_3-12 (δ_H 3.87) to C-3, from H-4 (δ_H 6.22) to C-3/C-5/C-6, and from H-6 (δ_H 4.35) to C-5 and C-8, showed the same settlement as with compound **10** (Figure 2). The main difference between **1** and **10** was found at C-10 of the side chain of the pyranone core, replaced by the fragment of methyl valerate group. This deduction was further identified by the HMBC correlations from H_3-11 (δ_H 3.66) and H_2-9 (δ_H 2.38) to C-10. Thus, the planar structure of **1** was established. Since there was only one chiral center in the molecule, the relative configuration was arbitrarily assigned as 6*R**. The absolute configuration of C-6 was subsequently assigned to be *R* by comparing the optical rotational value $[\alpha]_D^{25}$ +56.2 (*c* 0.11, MeOH) with $[\alpha]_D^{25}$ +96.0 (*c* 0.10, MeOH) of compound **10** [12]. Furthermore, the deduction was also confirmed by a time-dependent density functional theory–electronic circular dichroism (TDDFT-ECD) approach. As shown in Figure 3, the Boltzmann-averaged ECD spectrum of (6*R*)-**1** displayed a similar curve compared to the experimental one. Thus, the absolute configuration at C-6 in **1** was unambiguously assigned as 6*R* (Figure 1).

Table 1. ^{1}H NMR (δ_H, 600 MHz) and ^{13}C NMR (δ_C, 150 MHz) data for **1** in methanol-d_4.

No.	δ_H, Mult (J Hz)	δ_C, Type
1		167.0, C
2	5.56, d (2.1)	88.6, CH
3		173.8, C
4	6.22, d (2.1)	99.9, CH
5		168.7, C
6	4.35, dd (7.2, 4.8)	70.7, CH
7	1.83, m 1.68, m, overlap	35.2, CH_2
8	1.76, m 1.69, m, overlap	21.7, CH_2
9	2.38, t (6.7)	34.4, CH_2
10		175.6, C
11	3.66, s	52.0, CH_3
12	3.87, s	57.0, CH_3

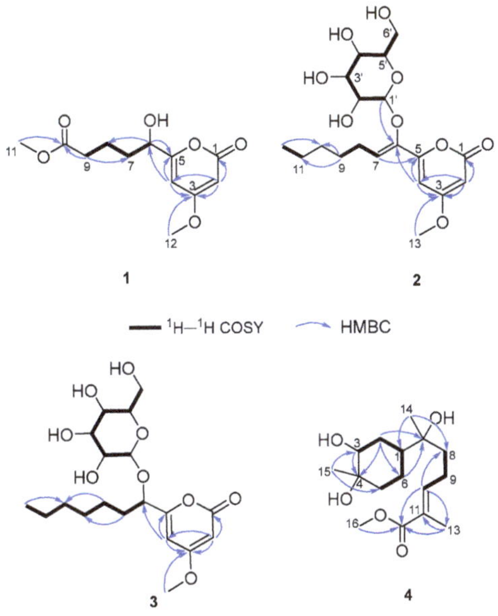

Figure 2. The key ^{1}H–^{1}H COSY and HMBC correlations of **1–4**.

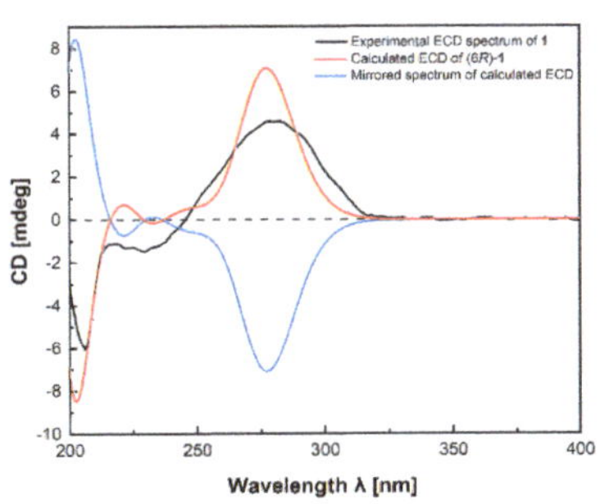

Figure 3. Experimental ECD spectrum of fallopiaxylarester A (**1**) (black); calculated ECD of (6*R*)–**1** (red); mirrored spectrum of calculated ECD (blue).

Compound **2**, a white solid, was determined to possess the molecular formula of $C_{19}H_{28}O_9$ with six degrees of unsaturation by using HRESIMS {*m/z* 423.1626 [M + Na]$^+$, (calcd for $C_{19}H_{28}O_9Na$, 423.1626)}. The spectra of **2** showed similar absorption bands, indicating the same presence of hydroxyl (3359 cm^{-1}) and carbonyl (1696 cm^{-1}) functionalities. The ^{1}H NMR spectra data showed signals of a typical sugar moiety at δ_H 5.12 (d, *J* = 3.7 Hz), 3.53 (dd, *J* = 10.0, 3.7 Hz), 3.78 (t, *J* = 9.3 Hz), 3.45 (t, *J* = 9.3 Hz), 3.90 (m), and 3.76 (m) (Table 2). Analysis of the ^{13}C NMR and DEPT spectra of **2** indicated the presence of 19 carbon signals, assignable to two methyl carbons (one methoxyl, δ_C 57.0 and 14.4), five methylene carbons (δ_C 27.1, 30.0, 32.8, 23.6 and 62.1), eight methine carbons (three olefinic, δ_C 89.4, 100.2, 125.0, 102.8, 73.4, 74.4, 71.0, and 75.5), one ester carbonyl carbon (δ_C 166.7), and three olefinic quaternary carbons (δ_C 174.0, 157.9 and 145.8). These substructures accounted for four out of five degrees of unsaturation, indicating one cyclic system in **2**. The ^{1}H–^{1}H COSY spectrum revealed three spin systems: (a) H-7/H-8/H-9; (b) H-11/H-12; and (c) H-1′/H-2′/H-3′/H-4′/H-5′/H-6′ (Figure 2). And those spin systems were connected by the key HMBC correlations from H-2 (δ_H 5.59) to C-1/C-3/C-4, from H_3-13 (δ_H 3.87) to C-3, from H-4 (δ_H 6.95) to C-3/C-5/C-6, from H-7 (δ_H 6.07) to C-5, and from H_2-9 (δ_H 1.48) to C-10/C-11.

Table 2. ^{1}H NMR (δ_H, 600 MHz) and ^{13}C NMR (δ_C, 150 MHz) data for **2**.

No.	δ_H, Mult (*J* Hz) [a]	δ_C, Type [a]	δ_H, Mult (*J* Hz) [b]	δ_C, Type [b]	δ_H, Mult (*J* Hz) [c]	δ_C, Type [c]
1		166.7, C		164.1, C		162.7, C
2	5.59, d (2.2)	89.4, CH	5.70, d (2.2)	89.7, CH	5.61, d (2.2)	88.6, CH
3		174.0, C		172.3, C		171.2, C
4	6.95, d (2.2)	100.2, CH	7.64, d (2.2)	99.7, CH	6.98, d (2.2)	98.3, CH
5		157.9, C		157.5 C		155.8, C
6		145.8, C		146.1, CH		144.3, C
7	6.07, t (7.5)	125.0, CH	6.24, t (7.5)	123.6, CH	5.91, t (7.5)	122.8, CH
8	2.42, m	27.1, CH_2	2.61, q (7.5)	26.9, CH_2	2.33, dd (15.0, 7.6)	25.3, CH_2
9	1.48, m	30.0, CH_2	1.33, m	29.7, CH_2	1.39, m	28.4, CH_2
10	1.36, m, overlap	32.8, CH_2	1.19, m, overlap	32.3, CH_2	1.28, m, overlap	31.1, CH_2
11	1.37, m, overlap	23.6, CH_2	1.18, m, overlap	23.2, CH_2	1.29, m, overlap	22.0, CH_2
12	0.92, t (6.9)	14.4, CH_3	0.75, t (7.0)	14.6, CH_3	0.87, t (6.9)	13.9, CH_3
13	3.87, s	57.0, CH_3	3.60, s	56.5, CH_3	3.81, s	56.4, CH_3
1′	5.12, d (3.7)	102.8, CH	5.68, d (3.7)	103.4, CH	4.96, d (3.7)	101.3, CH
2′	3.53, dd (10.0, 3.7)	73.4, CH	4.27, dd (10.5, 4.1)	74.0, CH	3.32, m, overlap	71.7, CH
3′	3.78, t (9.3)	74.4, CH	4.71, t (9.3)	75.2, CH	3.70, m	72.5, CH
4′	3.45, t (9.3)	71.0, CH	4.35, t (10.0)	71.9, CH	3.22, m	69.4, CH
5′	3.90, m	75.5, CH	4.66, dt (10.0, 3.5)	76.8, CH	3.54, m, overlap	74.7, CH
6′	3.76, m	62.1, CH_2	4.48, br s	62.9, CH_2	3.55, m, overlap	60.3, CH_2
2′-OH					5.52, d (4.9)	
3′-OH					5.10, m	
4′-OH					5.10, m	
6′-OH					4.53, t (5.8)	

[a] Measured in methanol-d_4, [b] measured in pyridine-d_5, [c] measured in DMSO-d_6.

Subsequently, the long-range HMBC correlation from H-1′ (δ_H 5.12) to the anomeric carbon C-6 suggested the sugar was attached at the C-6 of the side chain. Acid hydrolysis of **2** followed by HPLC analysis of the sugar derivative was applied to determine the type of sugar moiety, but without success. Then, a different deuterated solvent, pyridine-d_5, was used to analyze the proton signals on the sugar. In addition, the isolated anomeric proton signal was probed by a selective 1D-TOCSY experiment. In the 1D-TOCSY experiment, irradiation of the signal at δ_H 5.68 (1H, d, *J* = 3.7 Hz) enabled the identification of H-2′ (δ_H 4.27, dd, *J* = 10.5, 4.1 Hz), H-3′ (δ_H 4.71, t, *J* = 9.3 Hz), H-4′ (δ_H 4.35, t, *J* = 10.0 Hz), H-5′ (δ_H 4.66, dt, *J* = 10.0, 3.5 Hz), and H_2-6′ (δ_H 4.48, br s) in the same conjugated system (Figure 4, Table 2). The *J* values of $J_{H\text{-}5'/H\text{-}4'}$ (10.0 Hz) and $J_{H\text{-}5'/H\text{-}6'}$ (3.5 Hz) in ^{1}H NMR in pyridine-d_5 of **2** combined with the 1D-TOCSY experiment suggested an α-D-glucose. Furthermore, those chemical shift values of anomeric carbon at δ_C 102.8 (C-1′), four tertiary carbons at δ_C 73.4 (C-2′), 74.4 (C-3′), 71.0 (C-4′), 75.5 (C-5′), and a methylene oxide carbon at δ_C 62.1 (C-6′) in methanol-d_4 were highly similar to 6-Buty-AA-2G along with other derivatives of AA-2G in the literature [13], which also confirmed the inference that the sugar unit was α-D-glucose. In addition, the geometry of the double bond between C-6 and C-7 was inferred by the ROESY spectrum in DMSO-d_6. The ROESY correlations between H_2-8 (δ_H 2.33) and H-1′ (δ_H 4.96)/H-3′ (δ_H 3.70)/H-5′ (δ_H 3.54), and between H-4 (δ_H 6.98) and H-1′/H-3′, demonstrated the Z geometry of $\Delta^{6,7}$ (Figures 5 and S15). Thus, this undescribed **2** was established as shown in Figure 1 and named fallopiaxylarester B.

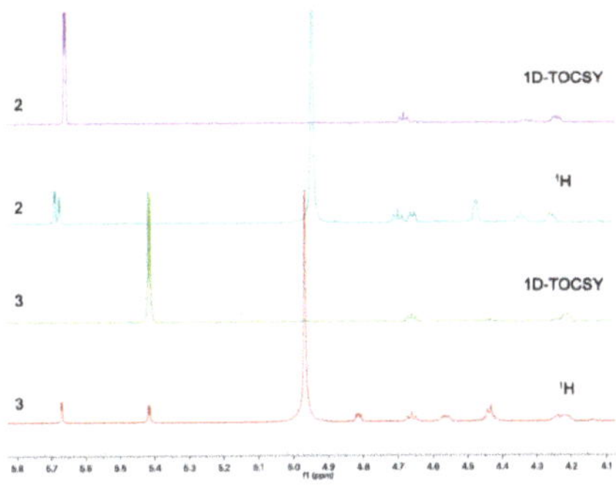

Figure 4. 1D-TOCSY spectra of compounds **2** (purple) and **3** (green).

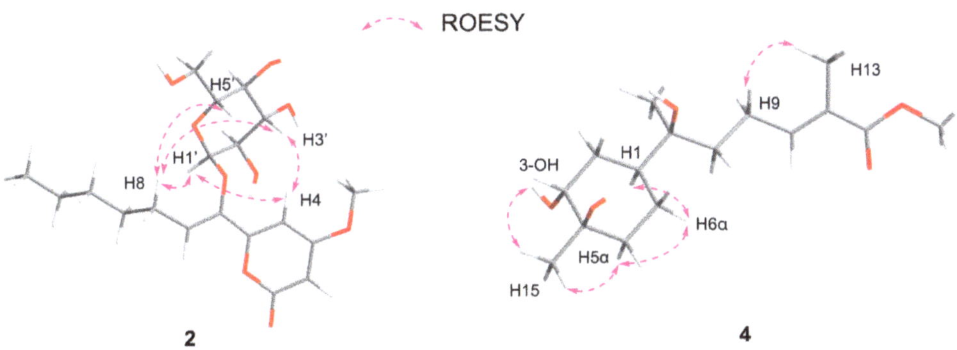

2 **4**

Figure 5. Energy-minimized structures of **2** and **4** with the key ROESY correlations.

Compound **3**, named fallopiaxylarester C, was isolated as a white solid. Its molecular formula of $C_{19}H_{30}O_9$ with five degrees of unsaturation was based on HRESIMS analysis {m/z 425.1780 $[M + Na]^+$, (calcd for $C_{19}H_{30}O_9Na$, 425.1782)}. Like **2**, the presence of α, β-unsaturated γ-lactone and hydroxyl groups in **3** was obvious by its IR absorption bands at υ_{max} 3380 and 1700 cm^{-1}. Furthermore, with the analysis of 1D NMR spectra, a sugar moiety in **3** was also quickly recognized. In addition, the HMBC correlation from H-6 (δ_H 4.52) to C-1′ suggested the same set as with **2** (Table 3). The main difference between these two compounds was the hydrogenation of the trisubstituted olefinic group at C-6/C-7 in **2**, with a 2 mass unit difference between **2** and **3**. Subsequently, an acid hydrolysis of **3** afforded the products, including a pyrone aglycone **3a** and a sugar moiety **3b**. The absolute configuration of C-6 in **3a** was assigned to be *R* form by comparing its optical value $[\alpha]_D^{25}$ +52.5 (*c* 0.11, MeOH) with +67.6 (*c* 0.25, MeOH) of nodulisporipyrones A [14]. According to the detailed analysis of ^{1}H NMR and 1D-TOCSY experiment of **3**, H-1′ (δ_H 5.42, d, J = 3.8Hz), H-2′ (δ_H 4.22, dd, J = 9.6, 3.8 Hz), H-3′ (δ_H 4.67, t, J = 9.6 Hz), H-4′

(δ_H 4.24, t, J = 9.6 Hz), H-5′ (δ_H 4.44, t, J = 9.6 Hz), H_2-6′ (δ_H 4.57, dd, J = 9.6, 5.6 Hz; δ_H 4.43, m), the sugar moiety was indicated as α-D-glucose (Figure 4, Table 3). Moreover, large similarities were observed by comparison of NMR data in DMSO-d_6 of **3** with 5-(α-D-glucopyranosyloxymethyl)-2-furancarboxylic acid and other analogs in the literature [15]. Thus, the structure of **3** was established as shown (Figure 1).

Table 3. ^{1}H NMR (δ_H, 600 MHz) and ^{13}C NMR (δ_C, 150 MHz) data for **3**.

No.	δ_H, Mult (J Hz) [a]	δ_C, Type [a]	δ_H, Mult (J Hz) [b]	δ_C, Type [b]	δ_H, Mult (J Hz) [c]	δ_C, Type [c]
1		167.1, C		164.5, C		164.2, C
2	5.57, d (2.2)	89.1, CH	5.68, d (2.3)	89.1, CH	5.57, d (2.2)	88.0, CH
3		173.6, C		171.9, C		170.9, C
4	6.55, d (2.2)	101.9, CH	6.99, d (2.3)	100.5, CH	6.50, d (2.2)	99.4, CH
5		165.5, C		165.8, C		163.3, C
6	4.52, t (6.2)	75.3, CH	4.82, dd (7.8, 4.4)	74.8, CH	4.38, dd (6.8, 4.8)	72.9, CH
7	1.85, dd (14.0, 7.6)	35.1, CH_2	1.89, m 1.84, m	35.2, CH_2	1.70, m	33.5, CH_2
8	1.43, m	26.2, CH_2	1.51, m	26.2, CH_2	1.32, m	24.4, CH_2
9	1.35, m 1.33, m	30.1, CH_2	1.18, m, overlap 1.10, m, overlap	29.7, CH_2	1.27, m, overlap	28.4, CH_2
10	1.31, m, overlap	32.8, CH_2	1.08, m, overlap	32.2, CH_2	1.23, m, overlap	31.1, CH_2
11	1.32, m, overlap	23.7, CH_2	1.16, m, overlap	23.3, CH_2	1.25, m, overlap	22.0, CH_2
12	0.90, t (6.8)	14.4, CH_3	0.78, t (7.3)	14.7, CH_3	0.85, t (6.8)	14.0, CH_3
13	3.87, s	57.0, CH_3	3.63, s	56.4, CH_3	3.81, s	56.4, CH_3
1′	4.80, d (3.8)	98.5, CH	5.42, d (3.8)	99.3, CH	4.66, d (3.8)	97.2, CH
2′	3.41, dd (9.8, 3.8)	73.2, CH	4.22, dd (9.6, 3.8)	74.0, CH	3.22, m	71.5, CH
3′	3.68, m, overlap	74.8, CH	4.67, t (9.6)	75.6, CH	3.45, m, overlap	73.7, CH
4′	3.29, m	71.7, CH	4.24, t (9.6)	72.6, CH	3.07, m	70.1, CH
5′	3.68, m, overlap	74.6, CH	4.44, t (9.6)	75.8, CH	3.45, m, overlap	73.0, CH
6′	3.68, m, overlap 3.83, m	62.7, CH_2	4.57, dd (9.6, 5.6) 4.43, m	63.3, CH_2	3.61, m 3.45, m, overlap	60.9, CH_2
2′-OH					5.04, d (5.9)	
3′-OH					4.92, d (4.1)	
4′-OH					4.99, d (5.2)	
6′-OH					4.52, t (5.4)	

[a] Measured in methanol-d_4, [b] measured in pyridine-d_5, [c] measured in DMSO-d_6.

Compound **4** was isolated as a colorless oil. Its molecular formula of $C_{16}H_{28}O_5$ with three degrees of unsaturation was also based on HRESIMS analysis {m/z 323.1829 $[M + Na]^+$, (calcd for $C_{16}H_{28}O_5Na$, 323.1829)}. The IR spectrum of **4** demonstrated characteristic absorption bands for hydroxyl (3425 cm^{-1}) and carbonyl (1687 cm^{-1}) groups. The 1D NMR and HSQC spectra of **4** revealed 16 carbon signals, including four methyl groups, five sp^3 methylene groups, one sp^2 methine, two sp^3 methine groups, and four quaternary carbons (three oxygenated carbons) (Table 4). The above information accounted for two degrees of unsaturation, indicating one cyclic system in compound**4**. The ^{1}H–^{1}H COSY spectrum revealed two spin systems: (a) H-5/H-6/H-1/H-2/H-3, and (b) H-8/H-9/H-10 (Figure 2). Furthermore, the HMBC correlations from H_3-15 (δ_H 1.04) to C-3/C-5, from H_2-2 (δ_H 1.59, 1.53)/H-3 (δ_H 3.36) to C-4, from H_2-2 (δ_H 1.59, 1.53)/H_2-6 (δ_H 1.29) to C-7, from H_3-14 (δ_H 0.96) to C-1/C-8, from H-10 (δ_H 6.71) to C-8/C-12/C-13, from H_3-13 (δ_H 1.77) to C-11/C-12, and from H_3-16 (δ_H 3.64) to C-12 made those two spin systems connected. Thus, the planar structure of **4** was established as shown (Figure 1) and named fallopiaxylarol A.

Table 4. ^{1}H NMR (δ_H, 600 MHz) and ^{13}C NMR (δ_C, 150 MHz) data for **4**.

No.	δ_H, Mult (J Hz) [a]	δ_C, Type [a]	δ_H, Mult (J Hz) [b]	δ_C, Type [b]
1	1.72, m	39.1, CH	1.59, m, overlap	38.8, CH
2	1.80, m	29.3, CH_2	1.59, m, overlap 1.53 m	29.2, CH_2
3	3.66, br s	74.0, CH	3.36, m, overlap	72.6, CH
4		70.9, C		69.5, C
5	1.74, m 1.55, m	33.6, CH_2	1.49, m 1.29, m, overlap	33.5, CH_2
6	1.49, m 1.40, m	22.1, CH_2	1.29, m, overlap	21.7, CH_2
7		74.0, C		72.0, C
8	1.60, m	38.8, CH_2	1.41, m	38.1, CH_2
9	2.26, m	22.9, CH_2	2.18, q (8.0)	22.7, CH_2
10	6.77, td (7.5, 1.2)	142.6, CH	6.71, td (7.5, 0.9)	143.5, CH
11		127.6, C		126.4, C
12		168.8, C	0.90, t (6.8)	167.7, C
13	1.84, s	12.4, CH_3	1.77, s	12.2, CH_3
14	1.14, s	23.7, CH_3	0.96, s	23.8, CH_3
15	1.26, s	27.6, CH_3	1.04, s	27.9, CH_3
16	3.73, s	51.8, CH_3	3.64, s	51.6, CH_3
3-OH			4.36, d (4.0)	
4-OH			3.96, s	
7-OH			3.88, s	

[a] Measured in chloroform-*d*, [b] measured in DMSO-d_6.

Initially, the ROESY correlation between H_3-13 and H_2-9 and the lack of correlation of H_3-13/H-10 assigned the *E*-geometry of $\Delta^{10,11}$, which was also supported by the *J* value of H-10 (7.5) (Figure 5). In addition, the ROESY correlations of H-1/3-OH/H_3-15/H-5α/H-6α suggested the *cis* orientation of the H-1, 3-OH, and H_3-15. Furthermore, the literature survey revealed that the NMR data of the six-membered ring and the optical rotation values of **4** were almost identical to those of (1*S*,3*R*,4*R*,7*S*)-3,4-dihydroxy-α-bisabolol [16]. Thus, the absolute configuration of **4** was tentatively determined as shown in Figure 1.

After the literature survey, as for the secondary metabolites produced by the genus *Xylaria*, the main structural differences between the co-isolated new pyranone derivatives in this case and the other analogues of the genus are the variation of substituents in the side chain attached pyranone core [7,12,16]. Although compound **5** has previously been reported as a natural product from fermentation extracts of endophytic fungi [17], this is the first report of its existence to be accompanied by a full suite of supporting spectroscopic data. The 14 known compounds, pestalotiopyrone M (**5**), 4-methoxy-6-nonyl-2-pyrone (**6**) [18], xylariaopyrone A (**7**) [19], xylariaopyrone H (**8**) [12], xylariaopyrone I (**9**) [12], xylapyrone D (**10**) [20], scirpyrone H (**11**) [21], 1α,10α-epoxy-3α,13-dihydroxyeremophil-7(11)-en-12,8β-olide (**12**) [4], 3α-hydroxymairetolide A (**13**) [4], mairetolide A (**14**) [22], (−)-5-methylmellein (**15**) [23], diaporthin (**16**) [24], mucorisocoumarin B (**17**) [25], and eucalyptene (**18**) [26], were also isolated from *Xylaria* sp. Z184. The structures of these compounds (**5–18**) were identified by comparing the spectral data to those reported in the respective references (Figures S80–S105).

The secondary metabolites generated by stains of *Xylaria* usually show obvious anti-inflammatory and antifungal activities [7,8]. In this case, compounds **2–10** and **15–18** and the crude extract were selected to evaluate the antimicrobial, anti-inflammatory and α-glucosidase-inhibition activities due to the limitation of samples. In antimicrobial assay, compounds **5**, **7**, and **8** displayed weak activity against *Staphylococcus areus* subsp. *aureus* with inhibition ratios of 25.9%, 31.5%, and 25.3% at a concentration of 100 μM. Unfortunately, in anti-inflammatory and α-glucosidase assay, only the crude extract potently inhibited LPS-induced NO production in RAW264.7 mouse macrophages, with an inhibition rate of 77.28 ± 0.82% at a concentration of 50 μg/mL. Although it was cytotoxic at this concentration, reducing the concentration to 6.25 μg/mL abrogated the cytotoxicity (Table 5).

Table 5. Inhibitory activities of compounds selected and crude extract on LPS-stimulated NO production.

Compounds	Concentration	NO Production Inhibition (%) [a]
2	50 μM	4.51 ± 0.35
3	50 μM	–3.93 ± 2.43
4	50 μM	1.85 ± 3.18
5	50 μM	–1.61 ± 0.53
12	50 μM	0.92 ± 2.97
13	50 μM	4.67 ± 2.36
14	50 μM	3.54 ± 1.26
18	50 μM	–0.92 ± 2.21
Crude extract	50 μg/mL	77.28 ± 0.82
	6.25 μg/mL	7.78 ± 3.29
L-NMMA [b]	50 μM	53.75 ± 1.28

[a] All compounds examined in a set of triplicated experiment. [b] Positive control.

3. Materials and Methods

3.1. General Experimental Procedures

Optical rotations were determined with a PerkinElmer 341 polarimeter (PerkinElmer, Waltham, MA, USA). UV absorptions were obtained by using a Waters UV-2401A spectrophotometer equipped with a DAD and a 1 cm path length cell. Methanolic samples were scanned from 190 to 400 nm in 1 nm steps. Measurements of IR spectra were performed using a Bruker Vertex 70 FT-IR spectrometer (Bruker, Karlsruhe, Germany). NMR spectra were recorded on Bruker AM-400 and AM-600 NMR spectrometers (Bruker, Karlsruhe, Germany) with TMS as internal standard, and NMR data were referenced to selected chemical shifts of methanol-d_4 (^{1}H: 3.31 ppm, ^{13}C: 49.0 ppm), chloroform-d (^{1}H: 7.26 ppm, ^{13}C: 77.0 ppm), and dimethyl sulfoxide-d_6 (^{1}H: 2.50 ppm, ^{13}C: 39.5 ppm), respectively. HRESIMS data were acquired on a Thermo Fisher LTQ XL LC/MS (Thermo Fisher, Palo Alto, CA, USA). Semi-preparative HPLC was performed on an Agilent 1220 instrument equipped with a UV detector with a semi-preparative column (RP-C_{18}, 5 μm, 250 × 10 mm, Welch Materials, Inc., Shanghai, China). Column chromatography was performed using Sephadex^{TM} LH-20 gel (40–70 μm; Merck KGaA, Darmstadt, Germany), and precoated silica gel plates (GF254, Qingdao Marine Chemical Co., Ltd., Qingdao, China) were used for TLC analyses. Spots were visualized by heating silica gel plates sprayed with 10% H_2SO_4 in EtOH. All HPLC solvents were purchased from Guangdong Guanghua Sci-Tech Co., Ltd. (Guangzhou, China). All solvents were of analytical grade (Guangzhou Chemical Regents Company, Ltd., Guangzhou, China).

3.2. Fungal Material

The fungus *Xylaria* sp. Z184 was isolated from the leaves of *Fallopia convolvulus* (L.) Á. Löve collected in Zhuyang Town, Henan province, P. R. China (N 34°14′12″ W 110°47′09″) in June 2022. Leaves of *F. convolvulus* (L.) Á. Löve were processed within 24 h and rinsed with sterile water. On a sterile workbench, after 30 min of ultraviolet light exposure, the leaves underwent sequential treatment with a 5% sodium hypochlorite solution, sterile water, and 75% ethanol, either soaked or rinsed, followed by drying with sterile filter paper. Leaves were trimmed into small squares with sterile scissors and placed into previously prepared PDA monoclonal agar plates, inoculating three petri plates in parallel. These plates were incubated at 30 °C for 3–7 days, until mycelial growth was observed extending from the inside of the tissue block to its surroundings. Distinct morphological colonies were subsequently transferred to new media for continued cultivation. This procedure was repeated until the fungal strains showed uniform growth, leading to the isolation of purified strains (Figure 6).

Figure 6. Photo of the fungus *Xylaria* sp. Z184.

To identify the strains, the standardized operating procedure was performed, which included genomic DNA extraction, 16*S*/18*S* amplification, PCR product detection and purification, and comparison of sequencing results with the NCBI-BLAST database (https://www.ncbi.nlm.nih.gov/) accessed on 10 December 2022, using ITS1 and ITS4 primers for both amplification and sequencing. The sequence data for this strain was submitted to the GenBank under accession No. KU645984. The fungal strain was deposited on 20% aqueous glycerol stock in a −80 °C freezer at the School of Pharmaceutical Sciences (Shenzhen), Shenzhen Campus of Sun Yat-sen University, Shenzhen, China.

3.3. Fermentation, Extraction, and Isolation

Xylaria sp. Z184 was cultured on potato dextrose agar for 5 days at 28 °C to prepare the seed culture. The cultured agar plates were cut into small pieces, which were then inoculated into 30 previously autoclaved Erlenmeyer flasks (350 mL), each containing 50 g of rice and 45 mL of distilled water. All flasks were incubated at 28 °C for 40 days. Cultural media was extracted with methanol four times, and the solvent was evaporated under reduced pressure at 45 °C. Then the extract was suspended in water and extracted four times with ethyl acetate. The combined ethyl acetate layers were concentrated under reduced pressure to yield a brown extract (7.5 g).

The crude extract was chromatographed on Sephadex LH-20 (MeOH) to give eight fractions (Fr.1–Fr.8). Fr. 3 (2.8 g) was separated with silica gel column chromatography (CC) with petroleum ether (PE)/EtOAc (20:1–0:1, *v*/*v*) to give seven subfractions (Fr. 3.1–Fr. 3.7). Fr. 3.3 (306.2 mg) was purified with silica gel CC using PE/EtOAc (15:1–1:1, *v*/*v*) to yield six further subfractions (Fr. 3.3.1–Fr. 3.3.6). Fr. 3.3.4 (25.8 mg) was purified by semi-preparative HPLC (MeOH/H_2O, 48:52, *v*/*v*, 3.0 mL/min) to yield compound **7** (4.3 mg, t_R 22.0 min). Fr. 4 (1.7 g) was separated by silica gel CC with CH_2Cl_2/MeOH (80:1–0:1, *v*/*v*) to obtain seven subfractions (Fr. 4.1–Fr. 4.7). Fr. 4.3 (277.5 mg) was purified by semi-preparative HPLC (MeCN/H_2O, 25:75, 0–14 min, then MeCN/H_2O, 50:50, 14.01–33 min, *v*/*v*, 3.0 mL/min) to yield compounds **13** (4.6 mg, t_R 12.1 min), **14** (1.6 mg, t_R 27.6 min), and **6** (4.1 mg, t_R 31.5 min). Fr. 4.5 (239.9 mg) was purified by semi-preparative HPLC (MeCN/H_2O, 20:80, 0–10 min, then MeCN/H_2O, 40:60, 10.01–21 min, *v*/*v*, 3.0 mL/min) to yield compounds **10** (8.7 mg, t_R 13.2 min) and **4** (6.0 mg, t_R 19.5 min). Fr. 4.6 (225.9 mg) was purified by semi-preparative HPLC (MeOH/H_2O, 40:60, *v*/*v*, 3.0 mL/min) to yield compounds **2** (3.9 mg, t_R 30.1 min) and **3** (12.1 mg, t_R 35.5 min). Fr. 5 (1.3 g) was separated with silica gel CC with CH_2Cl_2/MeOH (25:1–0:1, *v*/*v*) to give six subfractions (Fr. 5.1—Fr. 5.6). Fr. 5.1 (44.0 mg) was further purified by semi-preparative HPLC (MeOH/H_2O, 55:45, 0–23 min, then MeOH/H_2O, 65:35, 23.01–47 min, *v*/*v*, 3.0 mL/min) to yield compounds **15** (3.1 mg, t_R 21.8 min) and **18** (1.4 mg, t_R 45.5 min). Fr. 5.3 (399.7 mg) was purified by semi-preparative HPLC (MeOH/H_2O, 35:65, 0–19 min, then MeOH/H_2O, 54:46, 19.01–40 min, *v*/*v*, 3.0 mL/min) to yield compounds **1** (1.2 mg, t_R 6.0 min), **17** (8.3 mg, t_R 18.1 min), **16** (2.7 mg, t_R 31.9 min), and **11** (2.4 mg, t_R 38.2 min). Similarly, Fr. 5.4 (355.8 mg) was purified by semi-preparative HPLC ($MeCN/H_2O$, 10:90, *v*/*v*, 3.0 mL/min) to yield compounds **5** (3.5 mg, t_R 6.5 min), **8** (1.3 mg, t_R 8.1 min), **9** (8.3 mg, t_R 10.9 min), and **12** (2.1 mg, t_R 15.5 min).

3.4. Spectral and Physical Data of Compounds 1–5

Fallopiaxylarester A (**1**): White solid; $[\alpha]_D^{25}$ +56.2 (*c* 0.11, MeOH); UV (MeOH) λ_{max} (log ε): 279 (0.19), 204 (0.72) nm; IR (KBr) v_{max}: 3426, 2922, 1733, 1648, 1569, 1457, 1412, 1384, 1247, 1032, 832 cm^{-1}; ECD (MeOH) λ_{max} ($\Delta\varepsilon$): 279 (+3.0), 206 (–4.0) nm. ^1H and ^{13}C NMR data, see Table 1; HRESIMS (*m*/*z*): 279.0837 $[M + Na]^+$ (calcd for $C_{12}H_{16}O_6Na$, 279.0839).

Fallopiaxylarester B (**2**): White solid; $[\alpha]_D^{25}$ +102.9 (*c* 0.38, MeOH); UV (MeOH) λ_{max} (log ε): 310 (0.16), 260 (0.06), 219 (0.41) nm; IR (KBr) v_{max}: 3359, 2928, 2858, 1696, 1623, 1560, 1456, 1409, 1260, 1230, 1080, 1018, 817, 539 cm^{-1}; ECD (MeOH) λ_{max} ($\Delta\varepsilon$): 310 (+3.4), 231 (–4.1), 205 (–3.3) nm. ^1H and ^{13}C NMR data, see Table 2; HRESIMS (*m*/*z*): 423.1626 $[M + Na]^+$ (calcd for $C_{19}H_{28}O_9Na$, 423.1626).

Fallopiaxylarester C (**3**): White solid; $[\alpha]_D^{25}$ +124.2 (*c* 0.39, MeOH); UV (MeOH) λ_{max} (log ε): 281 (0.11), 204 (0.45) nm; IR (KBr) v_{max}: 3380, 2927, 2857, 1700, 1649, 1569, 1458, 1414, 1384, 1250, 1025, 836, 700 cm^{-1}; ECD (MeOH) λ_{max} ($\Delta\varepsilon$): 280 (+4.5), 232 (+0.9), 205 (+3.0) nm; ^1H and ^{13}C NMR data, see Table 3; HRESIMS (*m*/*z*): 425.1780 $[M + Na]^+$ (calcd for $C_{19}H_{30}O_9Na$, 425.1782).

Fallopiaxylarol A (**4**): Colorless oil; $[\alpha]_D^{25}$ –31.8 (*c* 0.10, MeOH); UV (MeOH) λ_{max} (log ε): 205 (0.36), 218 (0.48) nm; IR (KBr) v_{max}: 3546, 3426, 2945, 2930, 1688, 1287, 1150, 1036 cm^{-1}; ^1H and ^{13}C NMR data, see Table 4; HRESIMS (*m*/*z*): 323.1829 $[M + Na]^+$ (calcd for $C_{16}H_{28}O_5Na$, 323.1829).

Pestalotiopyrone M (**5**): White solid; UV (MeOH) λ_{max} nm (log ε) 206 (0.45), 293 (0.16); IR (KBr) ν_{max} 3311, 2961, 2928, 1711, 1565, 1365, 1014, 989 cm^{-1}; ^1H NMR (methanol-d_4, 600 MHz) δ_H: 4.54 (2H, s, H-7), 4.41 (2H, s, H-8), 4.18 (3H, s, H-10), 2.35 (3H, s, H-9); ^{13}C NMR (methanol-d_4, 150 MHz) δ_C: 171.6 (C, C-4), 167.4 (C, C-2), 163.0 (C, C-6), 115.2 (C, C-5), 110.1 (C, C-3), 63.1 (CH_3, C-10), 55.5 (CH_2, C-8), 55.1 (CH_2, C-7), 17.3 (CH_3, C-9); HRESIMS (*m*/*z*): 223.0578 $[M + Na]^+$ (calcd for $C_9H_{12}O_5Na$, 223.0577).

3.5. Computational Details (TDDFT-ECD) of 1

The conformational search of (6*R*)-**1** was performed by using torsional sampling (MCMM) conformational searches with an OPLS_2005 force field within an energy window of 21 kJ/mol. Conformers above 1% Boltzmann populations were re-optimized at the B3LYP/6-31G(d) level with the IEFPCM solvent model for methanol. The following TDDFT calculations of the re-optimized geometries were all performed at the B3LYP/6-311G(d,p) level with the IEFPCM solvent model for methanol. Frequency analysis was performed as well to confirm that the re-optimized geometries were at the energy minima. Finally, the SpecDis 1.62 [27] software was used to obtain the Boltzmann-averaged ECD spectra of **1** and visualize the result.

3.6. Biological Assays

3.6.1. Antimicrobial Activity

Compounds **2–10** and **15–18**, and the crude extract were evaluated for antimicrobial activities against *Staphylococcus aureus* subsp. *aureus* and fluconazole-resistant *Candida albicans*. The antimicrobial assay was conducted according to a previously described method [28]. Samples were added into a 96-well culture plate with a maximum test compound concentration of 100 μM. Bacterial liquid was added to each well until the final concentration was 5×10^5 CFU/mL. The plate was then incubated at 37 °C for 24 h, and the OD values at 595 nm were measured using a microplate reader. Blank bacterial medium served as control.

3.6.2. Anti-Inflammatory Activity

The RAW 264.7 cells (2×10^5 cells/well) were incubated in 96-well culture plates with or without 1 μg/mL lipopolysaccharide (LPS, Sigma Chemical Co., St. Louis, MO, USA) for 24 h in the presence or absence of the test compounds. Supernatant aliquots (50 μL) were then treated with 100 μL Griess reagent (Sigma Chemical Co., St. Louis, MO, USA). The absorbance was measured at 570 nm by using a Synergy TMHT microplate reader (BioTek Instruments Inc., Winooski, VT, USA). In this study, N^G-methyl-L-arginine acetate (L-NMMA, Sigma Chemical Co., USA) was used as a positive control. In the remaining medium, an MTT assay was carried out to determine whether the suppressive effect was related to cell viability. The inhibitory rate of nitric oxide (NO) production = (NO level of blank control − NO level of test samples)/NO level of blank control. The percentage of NO production was evaluated by measuring the amount of nitrite concentration in the supernatants with Griess reagent, as described previously [29].

3.6.3. Alpha-Glucosidase-Inhibition Activity

The α-glucosidase inhibition was assessed according to the slightly modified method of Ma et al. [30]. All samples were dissolved in DMSO at a concentration of 50 μM. The α-glucosidase (Sigma Chemical Co., St. Louis, MO, USA) and substrate (4–Nitrophenyl α-D-glucopyranoside, PNPG, Sigma Chemical Co., St. Louis, MO, USA) were dissolved in potassium phosphate buffer (0.1 M, pH 6.7). The samples were preincubated with α-glucosidase at 37 °C for 10 min. Then, PNPG was quickly added to the 96-well enzyme label plate to start the reaction, and the plate was incubated at 37 °C for 50 min. At the same time, a blank control without samples and a positive control of quercetin (10 mM) was set up. All samples were thoroughly mixed and analyzed in triplicate. The OD value was measured at 405 nm using a microplate reader. The inhibition percentage (%) was calculated by the following equation: Inhibition (%) = $(1 - OD_{\text{sample}})/OD_{\text{control blank}} \times 100$.

4. Conclusions

In this paper, three new pyranone derivatives (**1–3**) and a new bisabolane-type sesquiterpenoid (**4**) were discovered from the fungus *Xylaria* sp. Z184. Moreover, we co-isolated 14 previously reported compounds (**5–18**), and reported the first complete set of spectroscopic data for pyranone **5**. In vitro bioassays were performed on a number of the isolated compounds and crude fungal extract. Compounds **5**, **7**, and **8** were demonstrated to be weak growth inhibitors of *Staphylococcus aureus* subsp. *Aureus*, and the extract was shown to be a potent inhibitor of NO production in LPS-stimulated RAW 264.7 mouse macrophages, with an inhibition rate of 77.28 ± 0.82% at 50 μg/mL. Although the crude fungal extract showed certain inhibitory activity of NO production in LPS-stimulated RAW 264.7 mouse macrophages, unfortunately, in the subsequent isolated compounds, no such convenient activity was found. This suggests that there may still be other structural types of compounds in the extract that exhibit anti-inflammatory activity. Thus, in addition to revealing four novel compounds, this work enhances understanding of the structural diversity within the *Xylaria* metabolomes.

Supplementary Materials: The following supporting information can be downloaded at: https://www.mdpi.com/article/10.3390/molecules29081728/s1, Figures S1–S79: 1D, 2D NMR spectra, and HRESIMS of compounds **1–5**; Figures S80-S105: ^{1}H and ^{13}C NMR spectra of compounds **6–18**; Figure S106: The TLC and HPLC profiles of the crude extract of fungus *Xylaria* sp. Z184.

Author Contributions: Y.Z. led on the isolation and data curation. Y.J. conducted the writing of original draft. W.Y. was responsible for fungal isolation, fermentation, and identification of the samples. P.G. and Z.Z. supported on data curation and analysis. Z.L., G.Z., and M.W. contributed to the evaluation of antimicrobial activity. Y.X. led on funding acquisition and project administration. All authors have read and agreed to the published version of the manuscript.

Funding: This work was financially supported by the National Natural Science Foundation of China (No. 21977120 and 32270296), and the Key Basic Research Program of the Science, Technology, and Innovation Commission of Shenzhen Municipality (JCYJ20200109142215045).

Institutional Review Board Statement: Not applicable.

Informed Consent Statement: Not applicable.

Data Availability Statement: The authors declare that all relevant data supporting the results of this study are available within in the article and its Supplementary Materials file.

Acknowledgments: The authors thank Xiaonian Li, Jianchao Chen, and the Shenzhen Bay Laboratory for their helpful assistance in NMR measurement. And we express our sincere thanks to Kunming Institute of Botany, Chinese Academy of Sciences for providing support in quantum chemical calculation.

Conflicts of Interest: The authors declare no conflicts of interest.

References

1. Newman, D.J.; Cragg, G.M. Natural products as sources of new drugs over the nearly four decades from 01/1981 to 09/2019. *J. Nat. Prod.* **2020**, *83*, 770–803. [CrossRef] [PubMed]
2. Liu, H.; Tan, H.; Chen, Y.; Guo, X.; Wang, W.; Guo, H.; Liu, Z.; Zhang, W. Cytorhizins A-D, four highly structure-combined benzophenones from the endophytic fungus *Cytospora rhizophorae*. *Org. Lett.* **2019**, *21*, 1063–1067. [CrossRef] [PubMed]
3. Han, W.B.; Wang, G.Y.; Tang, J.J.; Wang, W.J.; Liu, H.; Gil, R.R.; Navarro-Vázquez, A.; Lei, X.X.; Gao, J.M. Herpotrichones A and B, two intermolecular [4+2] adducts with anti-neuroinflammatory activity from a *Herpotrichia* Species. *Org. Lett.* **2020**, *22*, 405–409. [CrossRef] [PubMed]
4. Yoiprommarat, S.; Unagul, P.; Suvannakad, R.; Klaysuban, A.; Suetrong, S.; Bunyapaiboonsri, T. Eremophilane sesquiterpenes from the mangrove fungus BCC 60405. *Phytochem. Lett.* **2019**, *34*, 84–90. [CrossRef]
5. Wu, W.; Dai, H.; Bao, L.; Ren, B.; Lu, J.; Luo, Y.; Guo, L.; Zhang, L.; Liu, H. Isolation and structural elucidation of proline-containing cyclopentapeptides from an endolichenic *Xylaria* sp. *J. Nat. Prod.* **2011**, *74*, 1303–1308. [CrossRef] [PubMed]
6. Xu, K.; Li, R.; Zhu, R.; Li, X.; Xu, Y.; He, Q.; Xie, F.; Qiao, Y.; Luan, X.; Lou, H. Xylarins A–D, two pairs of diastereoisomeric isoindoline alkaloids from the endolichenic fungus *Xylaria* sp. *Org. Lett.* **2021**, *23*, 7751–7754. [CrossRef] [PubMed]
7. Chen, W.; Yu, M.; Chen, S.; Gong, T.; Xie, L.; Liu, J.; Bian, C.; Huang, G.; Zheng, C. Structures and biological activities of secondary metabolites from *Xylaria* spp. *J. Fungi* **2024**, *10*, 190. [CrossRef]
8. Han, W.B.; Zhai, Y.J.; Gao, Y.; Zhou, H.Y.; Xiao, J.; Pescitelli, G.; Gao, J.M. Cytochalasins and an abietane-type diterpenoid with allelopathic activities from the endophytic fungus *Xylaria* species. *J. Agric. Food. Chem.* **2019**, *67*, 3643–3650. [CrossRef]
9. Jia, S.; Li, J.; Li, J.; Su, X.; Li, X.N.; Yao, Y.; Xue, Y. Anti-neuroinflammatory activity of new naturally occurring benzylated hydroxyacetophenone analogs from the endophytic fungus *Alternaria* sp. J030. *Chem. Biodivers.* **2022**, *19*, e202200751. [CrossRef] [PubMed]
10. Jia, S.; Su, X.; Yan, W.; Wu, M.; Wu, Y.; Lu, J.; He, X.; Ding, X.; Xue, Y. Acorenone C: A new spiro-sesquiterpene from a mangrove-associated fungus, *Pseudofusicoccum* sp. J003. *Front. Chem.* **2021**, *9*, 780304. [CrossRef] [PubMed]
11. Wu, Y.; Su, X.; Lu, J.; Wu, M.; Yang, S.Y.; Mai, Y.; Deng, W.; Xue, Y. In Vitro and in silico analysis of phytochemicals from *Fallopia dentatoalata* as dual functional cholinesterase inhibitors for the treatment of Alzheimer's disease. *Front. Pharmacol.* **2022**, *13*, 905708. [CrossRef] [PubMed]
12. Yang, W.W.; Lu, L.W.; Zhang, X.Q.; Bao, S.S.; Cao, F.; Guo, Z.Y.; Deng, Z.S.; Proksch, P. Xylariaopyrones E-I, five new α-pyrone derivatives from the endophytic fungus *Xylariales* sp. (HM-1). *Nat. Prod. Res.* **2022**, *36*, 2230–2238. [CrossRef] [PubMed]
13. Yamamoto, I.; Tai, A.; Fujinami, Y.; Sasaki, K.; Okazaki, S. Synthesis and characterization of a series of novel monoacylated ascorbic acid derivatives, 6-*O*-Acyl-2-*O*-α-D-glucopyranosyl-L-ascorbic acids, as skin antioxidants. *J. Med. Chem.* **2002**, *45*, 462–468. [CrossRef] [PubMed]
14. Zhao, Q.; Wang, C.X.; Yu, Y.; Wang, G.Q.; Zheng, Q.C.; Chen, G.D.; Lian, Y.Y.; Lin, F.; Guo, L.D.; Gao, H. Nodulisporipyrones A-D, new bioactive α-pyrone derivatives from *Nodulisporium* sp. *J. Asian Nat. Prod. Res.* **2015**, *17*, 567–575. [CrossRef] [PubMed]
15. Lichtenthaler, F.W.; Martin, D.; Weber, T.; Schiweck, H. Studies on Ketoses, 7–5-(α-D-Glucosyloxymethyl)furfural: Preparation from isomaltulose and exploration of its ensuing chemistry. *Eur. J. Org. Chem.* **1993**, *9*, 967–974. [CrossRef]
16. Miyazawa, M.; Nankai, H.; Kameoka, H. Biotransformation of (−)-α-bisabolol by plant pathogenic fungus, *Glomerella cingulata*. *Phytochemistry* **1995**, *40*, 69–72. [CrossRef]
17. Xu, J. The Preparation Method and Application of a Pyranoid Compound with Immunosuppressive Activity. CN 108913731 16 November 2021.
18. Cook, L.; Ternai, B.; Ghosh, P. Inhibition of human sputum elastase by substituted 2-pyrones. *J. Med. Chem.* **1987**, *30*, 1017–1023. [CrossRef] [PubMed]
19. Guo, Z.Y.; Lu, L.W.; Bao, S.S.; Liu, C.X.; Deng, Z.S.; Cao, F.; Liu, S.P.; Zou, K.; Proksch, P. Xylariaopyrones A-D, four new antimicrobial α-pyrone derivatives from endophytic fungus *Xylariales* sp. *Phytochem. Lett.* **2018**, *28*, 98–103. [CrossRef]

20. Zhang, H.; Deng, Z.; Guo, Z.; Peng, Y.; Huang, N.; He, H.; Tu, X.; Zou, K. Effect of culture conditions on metabolite production of *Xylaria* sp. *Molecules* **2015**, *20*, 7940–7950. [CrossRef] [PubMed]
21. Tian, J.F.; Yu, R.J.; Li, X.X.; Gao, H.; Guo, L.D.; Tang, J.S.; Yao, X.S. ^{1}H and ^{13}C NMR spectral assignments of 2-pyrone derivatives from an endophytic fungus of sarcosomataceae. *Magn. Reson. Chem.* **2015**, *53*, 866–871. [CrossRef] [PubMed]
22. Pérez-Castorena, A.L.; Arciniegas, A.; Guzmán, S.L.; Villaseñor, J.L.; Vivar, A.R. Eremophilanes from *Senecio mairetianus* and some reaction products. *J. Nat. Prod.* **2006**, *69*, 1471–1475. [CrossRef] [PubMed]
23. Arora, D.; Sharma, N.; Singamaneni, V.; Sharma, V.; Kushwaha, M.; Abrol, V.; Guru, S.; Sharma, S.; Gupta, A.P.; Bhushan, S.; et al. Isolation and characterization of bioactive metabolites from *Xylaria psidii*, an endophytic fungus of the medicinal plant *Aegle marmelos* and their role in mitochondrial dependent apoptosis against pancreatic cancer cells. *Phytomedicine* **2016**, *23*, 1312–1320. [CrossRef] [PubMed]
24. Hallock, Y.F.; Clardy, J.; Kenfield, D.S.; Strobel, G. De-*O*-methyldiaporthin, a phytotoxin from *Drechslera siccans*. *Phytochemistry* **1988**, *27*, 3123–3125. [CrossRef]
25. Feng, C.C.; Chen, G.D.; Zhao, Y.Q.; Xin, S.C.; Li, S.; Tang, J.S.; Li, X.X.; Hu, D.; Liu, X.Z.; Gao, H. New isocoumarins from a cold-adapted fungal strain *Mucor* sp. and their developmental toxicity to zebrafish embryos. *Chem. Biodivers.* **2014**, *11*, 1099–1108. [CrossRef]
26. Lin, Y.; Wu, X.; Feng, S.; Jiang, G.; Zhou, S.; Vrijmoed, L.L.P.; Jones, E.B.G. A novel *N*-cinnamoylcyclopeptide containing an allenic ether from the fungus *Xylaria* sp. (strain #2508) from the South China Sea. *Tetrahedron Lett.* **2001**, *42*, 449–451.
27. Bruhn, T.; Schaumloffel, A.; Hemberger, Y.; Bringmann, G. SpecDis: Quantifying the Comparison of Calculated and Experimental Electronic Circular Dichroism Spectra. *Chirality* **2013**, *25*, 243–249. [CrossRef] [PubMed]
28. Zhang, N.; Shi, Z.; Guo, Y.; Xie, S.; Qiao, Y.; Li, X.N.; Xue, Y.; Luo, Z.; Zhu, H.; Chen, C.; et al. The absolute configurations of hyperilongenols A–C: Rare 12,13-seco-spirocyclic polycyclic polyprenylated acylphloroglucinols with enolizable β, β'-tricarbonyl systems from *Hypericum longistylum* Oliv. *Org. Chem. Front.* **2019**, *6*, 1491–1502. [CrossRef]
29. Wu, Y.; Xie, S.; Hu, Z.; Wu, Z.; Guo, Y.; Zhang, J.; Wang, J.; Xue, Y.; Zhang, Y. Triterpenoids from whole plants of *Phyllanthus urinaria*. *Chin. Herb. Med.* **2017**, *9*, 193–196. [CrossRef]
30. Ma, Y.Y.; Zhao, D.G.; Zhou, A.Y.; Zhang, Y.; Du, Z.; Zhang, K. α-Glucosidase inhibition and antihyperglycemic activity of phenolics from the flowers of *Edgeworthia gardneri*. *J. Agric. Food. Chem.* **2015**, *63*, 8162–8169. [CrossRef]

molecules

MDPI

Review

Penicillides from *Penicillium* and *Talaromyces*: Chemical Structures, Occurrence and Bioactivities

Maria Michela Salvatore [1], Rosario Nicoletti [2,3,*], Filomena Fiorito [4,5] and Anna Andolfi [1,5]

1 Department of Chemical Sciences, University of Naples 'Federico II', 80126 Naples, Italy; mariamichela.salvatore@unina.it (M.M.S.); anna.andolfi@unina.it (A.A.)
2 Council for Agricultural Research and Economics, Research Centre for Olive, Fruit and Citrus Crops, 81100 Caserta, Italy
3 Department of Agricultural Sciences, University of Naples 'Federico II', 80055 Naples, Italy
4 Department of Veterinary Medicine and Animal Production, University of Naples 'Federico II', 80137 Naples, Italy; filomena.fiorito@unina.it
5 BAT Center-Interuniversity Center for Studies on Bioinspired Agro-Environmental Technology, University of Naples 'Federico II', 80138 Naples, Italy
* Correspondence: rosario.nicoletti@crea.gov.it

Abstract: Penicillide is the founder product of a class of natural products of fungal origin. Although this compound and its analogues have been identified from taxonomically heterogeneous fungi, they are most frequently and typically reported from the species of *Talaromyces* and *Penicillium*. The producing strains have been isolated in various ecological contexts, with a notable proportion of endophytes. The occurrence of penicillides in these plant associates may be indicative of a possible role in defensive mutualism based on their bioactive properties, which are also reviewed in this paper. The interesting finding of penicillides in fruits and seeds of *Phyllanthus emblica* is introductory to a new ground of investigation in view of assessing whether they are produced by the plant directly or as a result of the biosynthetic capacities of some endophytic associates.

Keywords: secondary metabolites; Eurotiomycetes; fungal extrolites; depsidones; bioactivity

1. Introduction

Fungi in the genus *Penicillium* (Eurotiomycetes, Aspergillaceae) represent one of the most exploited sources of chemodiversity, with a multitude of structural models and compound classes having been reported since the discovery of mycophenolic acid by Gosio [1]. Many species of these fungi have been described from all ecological contexts and geographical areas in the world, which are able to produce blockbuster drugs such as penicillin and compactin [2]. However, the recent affirmation of the principle 'one fungus, one name' in taxonomy [3] and the widespread employment of biomolecular markers for a more accurate identification have brought the separation of *Penicillium* species having symmetrical biverticillate conidiophores into the genus *Talaromyces* (Eurotiomycetes, Trichocomaceae) based on phylogenetic reconstructions [4].

Fifty years ago, a Japanese researcher reported on the finding of a new secondary metabolite from an isolate of *Penicillium* sp., which was named penicillide (11-hydroxy-3-[(1*S*)-1-hydroxy-3-methylbutyl]-4-methoxy-9-methyl-5H,7H-dibenzo[b,g][1,5]dioxocin-5-one) (Figure 1) [5]. This product is the founder of a class of bioactive products characterized by 2,4-dihydroxybenzilic alcohol and 2-hydroxy-4-methoxy-benzoic acid moieties linked together by ether and ester bonds, forming a typical 8-membered heterocycle in the place of the 7-membered one usually found in depsidones [6]. Rather than being one of a kind, more analogues of penicillide were later found by other independent research groups in the early 1990s; possibly due to difficulties in accessing the pertinent literature at that time, some of these compounds were given the unrelated names of purpactins [6] and vermixocins [7].

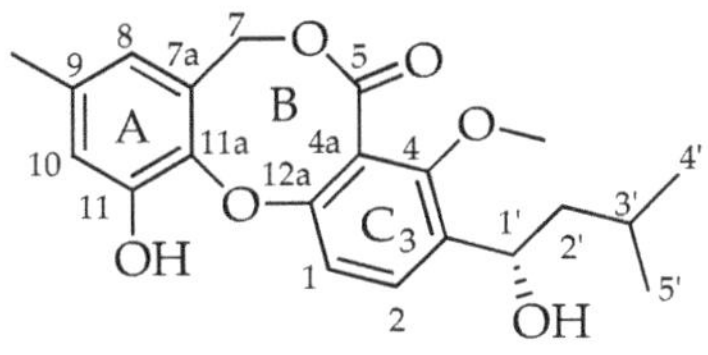

Figure 1. Structure of penicillide (**1**).

This paper offers an overview on the research activity which has been developed in the last five decades on penicillides, with reference to their occurrence in nature as secondary metabolites of *Penicillium* and *Talaromyces* species, as well as to their biological properties and possible biotechnological applications.

2. Chemical Structures

With reference to the general molecular scaffold outlined above, the A ring is invariable in all the known penicillides with the exception of hydroxypenicillide, while the C ring presents a side chain of five carbons at C-3. Substitutions in this chain vary among the ten analogues which have been identified from these fungi so far (Table 1). Besides searching databases, such as PubMed, Google Scholar and Web of Science, an accurate analysis of the available data based on the chemical structure was carried out using Scifinder.

Table 1. Penicillides from *Panicillium* and *Talaromyces*.

Code	Compounds	Chemical Structure	Nominal Mass	Formula
1	Penicillide (= vermixocin A)	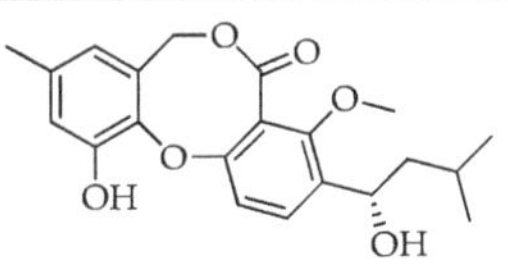	372	$C_{21}H_{24}O_6$
2	Purpactin A (= vermixocin B)	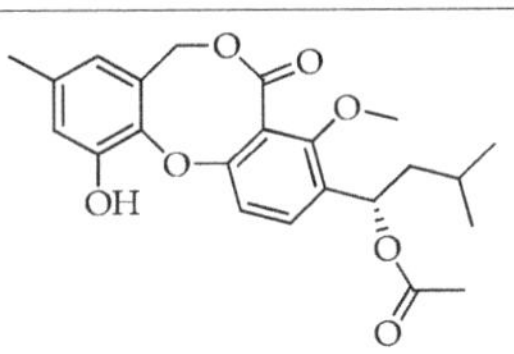	414	$C_{23}H_{26}O_7$
3	$\Delta^{1',3'}$-1-Dehydroxypenicillide	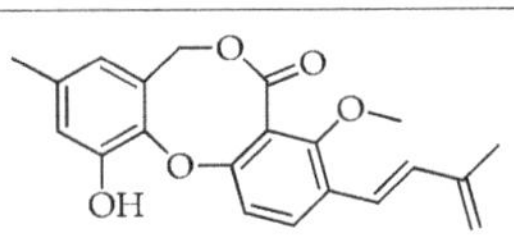	352	$C_{21}H_{20}O_5$
4	1′2′-Epoxy-3′,4′-didehydropenicillide	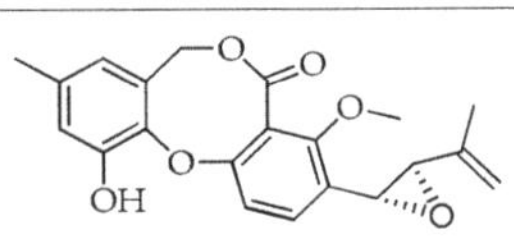	368	$C_{21}H_{20}O_6$
5	Dehydroisopenicillide (= MC-141, 1,2-dehydropenicillide)	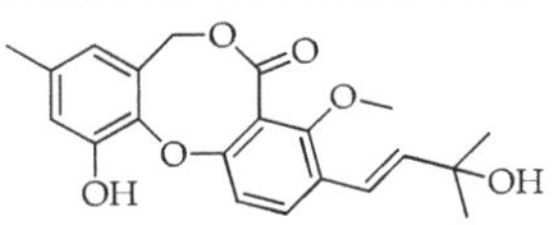	370	$C_{21}H_{22}O_6$

Table 1. *Cont.*

Code	Compounds	Chemical Structure	Nominal Mass	Formula
6	2′-Hydroxy-3′,4′-didehydropenicillide	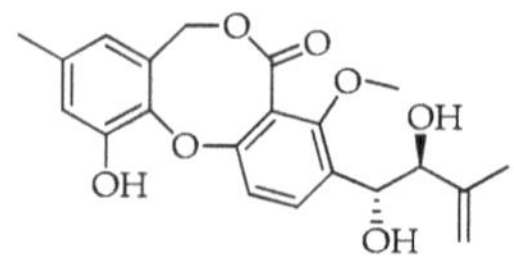	386	$C_{21}H_{22}O_7$
7	5′-Hydroxypenicillide	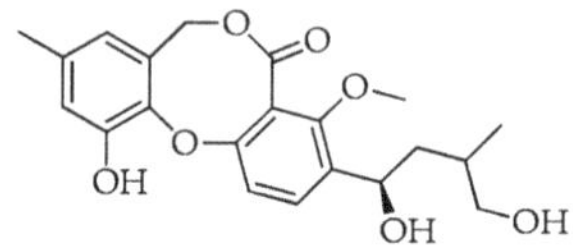	388	$C_{21}H_{24}O_7$
8	Hydroxypenicillide	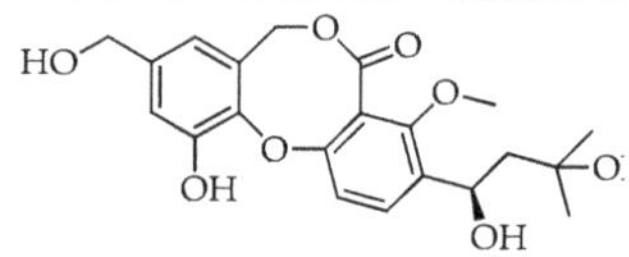	404	$C_{21}H_{24}O_8$
9	Isopenicillide	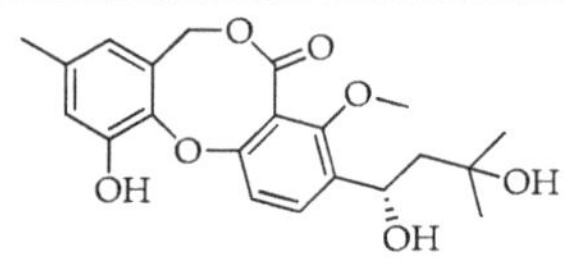	388	$C_{21}H_{24}O_7$
10	3′-*O*-Methyldehydroisopenicillide (= 3-methoxy-1′2-dehydropenicillide)	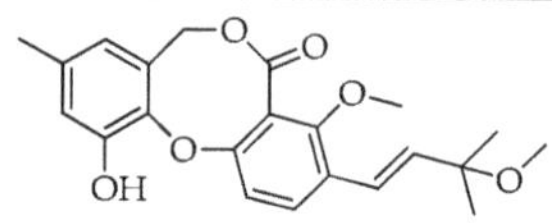	384	$C_{22}H_{24}O_6$
11	Prenpenicillide	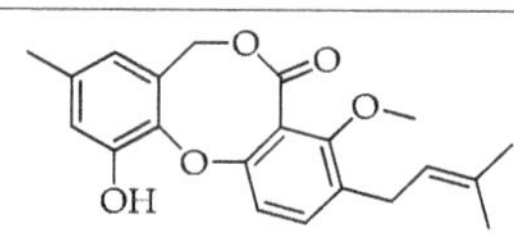	354	$C_{21}H_{22}O_5$

With reference to the structures in Table 1, it must be noted that isopenicillide (**9**) is not an isomeric form of penicillide (**1**) since it has an additional hydroxyl group on the side chain in ring C; rather, it is an isomer of 5′-hydroxypenicillide (**7**). Moreover, the denomination adopted for compounds **7** and **8** can be misleading, and it is incoherent because compound **8** has an additional hydroxyl group on the A ring.

3. Occurrence

Conforming to the above-introduced separation of symmetrically biverticillate species in *Talaromyces*, most of the strains producing penicillides are found to belong to this genus (Table 2). Similar to *Penicillium*, the latter taxon has been characterized as an outstanding source of chemodiversity [8–11], exhibiting some peculiar biosynthetic models which have been only or predominantly found in these fungi, such as the funicones [12,13].

Table 2. *Penicillium* and *Talaromyces* species/strains reported as producers of penicillides.

Species	Substrate	Location	Compounds	Ref.
Penicillium sp.	-	Japan	**1**	[5]
Penicillium sp.	endophytic in *Taxus cuspidata*	Gunma (Japan)	**4, 6, 5, 10**	[14]
Penicillium sp. F60760	soil	South Korea	**1**	[15]
Penicillium sp. H1	marine	China	**1**	[16]
Penicillium sp. MA-37 [1]	rhizosphere of *Bruguiera gymnorrhiza*	Hainan (China)	**1, 3, 5, 10**	[17]
Penicillium sp. PF9	endophytic in *Artemisia princeps*	South Korea	**1, 10**	[18]
Penicillium sp. SCS-KFD16	marine	China	**1**	[19]
Penicillium sp. ZLN29	marine sediment	Jiazhou Bay (China)	**1, 11**	[20]
P. chrysogenum MT-12	endophytic in *Huperzia serrata*	China	**1, 2, 8, 9**	[21]
P. montanense KYI635 [2]	soil	Kanagawa (Japan)	**1, 2**	[22]
P. simplicissimum IFM53375	unspecified	Japan	**1, 2**	[23]
Talaromyces sp.	moss (*Climacium dendroides*)	Gangneung (South Korea)	**1, 2**	[24]
Talaromyces sp. BTBU20213036	mud from intertidal zone	Qingdao (China)	**1**	[25]
Talaromyces sp. CS-258	mussel from cold seep	South China Sea	**1, 2, 5**	[26]
Talaromyces sp. HK1-18	mangrove soil	Hainan (China)	**1**	[27]
Talaromyces sp. IQ-567	endophytic in *Rhizophora mangle*	Tecomate Lagoon (Mexico)	**1**	[28]
Talaromyces sp. LF458	sponge (*Axinella verrucosa*)	Elba island (Italy)	**3, 10**	[29]
Talaromyces sp. M27416	sea water	Dongshan island (China)	**1**	[30]
Talaromyces sp. Wangcy005	gorgonian (*Subergorgia suberosa*)	Weizhou island (China)	**1, 2, 3**	[31]
Talaromyces sp. WHUF0362	mangrove soil	Hainan (China)	**2**	[32]
T. aculeatus IBT23209	soil	Araracuara (Colombia)	**1, 2**	[33]
T. aculeatus IBT23210			**2**	
T. aculeatus IBT23211			**2**	
T. aculeatus IBT23212			**2**	
T. aculeatus PSU-RSPG105	soil	Rajjaprabha Dam (Thailand)	**1, 2**	[34]
T. albobiverticillius IBT4466	imported pomegranate	Denmark	**2**	[35]
T. albobiverticillius CBS113167	air in cake factory	-	**2**	
T. amestolkiae CBS433.62	ground domestic waste	Verona (Italy)	**2**	[36]
T. amestolkiae CBS436.62	alum solution	-	**2**	
T. atroroseus IBT3933	-	-	**2**	[35]
T. atroroseus TRP-NRC	agricultural waste	Egypt	**2**	[37]
T. derxii NHL2981	soil	Kurashiki (Japan)	**1, 5**	[38]
T. flavus CCMF-276	soil	Jàchymov (Czechia)	**1, 2**	[39]
T. flavus ATCC74110	-	-	**1**	[40]
T. flavus TGGP34	endophytic in *Acanthus ilicifolius*	China	**1, 2, 5**	[41]
T. fuscoviridis CBS193.69	soil	the Netherlands	**2**	[42]
T. gwangjuensis CNUFC-WT19-1	freshwater	Yeosu (South Korea)	**2**	[43]
T. koreana CNUFC-YJW2-13	freshwater	Yeosu (South Korea)	**2**	
T. pinophilus AF-02	endophytic in *Allium fistulosum*	Lanzhou (China)	**1**	[44]
T. pinophilus FKI-5653	soil	Hachijo island (Japan)	**1, 5**	[45]

Table 2. *Cont.*

Species	Substrate	Location	Compounds	Ref.
T. pinophilus H608	mangrove sediment	Xiamen (China)	**1, 2, 3, 5, 7, 9**	[46]
T. pinophilus LT6	tobacco rhizosphere	Lecce area (Italy)	**1**	[47]
T. pinophilus SCAU037	rhizosphere of *Rhizophora stylosa*	Techeng isle (China)	**1, 9, 10**	[48]
T. pinophilus XS-20090E18	unidentified gorgonian	Xisha islands (China)	**1, 2, 8, 9**	[49]
T. purgamentorum CBS113145	forest leaf litter	Peña Roja (Colombia)	**1, 2**	[33]
T. purpureogenus CFRM02	unspecified	Karnataka (India)	**1**	[50]
T. purpureogenus FO-608	soil	Japan	**2**	[6]
T. purpureogenus ATCC44445	corn kernel	Georgia (USA)	**2**	
T. purpureogenus ATCC20204	-	Japan	**2**	
T. purpureogenus CBS286.36	-	Japan	**2**	
T. purpureogenus IBT17540	barley	Winnipeg (Canada)	**2**	
T. purpureogenus IBT11632	imported marjoram	Denmark	**2**	
T. purpureogenus IBT12779	marine	France	**2**	[36]
T. purpureogenus IMI112715	rhizosphere of *Trifolium alexandrinum*	Egypt	**2**	
T. purpureogenus IMI136126	molded corn	Wisconsin (USA)	**2**	
T. purpureogenus IMI136127	molded corn	Wisconsin (USA)	**2**	
T. purpureogenus NRRL3290	-	Georgia (USA)	**2**	
T. purpureogenus MM	cotton textile	Egypt	**1, 10**	[51]
T. ruber CBS237.93	unknown	Unknown	**2**	
T. ruber CBS113160	-	-	**2**	
T. ruber CBS132699	sandy soil	Sousse (Tunisia)	**2**	[36]
T. ruber FRR1503	preserved wood	Australia	**2**	
T. ruber IBT31167	-	-	**2**	
T. stellenboschiensis CBS135665	soil	Stellenbosch (South Africa)	**2**	[42]
T. stipitatus SK-4	leaf of *Acanthus ilicifolius*	Guangxi (China)	**1, 2**	[52]
T. thailandensis PSU-SPSF059	soil	Thailand	**2**	[53]
T. veerkampii CBS500.78	soil	Meta (Colombia)	**1, 2**	[42]
T. veerkampii CBS136668	soybean seed	Matou (Taiwan)	**1, 2**	[42]
T. verruculosus TGM14	mangrove (*Xylocarpus granatum*)	Hainan (China)	**1**	[54]

[1] This strain should be more correctly ascribed to *Talaromyces* based on an updated GenBank blast of its deposited ITS sequence. [2] This strain was originally identified as *Penicillium asperosporum*.

In light of the above, it is quite possible that the isolates provisionally identified as *Penicillium* sp. actually belong to *Talaromyces*. This can be directly verified for strain MA-37 through a blast in GenBank of the reported rDNA-ITS sequence [17], while the doubt remains in the other cases. However, apart from these incompletely identified strains, the fact that penicillides have been reported in three unrelated species, namely *P. chrysogenum* *P. montanense* and *P. simplicissimum*, represents an indication that these compounds may be common secondary metabolites in *Penicillium*, too. Yet, the connotation of penicillides as products characterizing the biosynthetic abilities of *Talaromyces* is well supported by the remark that as many as 18 species of this genus are listed in Table 1 as documented sources.

More infrequent are reports of these products from the taxonomically related genera *Aspergillus* (Aspergillaceae) and *Neosartorya* (Trichocomaceae), that is, penicillide [55–57], purpactin A, dehydropenicillide and $\Delta^{2'}$-1′-dehydropenicillide [58], while other occasional findings concern miscellaneous Ascomycetes species. This is the case of *Alternaria* (Dothideomycetes, Pleosporaceae), reported to produce penicillide [59]; *Guignardia* (= *Phyllosticta*: Dothideomycetes, Botryosphaeriaceae), producing purpactin A and prenpenicillide (identified as (*E*)-3-(3-methylbut-1-enyl)-11-hydroxy-4-methoxy-9-methyl-7H-dibenzo[b,g][1,5]dioxocin-5-one) [60]; *Scytalidium cuboideum* (Leotiomycetes, *incertae sedis*), producing purpactin A [61]; *Colpoma quercinum* (Leotiomycetes, Rhytismataceae), producing penicillide [62]; and *Pestalotiopsis* spp. (Sordariomycetes, Pestalotiopsidaceae), producing penicillide and purpactin A [63], dehydroisopenicillide and 3′-O-methyldehydroisopenicillide [64]. Fungi in the genus *Pestalotiopsis* have also been found to produce pestalotiollides A-B and sinopestalotiollides A-D; these are structural analogues with a modified side chain, which have not been reported from *Talaromyces* and *Penicillium* so far [64,65].

The assumption that penicillides are typical fungal products has been impaired by an intriguing report on the extraction in equable amounts of penicillide, purpactin A and the novel analogue 1′*S*-11-dehydroxypenicillide from fruits of the Indian gooseberry (*Phyllanthus emblica*) (Malpighiales, Phyllanthaceae) [66]. This finding has been followed by the detection of penicillide in seeds of the same plant by an independent research group [67]. Indeed, recent progresses in natural product research have disclosed the ability of taxonomically diverse endophytic fungi to synthesize compounds originally characterized from plants [68], which, in many cases, has postulated the transfer of gene clusters encoding for their synthesis [69,70]. By extension of this concept, horizontal gene transfer (HGT) could also have operated in the case of *P. emblica*, but in which direction? Did the plant acquire the genetic base from any endophytic fungus, or rather, did it occur that several endophytes borrowed this gene cluster from the plant and later spread to the various ecological contexts from which the penicillide producers have been reported? Alternatively, the extraction from *P. emblica* could be consequential to its production and accumulation in the plant tissues as resulting from the biosynthetic capacities of one or more endophytic associates, as has been demonstrated in the case of defensive mutualists of ryegrass and other plants [71]. Indeed, both *Talaromyces* and *Penicillium* species are at the forefront among endophytic fungi, having been reported from many and diverse plants in all environments [11,72], and the hypothesis that a systematic association established with *P. emblica* could lead to the accumulation of penicillides in its fruits and seeds deserves to be verified. The very recent report of the species *Talaromyces atroroseus* and *Penicillium choerospondiatis* from gooseberries [73] opens a new ground of investigation in this respect.

Finally, while this manuscript was in preparation, another paper was published reporting on the detection of purpactin A in the aqueous extract obtained from the Chinese vine *Sargentodoxa cuneata* (Ranunculales, Lardizabalaceae) [74]; undoubtedly, this last finding reinforces the need to more thoroughly examine plants as possible sources of penicillides.

4. Biological Properties

Biological properties have been essentially evaluated for the two most common products, penicillide and purpactin A; the available data are shown in Table 3.

It is generally agreed that the bioactivities of microbial secondary metabolites are related to their competitive interactions in the biocenosis. In this respect, both penicillide and purpactin A exhibit moderate antibacterial properties, which anyway could have an ecological impact considering the possible synergism with other antimicrobial compounds produced by *Penicillium* and *Talaromyces* species. Although various strain panels have been used in the antibacterial assays, the same MICs for both products generally resulted when they were concomitantly tested (e.g., against *Klebsiella pneumoniae, Pseudomonas aeruginosa* and *Vibrio parahaemolyticus*) [12]. The highest activities were detected for penicillide at 0.78 μg mL^{-1} against the methicillin and oxacillin resistant strain ATCC 43300 of *Staphylococcus aureus* [27], while purpactin A was active at a 5-fold higher concentration (4 μg mL^{-1}) against *E. coli* [26]. In the latter study, a positive correlation was observed between the acetylation of the side chain and antibacterial efficacy [26].

Antifungal activity was evaluated for penicillide against the opportunistic pathogenic Basidiomycete *Cryptococcus neoformans*, but effectiveness only resulted at a quite high concentration [20]. High active concentration also resulted in assays of this compound against brine shrimps [4], while it displayed antifouling properties against *Balanus amphitrite* at a lower concentration than purpactin A [34]. Conversely, antiplasmodial effects were observed at lower concentrations for purpactin A [20].

Assays for cytotoxic activity yielded quite variable results. In fact, on Hep G2 (hepatocellular carcinoma) cells, cytotoxicity was higher for penicillide, as assessed in two different laboratories [6,15], while incongruent values were obtained in two other laboratories for purpactin A on MCF-7 (breast cancer) cells [17,20]. In assays carried out at the same laboratory, it was slightly higher for penicillide against KB (epidermoid carcinoma) cells and slightly higher for purpactin A against Vero (kidney epithelium of African green monkey) cells [20]. Finally, the same values were measured for the two compounds in terms of incorporation of uridine, thymidine and valine in P388 (murine leukemia) cells [25].

Table 3. Biological activities of penicillide and purpactin A.

Biological Activity	Concentration	Results and Further Details	Ref.
	Penicillide (1)		
Antibacterial	100 μg mL^{-1}	*Acinetobacter baumannii* (40% inhibition)	[28]
	50 μg mL^{-1}	*Clostridium perfringens* (MIC)	[44]
	64 μg mL^{-1}	*Escherichia coli* (MIC)	[26]
	64 μg mL^{-1}	*Klebsiella pneumoniae* (MIC)	[26]
	50 μg mL^{-1}	*Micrococcus tetragenus* (MIC)	[44]
	32 μg mL^{-1}	*Pseudomonas aeruginosa* (MIC)	[26]
	100 μg mL^{-1}	*Staphylococcus aureus* (MIC)	[25]
	0.78 μg mL^{-1}	MRSA *S. aureus* (MIC)	[27]
	64 μg mL^{-1}	*Vibrio alginolyticus* (MIC)	[26]
	32 μg mL^{-1}	*Vibrio parahaemolyticus* (MIC)	[26]
Antifouling	2.6 μg mL^{-1}	*Balanus amphitrite* (EC_{50})	[49]
Antifungal	128 μg mL^{-1}	*Cryptococcus neoformans* (MIC)	[34]
Anti-inflammatory	11.5 μM	RAW264.7 (IC_{50})	[18]
Antimalarial	16.41 μM	*Plasmodium falciparum* (IC_{50})	[34]
Brine shrimp lethal	158.5 μM	LD_{50}	[17]
Cholesterol acyltransferase inhibition	22.9 μM	rabbit liver microsomes (IC_{50})	[22]
Cytotoxic	50 μg mL^{-1}	P388	[39]
	9.7 μM	Hep G2 (IC_{50})	[20]
	6.7 μM	HEp-2 (IC_{50})	[49]
	43.77 μM	KB (IC_{50})	[34]
	7.8 μM	RD (IC_{50})	[49]
	53.73 μM	Vero (IC_{50})	[34]
	50 μg mL^{-1}	incorporation of uridine, thymidine and valine in P388	[39]
m-Calpain inhibition	7.1 μM	SLLVY-AMC (IC_{50})	[15]
Oxytocin binding inhibition	67 μM	IC_{50}	[40]
α-Glucosidase inhibition	78.4 μM	IC_{50}	[48]

Table 3. *Cont.*

Biological Activity	Concentration	Results and Further Details	Ref.
	Purpactin A (2)		
Antibacterial	8 μg mL^{-1}	*Aeromonas hydrophila* (MIC)	[26]
	4 μg mL^{-1}	*E. coli* (MIC)	
	2.42 μmol L^{-1}	*Helicobacter pylori* 129 (MIC)	[32]
	4.83 μmol L^{-1}	*H. pylori* G27 (MIC)	
	64 μg mL^{-1}	*K. pneumoniae* (MIC)	[26]
	64 μg mL^{-1}	MRSA *S. aureus* (MIC)	
	8 μg mL^{-1}	*Micrococcus luteus* (MIC)	
	32 μg mL^{-1}	*P. aeruginosa* (MIC)	
	16 μg mL^{-1}	*Vibrio anguillarum* (MIC)	
	8 μg mL^{-1}	*Vibrio harveyi* (MIC)	
	32 μg mL^{-1}	*V. parahaemolyticus* (MIC)	
Antifouling	4.8 μg mL^{-1}	*B. amphitrite* (IC_{50})	[31]
	10 μg mL^{-1}	*B. amphitrite* (EC_{50})	[49]
Antimalarial	5.69 μM	*P. falciparum* (IC_{50})	[34]
Cholesterol acyltransferase inhibition	120 μM	rat microsomes (IC_{50})	[6,75]
	1.2 μM	J774 (IC_{50})	
	8.2 μM	rabbit liver microsomes (IC_{50})	[22]
Cytotoxic	15.1 μM	HCT-116 (IC_{50})	[31]
	38.95 μM	Hep G2 (IC_{50})	[29]
	50 μg mL^{-1}	incorporation of uridine, thymidine and valine in P388	[39]
	9.7 μM	J774 (IC_{50})	[6,75]
	52.5 μM	KB (IC_{50})	[34]
	16.4 μM	MCF-7 (IC_{50})	[31]
	75.28 μM	MCF-7 (IC_{50})	[34]
	41.21 μM	NIH 3 T3 (IC_{50})	[29]
	32.57 μM	Vero (IC_{50})	[34]
Elastase inhibition	37.2 μg mL^{-1}	IC_{50}	[76]
α-Glucosidase inhibition	80.9 μM	IC_{50}	[52]

Limited information has been gathered with reference to possible biotechnological applications related to miscellaneous enzyme inhibitory activities. These properties were approximately in the same range against α-glucosidase [33,39], while cholesterol acyltransferase inhibition was detected at lower concentrations for purpactin A, as measured in rabbit liver microsomes [8]. Anticholesterolemic activity of the latter compound was also documented in rat liver microsomes and J774 macrophages [36,41]. Moderate activity was detected for penicillide as a calpain inhibitor with possible applications for the treatment of muscular dystrophy and neurodegenerative diseases [2]; moreover, the reported activity of penicillide as an oxytocin antagonist [26] could be exploited in gynecology. Finally, purpactin A was found to consistently act as an elastase inhibitor with possible application in the treatment of chronic obstructive pulmonary disease [42] and as an inhibitor of TMEM16A, a Ca^{2+}-activated Cl^{-} channel protein involved in mucus secretion in inflamed airways, which has been proposed as a drug target for diseases associated with mucus hypersecretion, including asthma. The compound prevented Ca^{2+}-induced mucin release in cytokine-treated airway cells, while it did not affect cell viability, epithelial barrier integrity and activities of membrane transport proteins essential for maintaining airway hydration [77].

5. Biosynthesis of Penicillide and Purpactin A

Penicillides are structurally related compounds biosynthetically derived from the polyketide pathway. It has been proposed that the skeleton of these natural products, characterized by 2,4-dihydroxybenzilic alcohol acid and 2,4-dihydroxybenzoic acid moieties linked by ether and ester bonds, derives from chrysophanol anthrone after the decarboxylation of a single octaketide chain. The subsequent oxidative cleavage of the B ring of chrysophanol anthrone is assumed to generate a benzophenone intermediate which is then oxidized to produce the tricyclic skeleton (spirobenzofuran-l,2′-cyclohexa-3′,5′-diene-2′,3-dione). Further structural modifications (i.e., prenylation, acetylation and methylation) of this latter compound generate purpactin B which is synthesized as an intermediate; then, the oxidation of its hydroxymethyl group leads to purpactin A [78,79].

The biosynthesis of penicillide can be deduced starting from the intermediate spirobenzofuran-l,2′-cyclohexa-3′,5′-diene-2′,3-dione (Figure 2). Slight modifications of the side chain or the addition of functional groups on the tricyclic skeleton (dibenzo[b,g][1,5]dioxocin-5(7H)-one) may generate the other known compounds of this class.

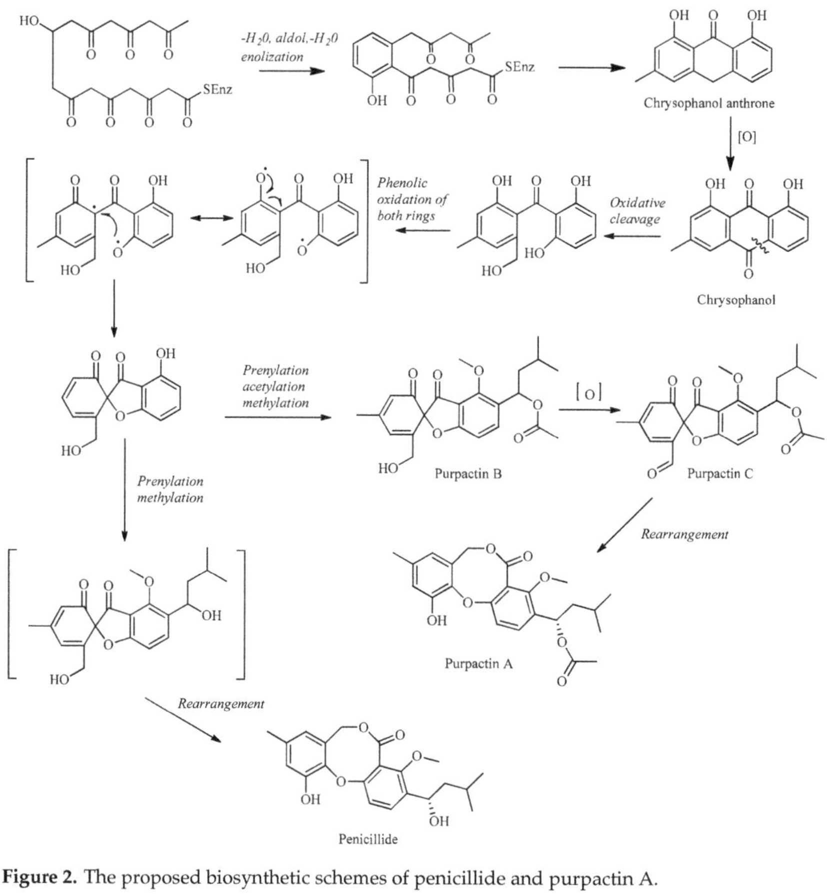

Figure 2. The proposed biosynthetic schemes of penicillide and purpactin A.

6. Related Products

Several products of *Penicillium* and *Talaromyces* are structurally related to penicillides, but they cannot be strictly considered members of this class of natural products because of the lack of the typical structural features (Figure 3). This is the case of purpactins B and C which, despite having been named as purpactin A homologues, are characterized by spirobenzofuran-1,2′-cyclohexa-3′,5′-diene-2′,3-dione instead of dibenzo[b,g][1,5]dioxocin-5(7H)-one. Considering the biosynthetic pathway in Figure 2, purpactin B is an intermediate in the biosynthesis of purpactin A, which is formed by the acetylation and rearrangement of the former compound. The common biosynthetic origin of purpactins could explain why these compounds were frequently reported from the same fungal sources [6,37,52].

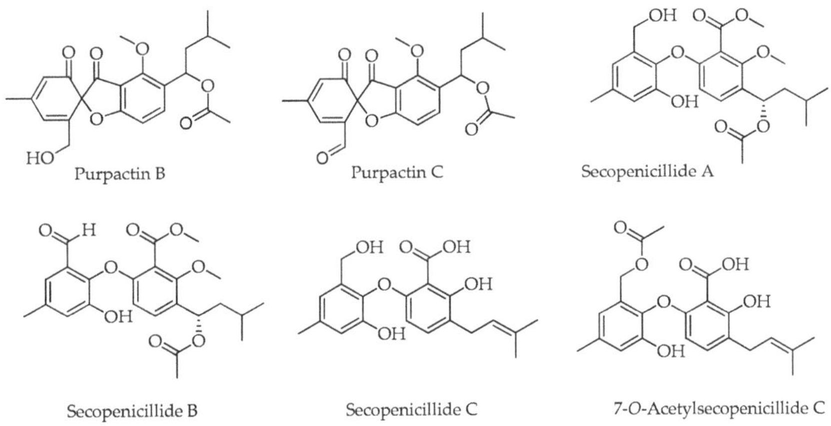

Figure 3. Compounds structurally and/or biosynthetically related to penicillides reported from *Penicillium* and *Talaromyces*.

It has been proposed that secopenicillides A, B and C could share the same biosynthetic route of penicillide and purpactin A [46]. However, these compounds, and the related 7-*O*-acetylsecopenicillide C [17], are characterized by the presence of a diphenylether moiety which does not conform to the genuine structural model of penicillides. In particular, Wu and coworkers proposed that secopenicillide A originates from purpactin C which is converted to purpactin A via reduction of the aldehydic group to form an alcohol intermediate, followed by esterification [46]. Secopenicillides A and B were isolated for the first time from *P. simplicissimum* [23], while secopenicillide C was detected as a new compound from *T. pinophilus,* and its production was found to be enhanced in co-culture with *Trichoderma harzianum* [45]. It must be noted that all these metabolites were detected along with other known penicillides.

7. Conclusions

As can be concluded from the available information on their occurrence and structures examined in this review, species of *Penicillium* and *Talaromyces* from various ecological contexts represent the main natural source of penicillides. However, the reported extraction of these depsidones from two unrelated plant species introduces the opportunity for further investigations in the aim to assess whether, or not, this biosynthetic aptitude results from any endophytic associates. In the latter alternative, further studies are also advisable to investigate if the inter-organism spread derives from the transfer of a pertinent gene cluster through HGT, which may eventually confer an ecological advantage depending on the bioactive properties of these compounds.

The broad-range laboratory investigations carried out so far have pointed out the multifaceted bioactivities of penicillides. Particularly, the best antibiotic effects have been documented against methicillin and oxacillin resistant *S. aureus* [27], deserving further assessments in the perspective to integrate the available panel of these highly demanded drugs. In this respect, a relevant contribution is expected by the pharmaceutical industry through the realization of more potent semi-synthetic derivatives, as a follow-up of the pioneering research activity carried out in some laboratories [80,81]. Considering the increasing impact of pulmonary diseases, a notable contribution in the achievement of new drugs is also expected from further investigations on the above-outlined effects on mucus hypersecretion [42]. Finally, following the preliminary evidence obtained for depsidones [82,83], more valuable opportunities could result from the exploration of the antiviral effects of these compounds.

Author Contributions: Conceptualization, R.N. and A.A.; resources, F.F.; writing—original draft preparation, M.M.S.; writing—review and editing, M.M.S., R.N. and A.A.; funding, A.A. and F.F. All authors have read and agreed to the published version of the manuscript.

Funding: This research was carried out within the program 'Finanziamento della Ricerca di Ateneo (FRA) 2022 dell'Università degli Studi di Napoli Federico II' and received funding from progetto PRIN: Progetti di Ricerca di Rilevante Interesse Nazionale: Bando 2022 PNRR Prot. P2022WXE4T, financed by the European Union Next-Generation EU (Piano Nazionale di Ripresa e Resilienza, PNRR).

Conflicts of Interest: The authors declare no conflicts of interest.

References

1. Gosio, B. Richerche batteriologiche e chemiche sulle alterazoni del mais. *Riv. Igiene Sanità Pub.* **1896**, *7*, 825–849.
2. Frisvad, J.C. Taxonomy, chemodiversity and chemoconsistency of *Aspergillus*, *Penicillium* and *Talaromyces* species. *Front. Microbiol.* **2015**, *5*, 773. [CrossRef]
3. Hawksworth, D.L.; Crous, P.W.; Redhead, S.A.; Reynolds, D.R.; Samson, R.A.; Seifert, K.A.; Taylor, J.W.; Wingfield, M.J.; Abaci, Ö.; Aime, C.; et al. The Amsterdam declaration on fungal nomenclature. *IMA Fungus* **2011**, *2*, 105–111. [CrossRef]
4. Samson, R.A.; Yilmaz, N.; Houbraken, J.; Spierenburg, H.; Seifert, K.A.; Peterson, S.W.; Varga, J.; Frisvad, J.C. Phylogeny and nomenclature of the genus *Talaromyces* and taxa accommodated in *Penicillium* subgenus *Biverticillium*. *Stud. Mycol.* **2011**, *70*, 159–183. [CrossRef]
5. Sassa, T. Structure of penicillide, a new metabolite produced by a *Penicillium* species. *Tetrahedron Lett.* **1974**, *15*, 3941–3942. [CrossRef]
6. Tomoda, H.; Nishida, H.; Masuma, R.; Cao, J.; Okuda, S.; Omura, S. Purpactins, new inhibitors of acyl-CoA: Cholesterol acyltransferase produced by *Penicillium purpurogenum* I. Production, isolation and physico-chemical and biological properties. *J. Antibiot.* **1991**, *44*, 136–143. [CrossRef]
7. Proksa, B.; Uhrínová, S.; Adamcová, J.; Fuska, J. Hydrogenation of vermistatin. *Monatshefte Chem. Chem. Mon.* **1992**, *123*, 251–256. [CrossRef]
8. Lan, D.; Wu, B. Chemistry and bioactivities of secondary metabolites from the genus *Talaromyces*. *Chem. Biodivers.* **2020**, *17*, e2000229. [CrossRef]
9. Lei, L.R.; Gong, L.Q.; Jin, M.Y.; Wang, R.; Liu, R.; Gao, J.; Liu, M.D.; Huang, L.; Wang, G.Z.; Wang, D.; et al. Research advances in the structures and biological activities of secondary metabolites from *Talaromyces*. *Front. Microbiol.* **2022**, *13*, 984801. [CrossRef]
10. Nicoletti, R.; Salvatore, M.M.; Andolfi, A. Secondary metabolites of mangrove-associated strains of *Talaromyces*. *Mar. Drugs* **2018**, *16*, 12. [CrossRef]
11. Nicoletti, R.; Andolfi, A.; Salvatore, M.M. Endophytic fungi of the genus *Talaromyces* and plant health. In *Microbial Endophytes and Plant Growth*; Accademic Press: London, UK, 2023; pp. 183–213.
12. Nicoletti, R.; Manzo, E.; Ciavatta, M.L. Occurence and bioactivities of funicone-related compounds. *Int. J. Mol. Sci.* **2009**, *10*, 1430–1444. [CrossRef]
13. Salvatore, M.M.; DellaGreca, M.; Andolfi, A. New insights into chemical and biological properties of funicone-like compounds. *Toxins* **2022**, *14*, 466. [CrossRef] [PubMed]
14. Kawamura, H.; Kaneko, T.; Koshino, H.; Esumi, Y.; Uzawa, J.; Sugawara, F. Penicillides from *Penicillium* sp. isolated from *Taxus cuspidata*. *Nat. Prod. Lett.* **2000**, *14*, 477–484. [CrossRef]
15. Chung, M.C.; Lee, H.J.; Chun, H.K.; Kho, Y.H. Penicillide, a nonpeptide calpain inhibitor, produced by *Penicillium* sp. F60760. *J. Microbiol. Biotechnol.* **1998**, *8*, 188–190.
16. Wang, C.; Gao, Y.K.; Lei, F.H.; Tan, X.C.; Shen, L.Q.; Gao, C.H.; Yi, X.X.; Li, X.Y. A new glycosyl ester isolated from marine-derived *Penicillium* sp. *Chin. Tradit. Herb. Drugs* **2019**, *50*, 2518–2523.
17. Zhang, Y.; Li, X.M.; Shang, Z.; Li, C.S.; Ji, N.Y.; Wang, B.G. Meroterpenoid and diphenyl ether derivatives from *Penicillium* sp. MA-37, a fungus isolated from marine mangrove rhizospheric soil. *J. Nat. Prod.* **2012**, *75*, 1888–1895. [CrossRef]

18. Jeon, H.; Shim, S.H. Chemical constituents of the endophyte *Penicillium* sp. isolated from *Artemisia princeps*. *Chem. Nat. Compd.* **2020**, *56*, 122–124. [CrossRef]
19. Kong, F.D. Secondary metabolites from marine fungus *Penicillium* sp. SCS-KFD16. *Chin. Tradit. Herb. Drugs* **2018**, *24*, 5029–5033.
20. Gao, H.; Zhou, L.; Li, D.; Gu, Q.; Zhu, T.J. New cytotoxic metabolites from the marine-derived fungus *Penicillium* sp. ZLN29. *Helv. Chim. Acta* **2013**, *96*, 514–519. [CrossRef]
21. Zhang, X.; Qi, B.W.; Yang, H.Y.; Jiang, F.F.; Ding, N.; Wu, Y.; Liu, X.; Tu, P.F.; Shi, S.P. Two new diphenyl ether derivatives from *Penicillium chrysogenum* MT-12, an endophytic fungus isolated from *Huperzia serrata*. *Chin. Tradit. Herb. Drugs* **2018**, *49*, 2496–2501.
22. Kuroda, K.; Morishita, Y.; Saito, Y.; Ikuina, Y.; Ando, K.; Kawamoto, I.; Matsuda, Y. As-186 Compounds, new inhibitors of Acyl-Coa: Cholesterol acyltransferase from *Penicillium asperosporum* KYI635. *J. Antibiot.* **1994**, *47*, 16–22. [CrossRef] [PubMed]
23. Komai, S.I.; Hosoe, T.; Itabashi, T.; Nozawa, K.; Yaguchi, T.; Fukushima, K.; Kawai, K.I. New penicillide derivatives isolated from *Penicillium simplicissimum*. *J. Nat. Med.* **2006**, *60*, 185–190. [CrossRef]
24. Hwang, H.; Kwon, C.; Kwon, J. Chemical constituents isolated from the Moss-derived fungus *Talaromyces* sp. *J. Korean Magn. Reson. Soc.* **2020**, *24*, 123–128.
25. Song, F.; Dong, Y.; Wei, S.; Zhang, X.; Zhang, K.; Xu, X. New antibacterial secondary metabolites from a marine-derived *Talaromyces* sp. strain BTBU20213036. *Antibiotics* **2022**, *11*, 222. [CrossRef]
26. Wu, Z.; Li, X.-M.; Yang, S.-Q.; Wang, B.-G.; Li, X. Antibacterial polyketides from the deep-sea cold-seep-derived fungus *Talaromyces* sp. CS-258. *Mar. Drugs* **2024**, *22*, 204. [CrossRef]
27. Hui, Z.; Baocong, H.; Yaoyao, Z.; Jinqiu, M.; Xiahao, Z.; Chengyun, W. Secondary metabolites and antibacterial activity of the marine-derived fungus, *Talaromyces* sp. HK1-18. *Chin. Mar. Med.* **2022**, *42*, 67–72.
28. Carrillo-Jaimes, K.; Fajardo-Hernández, C.A.; Hernández-Sedano, F.; Cano-Sánchez, P.; Morales-Jiménez, J.; Quiroz-García, B.; Rivera-Chávez, J. Antibacterial activity and AbFtsZ binding properties of fungal metabolites isolated from mexican mangroves. *Rev. Bras. Farmacogn.* **2024**, *34*, 564–576. [CrossRef]
29. Wu, B.; Ohlendorf, B.; Oesker, V.; Wiese, J.; Malien, S.; Schmaljohann, R.; Imhoff, J.F. Acetylcholinesterase inhibitors from a marine fungus *Talaromyces* sp. strain LF458. *Mar. Biotechnol.* **2015**, *17*, 110–119. [CrossRef] [PubMed]
30. Tang, L.; Xia, J.; Chen, Z.; Lin, F.; Shao, Z.; Wang, W.; Hong, X. Cytotoxic and antibacterial meroterpenoids isolated from the marine-derived fungus *Talaromyces* sp. M27416. *Mar. Drugs* **2024**, *22*, 186. [CrossRef]
31. Chen, M.; Han, L.; Shao, C.L.; She, Z.G.; Wang, C.Y. Bioactive diphenyl ether derivatives from a gorgonian-derived fungus *Talaromyces* sp. *Chem. Biodivers.* **2015**, *12*, 443–450. [CrossRef]
32. Lv, H.; Su, H.; Xue, Y.; Jia, J.; Bi, H.; Wang, S.; Zhang, J.; Zhu, M.; Emam, M.; Wang, H.; et al. Polyketides with potential bioactivities from the mangrove-derived fungus *Talaromyces* sp. WHUF0362. *Mar. Life Sci. Technol.* **2023**, *5*, 232–241. [CrossRef]
33. Yilmaz, N.; López-Quintero, C.A.; Vasco-Palacios, A.M.; Frisvad, J.C.; Theelen, B.; Boekhout, T.; Samson, R.A.; Houbraken, J. Four novel *Talaromyces* species isolated from leaf litter from Colombian Amazon rain forests. *Mycol. Progr.* **2016**, *15*, 1041–1056. [CrossRef]
34. Daengrot, C.; Rukachaisirikul, V.; Tadpetch, K.; Phongpaichit, S.; Bowornwiriyapan, K.; Sakayaroj, J.; Shen, X. Penicillanthone and penicillidic acids A-C from the soil-derived fungus *Penicillium aculeatum* PSU-RSPG105. *RSC Adv.* **2016**, *6*, 39700–39709. [CrossRef]
35. Frisvad, J.C.; Yilmaz, N.; Thrane, U.; Rasmussen, K.B.; Houbraken, J.; Samson, R.A. *Talaromyces atroroseus*, a new species efficiently producing industrially relevant red pigments. *PLoS ONE* **2013**, *8*, e84102. [CrossRef]
36. Yilmaz, N.; Houbraken, J.; Hoekstra, E.S.; Frisvad, J.C.; Visagie, C.M.; Samson, R.A. Delimitation and characterisation of *Talaromyces purpurogenus* and related species. *Persoonia* **2012**, *29*, 39–54. [CrossRef] [PubMed]
37. Salim, R.G.; Fadel, M.; Youssef, Y.A.; Taie, H.A.A.; Abosereh, N.A.; El-Sayed, G.M.; Marzouk, M. A local *Talaromyces atroroseus* TRP-NRC isolate: Isolation, genetic improvement, and biotechnological approach combined with LC/HRESI-MS characterization, skin safety, and wool fabric dyeing ability of the produced red pigment mixture. *J. Genet. Eng. Biotechnol.* **2022**, *20*, 62. [CrossRef]
38. Suzuki, K.; Nozawa, K.; Udagawa, S.; Nakajima, S.; Kawai, K. Penicillide and dehydroisopenicillide from *Talaromyces derxii*. *Phytochemistry* **1991**, *30*, 2096–2098. [CrossRef]
39. Proksa, B.; Uhrin, D.; Adamcova, J.; Fuska, J. Vermixocins A and B, two novel metabolites from *Penicillium vermiculatum*. *J. Antibiot.* **1992**, *45*, 1268–1272. [CrossRef]
40. Salituro, G.M.; Pettibone, D.J.; Clineschmidt, B.V.; Williamson, J.M.; Zink, D.L. Potent, non-peptidic oxytocin receptor antagonists from a natural source. *Bioorgan. Med. Chem. Lett.* **1993**, *3*, 337–340. [CrossRef]
41. Cai, J.; Wang, L.; Zhang, Z. Study on the secondary metabolites and bioactivities of a medicinal mangrove endophytic fungus *Talaromyces flavus* TGGP34. *Chin. J. Mar. Drugs* **2021**, *40*, 37–43.
42. Visagie, C.M.; Yilmaz, N.; Frisvad, J.C.; Houbraken, J.; Seifert, K.A.; Samson, R.A.; Jacobs, K. Five new *Talaromyces* species with ampulliform-like phialides and globose rough walled conidia resembling *T. verruculosus*. *Mycoscience* **2015**, *56*, 486–502. [CrossRef]
43. Nguyen, T.T.T.; Frisvad, J.C.; Kirk, P.M.; Lim, H.J.; Lee, H.B. Discovery and extrolite production of three new species of *Talaromyces* belonging to sections helici and purpurei from freshwater in Korea. *J. Fungi* **2021**, *7*, 722. [CrossRef]
44. Zhai, M.M.; Niu, H.T.; Li, J.; Xiao, H.; Shi, Y.P.; Di, D.L.; Crews, P.; Wu, Q.X. Talaromycolides A-C, novel phenyl-substituted phthalides isolated from the green chinese onion-derived fungus *Talaromyces pinophilus* AF-02. *J. Agric. Food Chem.* **2015**, *63*, 9558–9564. [CrossRef] [PubMed]

45. Nonaka, K.; Abe, T.; Iwatsuki, M.; Mori, M.; Yamamoto, T.; Shiomi, K.; Ômura, S.; Masuma, R. Enhancement of metabolites productivity of *Penicillium pinophilum* FKI-5653, by co-culture with *Trichoderma harzianum* FKI-5655. *J. Antibiot.* **2011**, *64*, 769–774. [CrossRef]
46. Wu, C.; Zhao, Y.; Chen, R.; Liu, D.; Liu, M.; Proksch, P.; Guo, P.; Lin, W. Phenolic metabolites from mangrove-associated *Penicillium pinophilum* fungus with lipid-lowering effects. *RSC Adv.* **2016**, *6*, 21969–21978. [CrossRef]
47. Salvatore, M.M.; DellaGreca, M.; Nicoletti, R.; Salvatore, F.; Vinale, F.; Naviglio, D.; Andolfi, A. Talarodiolide, a new 12-membered macrodiolide, and GC/MS investigation of culture filtrate and mycelial extracts of *Talaromyces pinophilus*. *Molecules* **2018**, *23*, 950. [CrossRef]
48. He, F.; Li, X.; Yu, J.H.; Zhang, X.; Nong, X.; Chen, G.; Zhu, K.; Wang, Y.Y.; Bao, J.; Zhang, H. Secondary metabolites from the mangrove sediment-derived fungus *Penicillium pinophilum* SCAU037. *Fitoterapia* **2019**, *136*, 104177. [CrossRef]
49. Zhao, D.L.; Shao, C.L.; Zhang, Q.; Wang, K.L.; Guan, F.F.; Shi, T.; Wang, C.Y. Azaphilone and diphenyl ether derivatives from a Gorgonian-derived strain of the fungus *Penicillium pinophilum*. *J. Nat. Prod.* **2015**, *78*, 2310–2314. [CrossRef]
50. Pandit, S.G.; Puttananjaiah, M.H.; Serva Peddha, M.; Dhale, M.A. Safety efficacy and chemical profiling of water-soluble *Talaromyces purpureogenus* CFRM02 pigment. *Food Chem.* **2020**, *310*, 125869. [CrossRef]
51. Shaaban, M.; El-Metwally, M.M.; Laatsch, H. New bioactive metabolites from *Penicillium purpurogenum* MM. *Zeit. Naturforsch.—Sect. B J. Chem. Sci.* **2016**, *71*, 287–295. [CrossRef]
52. Cai, R.; Chen, S.; Long, Y.; Li, C.; Huang, X.; She, Z. Depsidones from *Talaromyces stipitatus* SK-4, an endophytic fungus of the mangrove plant *Acanthus ilicifolius*. *Phytochem. Lett.* **2017**, *20*, 196–199. [CrossRef]
53. Ningsih, B.N.S.; Rukachaisirikul, V.; Phongpaichit, S.; Muanprasat, C.; Preedanon, S.; Sakayaroj, J.; Intayot, R.; Jungsuttiwong, S. Talarostatin, a vermistatin derivative from the soil-derived fungus *Talaromyces thailandensis* PSU-SPSF059. *Nat. Prod. Res.* **2024**, *38*, 2535–2542. [CrossRef]
54. Wang, B.; Chen, Q.H.; Jiang, T.; Cai, Y.W.; Huang, G.L.; Sun, X.P.; Zheng, C.J. Secondary metabolites from the mangrove-derived fungus *Penicillium verruculosum* and their bioactivities. *Chem. Nat. Compd.* **2021**, *57*, 588–591. [CrossRef]
55. Yu, M.; Chen, X.; Jiang, M.; Li, X. Two marine natural products, penicillide and verrucarin J, are identified from a chemical genetic screen for neutral lipid accumulation effectors in *Phaeodactylum tricornutum*. *Appl. Microbiol. Biotechnol.* **2020**, *104*, 2731–2743. [CrossRef] [PubMed]
56. Chaiyosang, B.; Kanokmedhakul, K.; Yodsing, N.; Boonlue, S.; Yang, J.X.; Wang, Y.A.; Andersen, R.J.; Yahuafai, J.; Kanokmedhakul, S. Three new indole diterpenoids from *Aspergillus aculeatus* KKU-CT2. *Nat. Prod. Res.* **2022**, *36*, 4973–4981. [CrossRef]
57. de Sá, J.D.; Pereira, J.A.; Dethoup, T.; Cidade, H.; Sousa, M.E.; Rodrigues, I.C.; Costa, P.M.; Mistry, S.; Silva, A.M.; Kijjoa, A. Anthraquinones, diphenyl ethers, and their derivatives from the culture of the marine sponge-associated fungus *Neosartorya spinosa* KUFA 1047. *Mar. Drugs* **2021**, *19*, 457. [CrossRef]
58. Song, F.; Lin, R.; Yang, N.; Jia, J.; Wei, S.; Han, J.; Li, J.; Bi, H.; Xu, X. Antibacterial secondary metabolites from marine-derived fungus *Aspergillus* sp. IMCASMF180035. *Antibiotics* **2021**, *10*, 377. [CrossRef]
59. Tapfuma, K.I.; Uche-Okereafor, N.; Sebola, T.E.; Hussan, R.; Mekuto, L.; Makatini, M.M.; Green, E.; Mavumengwana, V. Cytotoxic activity of crude extracts from *Datura stramonium*'s fungal endophytes against A549 lung carcinoma and UMG87 glioblastoma cell lines and LC-QTOF-MS/MS based metabolite profiling. *BMC Complement. Altern. Med.* **2019**, *19*, 330. [CrossRef] [PubMed]
60. Xia, X.K.; Liu, F.; She, Z.G.; Yang, L.G.; Li, M.F.; Vrijmoed, L.L.P.; Lin, Y.C. 1H and 13C NMR assignments for 6-demethylvermistatin and two penicillide derivatives from the mangrove fungus *Guignardia* sp. (No. 4382) from the South China Sea. *Magn. Reson. Chem.* **2008**, *46*, 693–696. [CrossRef]
61. Sy-Cordero, A.A.; Figueroa, M.; Raja, H.A.; Meza Aviña, M.E.; Croatt, M.P.; Adcock, A.F.; Kroll, D.J.; Wani, M.C.; Pearce, C.J.; Oberlies, N.H. Spiroscytalin, a new tetramic acid and other metabolites of mixed biogenesis from *Scytalidium cuboideum*. *Tetrahedron* **2015**, *71*, 8899–8904. [CrossRef]
62. Primahana, G.; Narmani, A.; Surup, F.; Teponno, R.B.; Arzanlou, M.; Stadler, M. Five tetramic acid derivatives isolated from the iranian fungus *Colpoma quercinum* CCTU A372. *Biomolecules* **2021**, *11*, 783. [CrossRef] [PubMed]
63. Arunpanichlert, J.; Rukachaisirikul, V.; Phongpaichit, S.; Supaphon, O.; Sakayaroj, J. Meroterpenoid, isocoumarin, and phenol derivatives from the seagrass-derived fungus *Pestalotiopsis* sp. PSU-ES194. *Tetrahedron* **2015**, *71*, 882–888. [CrossRef]
64. Xia, X.; Kim, S.; Liu, C.; Shim, S.H. Secondary metabolites produced by an endophytic fungus *Pestalotiopsis sydowiana* and their 20S proteasome inhibitory activities. *Molecules* **2016**, *21*, 944. [CrossRef] [PubMed]
65. Xiao, J.; Hu, J.Y.; Sun, H.D.; Zhao, X.; Zhong, W.T.; Duan, D.Z.; Wang, L.; Wang, X.L. Sinopestalotiollides A–D, cytotoxic diphenyl ether derivatives from plant endophytic fungus *Pestalotiopsis palmarum*. *Bioorgan. Med. Chem. Lett.* **2018**, *28*, 515–518. [CrossRef] [PubMed]
66. Zhang, Y.; Zhao, L.; Guo, X.; Li, C.; Li, H.; Lou, H.; Ren, D. Chemical constituents from *Phyllanthus emblica* and the cytoprotective effects on H_2O_2-induced PC12 cell injuries. *Arch. Pharm. Res.* **2016**, *39*, 1202–1211. [CrossRef] [PubMed]
67. Bhardwaj, N.; Sharma, P.; Singh, K.; Rana, D.; Kumar, V. *Phyllanthus emblica* seed extract as corrosion inhibitor for stainless steel used in petroleum industry (SS-410) in acidic medium. *Chem. Phys. Impact* **2021**, *3*, 100038. [CrossRef]
68. Singh, A.; Singh, D.K.; Kharwar, R.N.; White, J.F.; Gond, S.K. Fungal endophytes as efficient sources of plant-derived bioactive compounds and their prospective applications in natural product drug discovery: Insights, avenues, and challenges. *Microorganisms* **2021**, *9*, 197. [CrossRef] [PubMed]

69. Tiwari, P.; Bae, H. Horizontal gene transfer and endophytes: An implication for the acquisition of novel traits. *Plants* **2020**, *9*, 305. [CrossRef]
70. Bielecka, M.; Pencakowski, B.; Nicoletti, R. Using next-generation sequencing technology to explore genetic pathways in endophytic fungi in the syntheses of plant bioactive metabolites. *Agriculture* **2022**, *12*, 187. [CrossRef]
71. Caradus, J.R.; Johnson, L.J. *Epichloë* fungal endophytes—From a biological curiosity in wild grasses to an essential component of resilient high performing ryegrass and fescue pastures. *J. Fungi* **2020**, *6*, 322. [CrossRef]
72. Mamangkey, J.; Mendes, L.W.; Mustopa, A.Z.; Hartanto, A. Endophytic *Aspergillii* and *Penicillii* from medicinal plants: A focus on antimicrobial and multidrug resistant pathogens inhibitory activity. *Biotechnologia* **2024**, *105*, 83–95. [CrossRef] [PubMed]
73. Dutta, M.; Hazra, A.; Bhattacharya, E.; Bose, R.; Mandal Biswas, S. Characterization and metabolomic profiling of two pigment producing fungi from infected fruits of Indian gooseberry. *Arch. Microbiol.* **2023**, *205*, 141. [CrossRef] [PubMed]
74. Xu, F.; Yu, P.; Wu, H.; Liu, M.; Liu, H.; Zeng, Q.; Wu, D.; Wang, X. Aqueous extract of *Sargentodoxa cuneata* alleviates ulcerative colitis and its associated liver injuries in mice through the modulation of intestinal flora and related metabolites. *Front. Microbiol.* **2024**, *15*, 1295822. [CrossRef]
75. Nishida, H.; Tomoda, H.; Cao, J.; Araki, S.; Okuda, S.; Omura, S. Purpactins, new inhibitors of Acyl-Coa: Cholesterol acyltransferase produced by *Penicillium purpurogenum* III. Chemical modification of purpactin A. *J. Antibiot.* **1991**, *44*, 152–159. [CrossRef]
76. Sturdikova, M.; Proksa, B.; Fuska, J.; Stancikova, M. Vermilutin, an elastase inhibitor produced by *Penicillium vermiculatum*. *Biologia* **1995**, *50*, 233–236.
77. Yimnual, C.; Satitsri, S.; Ningsih, B.N.S.; Rukachaisirikul, V.; Muanprasat, C. A fungus-derived purpactin A as an inhibitor of TMEM16A chloride channels and mucin secretion in airway epithelial cells. *Biomed. Pharmacother.* **2021**, *139*, 111583. [CrossRef]
78. Nishida, H.; Tomoda, H.; Okuda, S.; Omura, S. Biosynthesis of purpactin A. *J. Org. Chem.* **1992**, *57*, 1271–1274. [CrossRef]
79. Dewick, P.M. *Medicinal Natural Products: A Biosyntheic Approach*, 3rd ed.; John Wiley & Sons: Hoboken, NJ, USA, 2016.
80. Zhang, Q.; Deng, C.; Fang, L.; Xu, W.; Zhao, Q.; Zhang, J.; Wang, Y.; Lei, X. Synthesis and evaluation of the analogues of penicillide against cholesterol ester transfer protein. *Chin. J. Chem.* **2013**, *31*, 355–370. [CrossRef]
81. Tomoda, H.; Omura, S. Potential therapeutics for obesity and atherosclerosis: Inhibitors of neutral lipid metabolism from microorganisms. *Pharmacol. Ther.* **2007**, *115*, 375–389. [CrossRef] [PubMed]
82. Zeukang, R.D.; Siwe-Noundou, X.; Fotsing, M.T.; Kuiate, T.T.; Mbafor, J.T.; Krause, R.W.M.; Choudhary, M.I.; de Théodore Atchadé, A. Cordidepsine is a potential new anti-HIV depsidone from *Cordia millenii*, Baker. *Molecules* **2019**, *24*, 3202. [CrossRef]
83. Khayat, M.T.; Ghazawi, K.F.; Samman, W.A.; Alhaddad, A.A.; Mohamed, G.A.; Ibrahim, S.R.M. Recent advances on natural depsidones: Sources, biosynthesis, structure-activity relationship, and bioactivities. *PeerJ* **2023**, *11*, e15394. [CrossRef] [PubMed]

molecules

MDPI

Article

Bioactive Secondary Metabolites from an Arctic Marine-Derived Strain, *Streptomyces* sp. MNP-1, Using the OSMAC Strategy

Mengna Wu [1], Zijun Liu [1], Jiahui Wang [1], Wentao Hu [2] and Huawei Zhang [1,*]

[1] School of Pharmaceutical Sciences, Zhejiang University of Technology, Hangzhou 310014, China; 211123070049@zjut.edu.cn (M.W.); 17857693513@163.com (Z.L.); 211123070027@zjut.edu.cn (J.W.)
[2] College of Pharmaceutical Science & Collaborative Innovation Center of Yangtze River Delta Region Green Pharmaceuticals, Zhejiang University of Technology, Hangzhou 310014, China; liq2363@gmail.com
* Correspondence: hwzhang@zjut.edu.cn; Fax: +86-571-88320913

Abstract: An Arctic marine-derived strain, MNP-1, was characterized by a combined methodological approach, incorporating a variety of analytical techniques including morphological features, biochemical characteristics, and 16S ribosomal RNA (rRNA) sequence analysis. The chemical investigation of *Streptomyces* sp. MNP-1 using the OSMAC (one strain many compounds) strategy yielded the isolation of twenty known compounds (**1–20**), which were unambiguously identified by various spectroscopic approaches including ^{1}H and ^{13}C NMR and ESI-MS (previously reported data). Bioassay results indicated that compounds **2**, **3**, **5**, **9**, **14**, **15**, and **20** had antimicrobial activity against human pathogenic strains including *Staphylococcus aureus*, *Escherichia coli*, and *Candida albicans* with MIC values ranging from 4 to 32 μg/mL, and compounds **3** and **14** exhibited moderate inhibitory activity on A549, MCF-7, and HepG2 tumor lines showing IC_{50} values within the range of 19.88 to 35.82 μM. These findings suggest that *Streptomyces* sp. MNP-1 is one of the prolific manufacturers of bioactive secondary metabolites with therapeutic potential.

Keywords: extremophile; *Streptomyces*; secondary metabolite; OSMAC strategy; antimicrobial activity; anticancer

1. Introduction

Extremophilic microbes in polar marine ecosystems have evolved over million years and developed various unique metabolisms to survive in inhospitable environments, such as oligotrophic conditions, low temperature, and high salinity and osmotic pressure, and have been discovered to be an unexplored treasure of natural products [1,2]. Actinomycetes, particularly *Streptomyces* sp., represent a significant source of antibiotics, with these microbes producing a diversity of β-lactam, tetracycline, macrolide, aminoglycoside, and glycopeptide antibiotics that have proven clinically effective [3]. *Streptomyces* from extreme environmental sources have the potential to produce compounds that overcome multi-drug resistance and new activities, and represent a crucial source of novel antibiotics [4]. Numerous experiments have indicated that most biosynthetic gene clusters (BGCs), which are directly involved in microbial secondary metabolite (SM) production activities, are silent under conventional circumstances [5–8]. It is fortunate that the "one strain many compounds" (OSMAC) strategy has emerged as the simplest and the most effective method to activate these cryptic BGCs [9,10]. For instance, one desert-derived strain, *Streptomyces* sp. C34, was found to make three novel compounds, chaxalactins A–C, which demonstrated high activity in fighting Gram-positive bacteria, exhibiting a minimal inhibitory concentration (MIC) of less than 1 μg/mL, by varying different culture

https://doi.org/10.3390/molecules30081657

media [11]. Three new cyclopentenone chemicals, aspergispones A–C, were identified from the sea-derived fungus *Aspergillus* sp. SCSIO 41501 by means of diversity medium cultures [12].

Herein, the chemical analysis of the Arctic marine-derived isolate MNP-1 using the OSMAC strategy resulted in the separation of twenty known chemicals (**1–20**), of which compound **11** was firstly characterized from microbes and **3** and **10** were produced by the *Streptomyces* strain for the first time. In vitro bioactivity tests suggested that compounds **3** and **10** have predominant activity on the A549, MCF-7, and HepG2 tumor cell lines (see Figure 1 below).

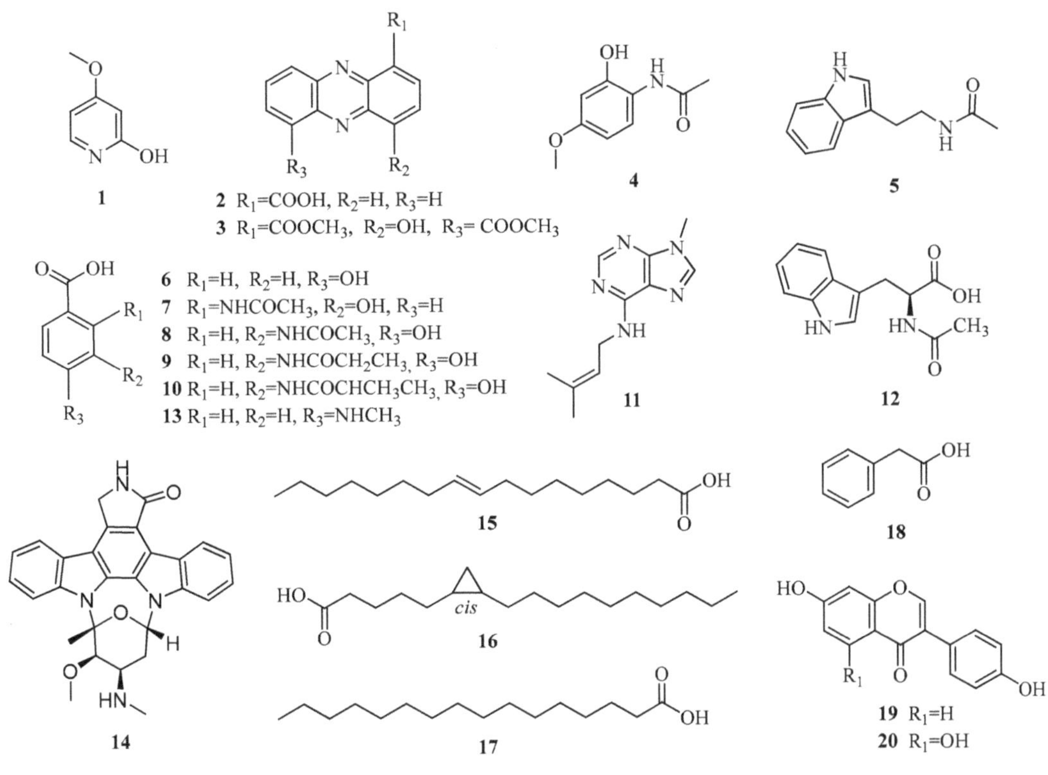

Figure 1. Chemical structures of compounds **1–20** from strain MNP-1.

2. Results

2.1. Strain Classification

The Arctic ore-derived isolate MNP-1 was identified by a combination of various approaches including an analysis of morphological features, biochemical characteristics, and 16S rRNA sequence analysis. The findings indicated that its single colony was round and grayish-white, and Gram's staining test was positive (see Figure 2c below). The 16S rRNA sequence-based phylogenetic analysis (see Figure 3, Table S1) indicated that strain MNP-1 was the closest interspecies relative to *Streptomyces acrimycini* and *S. pratensis*, suggesting it belonged to the genus *Streptomyces*.

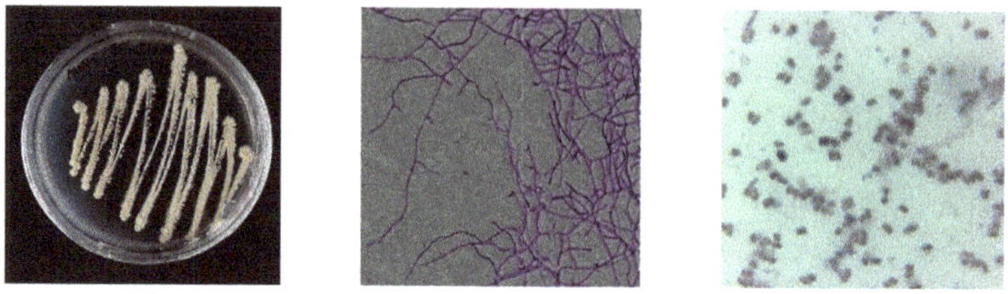

(**a**) (**b**) (**c**)

Figure 2. (**a**) Morphology of strain MNP-1. (**b**) Microscopic image of the mycelium and spore of the strain MNP-1 (25 × 100). (**c**) Gram staining results of the strain MNP-1 (25 × 100).

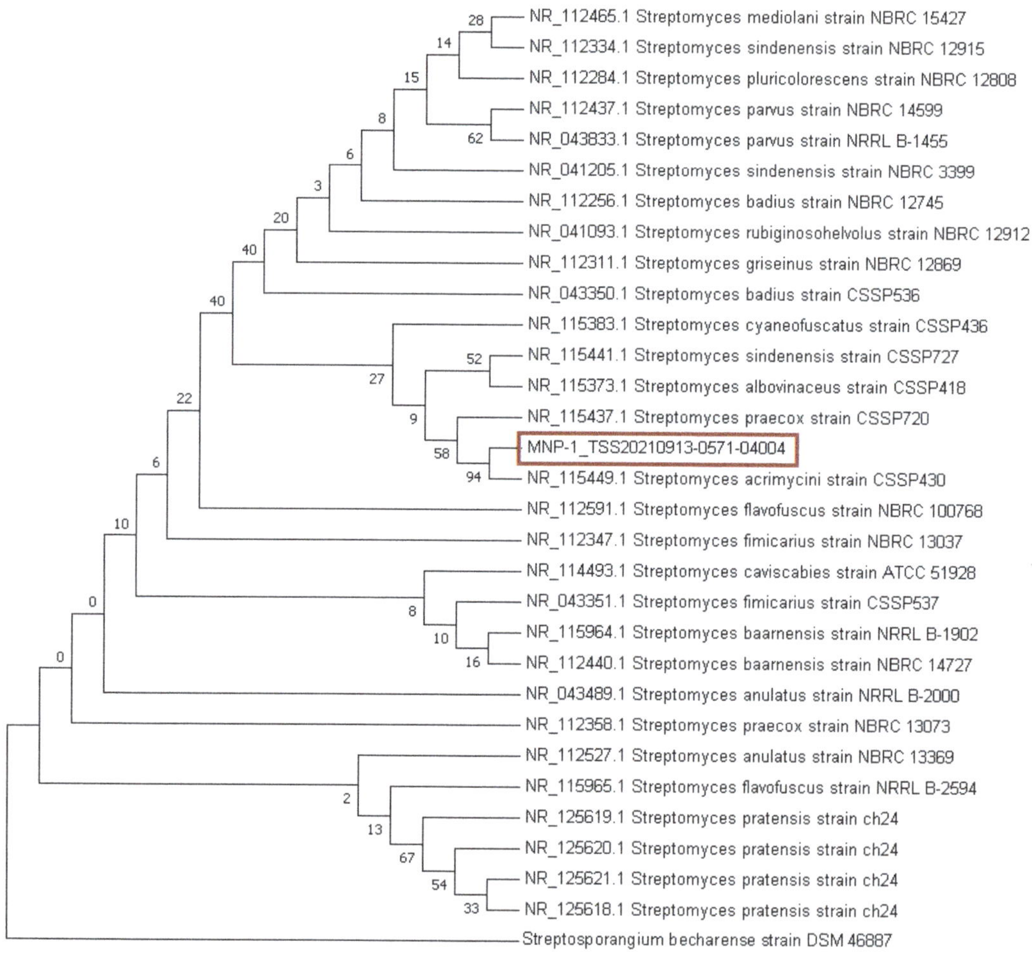

Figure 3. Phylogenetic tree of the strain MNP-1 (the sequence of MNP-1 is marked in red).

2.2. Medium Evaluation

The culture conditions of *Streptomyces* sp. MNP-1 were optimized by comparing the differences of metabolites under different culture conditions using the integrated OSMAC strategy with the triple orientation of fermentation crude extract yield, compound structure prediction based on hydrogen spectroscopy characterization and LC-MS/MS metabolomics, and metabolite activity evaluation (see Tables S2–S9, Figures S1–S10). Based on the results of the comprehensive evaluation of 18 solid/liquid media (see Table S2), three optimal medium compositions were screened to obtain a rice solid medium supplemented with 20 mg/L $CuSO_4 \cdot 5H_2O$ (#2), Gauze's Synthetic Medium No. 1 with supplemented with 50 μM 5-Aza-C (#7), and MC1 liquid medium (#16).

2.3. Structure Elucidation

Twenty chemicals were identified in the crude extraction of *Streptomyces* sp. MNP-1 using a combination of multiple analytical chromatographic methods (see Figure 1). As demonstrated by ^{1}H-NMR spectra, ^{13}C-NMR spectra, and MS analyses, in a comparative analysis with previous reports, these chemicals were definitively identified as 4-methoxypyridin-2-ol (**1**) [13], 1-phenazinecarboxylic acid (**2**) [14], 4-hydroxyphenazine-1, 6-dicarboxylic acid dimethyl ester (**3**) [15], *N*-(2-hydroxy-4-methoxyphenyl)-acetamide (**4**) [16], *N*-(2-(1H-indol-3-yl)ethyl) acetamide (**5**) [17], 4-hydroxybenzoic acid (**6**) [18], 2-(acetylamino)-3-hydroxy-benzoic acid (**7**) [19], 3-(acetylamino)-4-hydroxy-benzoic acid (**8**) [20], 4-hydroxy-3-propanamidobenzoic acid (**9**) [21], 4-hydroxy-3-[(2-methyl-1-oxopropyl) amino]-benzoic acid (**10**) [22], 9-methyl-*N*-(3-methyl-2-buten-1-yl)-9H-purin-6-amine (**11**) [23], *N*-acetyl-*L*-tryptophan (**12**) [24], 4-(methylamino)benzoic acid (**13**) [25], staurosporine (**14**) [26], (9*Z*)-9-heptadecenoic acid (**15**) [27], (6*S*,7*R*)-6,7-methyleneheptadecanoic acid (**16**) [28], palmitic acid (**17**) [29], phenyl acetic acid (**18**) [30], daidzein (**19**) [31], and genistein (**20**) [32].

(**1**): White powder; ^{1}H-NMR (CD_3OD) δ_H: 6.83 (1H, d, *J* = 7.8, H-3), 7.55 (1H, dd, *J* = 6.6, 1.8, H-5), 7.56 (1H, d, *J* = 6.6, H-6). ^{13}C-NMR (CD_3OD) δ_C: 148.6 (C-2), 113.9 (C-3), 125.1 (C-4), 115.7 (C-5), 152.2 (C-6), 56.4 (4-OCH_3). ESI-MS (*m/z*): 249.2 $[2M - H]^-$.

(**2**): Yellow-green crystals; ^{1}H-NMR ($CDCl_3$) δ_H: 8.98 (1H, dd, *J* = 7.1, 1.4), 8.01 (1H, m), 8.53 (1H, dd, *J* = 8.7, 1.4), 8.28 (1H, dd, *J* = 8.5, 1.5), 8.01 (1H, m), 8.01 (1H, m), 8.34 (1H, dd, *J* = 8.5, 1.5), 15.53 (1H, s). ^{13}C-NMR($CDCl_3$) δ_C: 125.1 (C-1), 137.6 (C-2), 130.4 (C-3), 135.2 (C-4), 143.6 (C-4a), 144.3 (C-5a), 128.1 (C-6), 133.3 (C-7), 131.9 (C-8), 130.4 (C-9), 140.0 (C-9a), 140.2 (C-10a), 166.0 (1-COOH). ESI-MS (*m/z*): 247.0 $[M + Na]^+$.

(**3**): Purple powder; ^{1}H-NMR ($CDCl_3$) δ_H: 7.29 (1H, d, *J* = 8.3, H-3), 7.94 (1H, dd, *J* = 7.9, H-8), 8.41 (1H, d, *J* = 7.1, H-7), 8.54 (1H, d, *J* = 6.9, H-2), 8.57 (1H, d, *J* = 8.8, H-9), 4.07 (3H, s, H-2′), 4.10 (3H, s, H-2″). HRESI-MS (*m/z*): 335.07 $[M + Na]^+$ (calcd. for $C_{16}H_{12}N_2O_5$).

(**4**): Brown powder; ^{1}H-NMR (CD_3OD) δ_H: 6.46 (1H, d, *J* = 2.5, H-3), 6.33 (1H, dd, *J* = 8.6, 2.5, H-5), 7.48 (1H, d, *J* = 8.6, H-6), 2.11 (3H, s, H-3′), 3.81 (3H,s). ^{13}C-NMR (CD_3OD) δ_C: 119.5 (C-1), 153.9 (C-2), 100.2 (C-3), 157.1 (C-4), 107.4 (C-5), 126.0 (C-6), 172.9 (C-2′), 23.3 (C-3′), 56.1 (4-OCH_3). HRESI-MS (*m/z*): 204.06 $[M + Na]^+$ (calculated for $C_9H_{11}NO_3$).

(**5**): Brown powder; ^{1}H-NMR (CD_3OD, 600 MHz) δ_H: 7.00 (1H, m, H-6), 7.07 (1H, s, H-2), 7.09 (1H, m, H-7), 7.33 (1H, d, *J* = 8.2, H-8), 7.55 (1H, d, *J* = 7.9, H-5), 2.94 (2H, t, *J* = 7.3, H-2′), 3.47 (2H, t, *J* = 7.4, H-1′), 1.92 (3H, s, H-5′). HRESI-MS (*m/z*): 225.10 $[M + Na]^+$ (calculated for $C_{12}H_{14}N_2O$).

(**6**): White powder; ^{1}H-NMR (CD_3OD) δ_H: 6.81 (2H, d, *J* = 8.7, H-2, 6), 7.87 (2H, d, *J* = 8.7, H-3, 5). ^{13}C-NMR (CD_3OD) δ_C: 163.0 (C-4), 123.8 (C-1), 115.9 (C-2, 6), 132.9 (C-3, 5), 163.0 (C-1). HRESI-MS (*m/z*): 139.04 $[M + H]^+$ (calculated for $C_7H_6O_3$).

(**7**): Brown powder; ^{1}H-NMR (CD_3OD) δ_H: 7.04 (1H, d, *J* = 7.9, H-4), 6.94 (1H, dd, *J* = 8.6, 1.5, H-5), 7.54 (1H, d, *J* = 7.7, H-6), 2.25 (3H, s, H-3′). HRESI-MS (*m/z*): 218.04 $[M + Na]^+$ (calculated for $C_9H_9NO_4$).

(**8**): White powder; ^{1}H-NMR (CD_3OD) δ_H: 6.91 (1H, d, *J* = 8.4, H-5), 7.71 (1H, dd, *J* = 8.4, 2.0, H-6), 8.40 (1H, d, *J* = 2.0), 2.20 (3H, s, H-3′). ^{13}C-NMR (CD_3OD) δ_C: 122.9 (C-1), 125.5 (C-2), 126.5 (C-3), 153.7 (C-4), 116.0 (C-5), 128.4 (C-6), 171.9 (C-2′), 23.1 (C-3′), 169.5 (1-COOH). HRESI-MS (*m/z*): 196.06 $[M + H]^+$ (calculated for $C_9H_9NO_4$).

(**9**): Yellow powder; ^{1}H-NMR (CD_3OD) δ_H: 8.41 (1H, d, *J* = 1.8), 6.93 (1H, d, *J* = 8.5), 7.72 (1H, dd, *J* = 8.3, 2.0), 2.50 (2H, q, *J* = 8.2, 7.7), 1.25 (3H, t, *J* =7.4). ^{13}C-NMR (CD_3OD) δ_C: 123.9 (C-1), 125.8 (C-2), 127.0 (C-3), 153.9 (C-4), 116.5 (C-5), 128.7 (C-6), 176.0 (C-2′), 30.7 (C-3′), 10.2 (C-4′), 170.2 (1-COOH). HRESI-MS (*m*/*z*): 232.06 [M + Na]$^+$ (calculated for $C_{10}H_{11}NO_4$).

(**10**): Brown powder; ^{1}H-NMR (CD_3OD) δ_H: 6.92 (1H, d, *J* = 8.5, H-5), 7.73 (1H, d, *J* = 8.4, H-6), 8.37 (1H, s, H-2), 2.79 (H, m, H-3′), 1.26 (6H, d, *J* = 7.1). ^{13}C-NMR(CD_3OD) δ_C: 125.8 (C-1), 127.3 (C-2), 129.3 (C-3), 152.4 (C-4), 116.7 (C-5), 130.6 (C-6), 36.8 (C-3′), 20.0 (C-4′,5′). HRESI-MS (*m*/*z*): 246.07 [M + Na]$^+$ (calculated for $C_{11}H_{13}NO_4$).

(**11**): Amorphous yellow solid; ^{1}H-NMR (CD_3OD) δ_H: 8.26 (1H, s, H-4), 8.01 (1H, s, H-8), 4.18 (2H, d, *J* = 7.5, H-2′), 5.39 (H, t, *J* = 6.9, H-3′), 3.82 (1H, s, 9-CH_3). HRESI-MS (*m*/*z*): 218.14 [M + H]$^+$ (calculated for $C_{11}H_{15}N_5$).

(**12**): White powder; ^{1}H-NMR (CD_3OD) δ_H: 7.01 (1H, ddd, *J* = 8.0, 7.0, 1.0, H-6), 7.08 (1H, s, H-2), 7.08 (1H, ddd, *J* = 8.1, 7.0, 1.2, H-7), 7.32 (1H, d, *J* = 8.2, H-8), 7.56 (1H, d, *J* = 7.9, H-5), 3.16 (1H, dd, *J* = 15.1, 8.4, H-1′), 3.34 (1H, dd, *J* = 15.2, 4.7, H-1′), 4.72 (1H, dd, *J* = 8.0, 5.2, H-2′), 1.90 (1H, s, H-5′). ^{13}C-NMR (CD_3OD) δ_C: 124.3(C-2), 111.1(C-3), 128.9(C-4), 119.2(C-5), 119.8(C-6), 122.4(C-7), 112.2(C-8), 138.1(C-9), 175.2(2′-COOH), 28.5 (C-1′), 54.8 (C-2′), 173.2(C-4′), 22.4(C-5′). ESI-MS (*m*/*z*): 247.4 [M + H]$^+$.

(**13**): Amorphous yellow powder; ^{1}H-NMR (CD_3OD) δ_H: 7.79 (2H, m, H-2, 6), 6.56 (2H, m, H-3, 5), 2.81 (3H, s, H-2′). ^{13}C-NMR (CD_3OD) δ_C: 118.8 (C-4), 132.7 (C-2,6), 111.8 (C-3,5), 155.3 (C-4), 29.9 (C-2′), 171.3 (C-2′, 1-COOH). HRESI-MS (*m*/*z*): 152.07 [M + H]$^+$ (calculated for $C_8H_9NO_2$).

(**14**): Yellow-green powder; ^{1}H-NMR (DMSO-d6) δ_H: 7.58 (1H, d, *J* = 8.2,H-1), 7.45 (1H, t, *J* = 7.6,H-2), 7.29 (1H, m,H-3), 9.28 (1H, d, *J* = 7.9, H-4), 8.49 (1H, s, H-6), 4.94 (2H, s, H-7), 7.96 (1H, d, *J* = 7.8, H-8), 7.29 (1H, m,H-9), 7.41 (1H, t, *J* = 7.8, H-10), 7.96 (1H, d, *J* = 8.5,H-11), 4.06 (1H, d, *J* = 3.5, H-3′), 3.26 (1H, m, H-4′), 6.70 (1H, d, *J* = 3.9, H-6′), 3.32 (3H, s,), 2.30 (3H, s,), 1.46 (3H, s). ^{13}C-NMR (DMSO-d6) δ_C: 172.2 (C-5), 45.4 (C-7), 29.4 (C-5′), 82.8 (C-3′), 79.9 (C-6′), 91.1 (C-2′), 50.1 (C-4′), 57.3 (3′-OCH_3), 29.7 (2′-CH_3), 33.3 (C-2″). ESI-MS (*m*/*z*): 467.2 [M + H]$^+$.

(**15**): Yellowish oil; ^{1}H-NMR ($CDCl_3$) δ_H: 5.34 (2H, m, H-9, 10), 1.31 (18H, m), 1.63 (2H, m), 2.01 (4H, m), 2.34 (2H, t, *J* = 7.5), 0.88 (3H, t, *J* = 6.9, H-17). ^{13}C-NMR ($CDCl_3$) δ_C: 130.2(C-9), 129.9(C-10), 180.2(C-1), 14.2 (C-17). HRESI-MS (*m*/*z*): 267.20 [M − H]$^-$ (calculated for $C_{17}H_{32}O_2$).

(**16**): Colorless oil; ^{1}H-NMR ($CDCl_3$) δ_H: -0.33 (1H, ddd, *J* = 5.1, 5.1, 4.6, H-18), 0.56 (1H, m, H-18), 0.65 (2H, m, H-6, 7), 1.13 (2H, m), 1.32 (20H, m), 1.63 (2H, m), 2.34 (2H, t, *J* = 7.5), 0.89 (3H, t, *J* = 7.0), 1.32 (1H, m, 1-OH). ^{13}C-NMR ($CDCl_3$) δ_C: 11.1 (C-18), 15.9 (C-6, 7), 180.1 (C-1), 14.2 (C-17). HRESI-MS (*m*/*z*): 283.26 [M + H]$^+$ (calculated for $C_{18}H_{34}O_2$).

(**17**): White flaky solid; ^{1}H-NMR ($CDCl_3$) δ_H: 1.26 (24H, m), 1.63 (2H, m, H-3), 2.35 (2H, t, *J* = 7.5, H-2), 0.89 (3H, t, *J* = 7.0, H-16), 1.26 (1H, m,1-OH). ^{13}C-NMR ($CDCl_3$) δ_C: 180.1 (C-1), 34.2 (C-2), 24.9 (C-3), 29.2 (C-4), 29.5 (C-5), 29.6 (C-6), 29.7 (C-7), 29.8 (C8-C12), 29.3 (C-13), 32.1 (C-14), 22.8 (C-15), 14.3 (C-16). HRESI-MS (*m*/*z*): 255.23 [M − H]$^-$ (calculated for $C_{16}H_{32}O_2$).

(**18**): White powder; ^{1}H-NMR (CD_3OD, 600 MHz) δ_H: 7.27 (5H, m), 3.59 (2H, s, H-1′). ^{13}C-NMR (CD_3OD, 150 MHz) δ_C: 41.9 (C-1′), 175.6 (1′-COOH), 136.0 (C-1), 130.3 (C-2,6), 129.4 (C-3,5), 127.9 (C-4). ESI-MS (*m*/*z*): 134.9 [M − 2H]$^-$.

(**19**): Yellow solid; ^{1}H-NMR (CD_3OD) δ_H: 8.06 (1H, d, *J* = 8.0, H-5), 6.94 (1H, m, H-6), 6.85 (1H, s, H-8), 7.37 (2H, d, *J* = 6.8, H-2′, 6′), 6.84 (2H, d, *J* = 7.0, H-3′, 5′), 8.13 (1H, s, H-2). ^{13}C-NMR (CD_3OD) δ_C: 126.0 (C-3), 128.5 (C-5), 116.4 (C-6), 103.2 (C-8), 118.2 (C-10), 124.3 (C-1′), 131.4 (C-2′, 6′), 116.2 (C-3′, 5′), 154.7 (C-2), 164.6 (C-7), 159.8 (C-9), 158.7 (C-4′), 178.2 (C-4). ESI-MS (*m*/*z*): 277.0 [M + Na]$^+$.

(**20**): Yellow solid; ^{1}H-NMR (CD_3OD) δ_H: 8.04 (1H, s), 6.33 (1H, s), 6.22 (1H, s), 7.36 (2H, d, J = 9.0), 6.84 (2H, d, J = 7.0). ^{13}C-NMR (CD_3OD) δ_C:154.8 (C-2), 123.3 (C-3), 182.3 (C-4), 163.9 (C-5), 100.1 (C-6), 165.9 (C-7), 94.8 (C-8), 158.8 (C-9), 106.3 (C-10), 124.7 (C-1′), 131.4 (2′, 6′), 116.3 (3′, 5′), 159.7 (C-4′). ESI-MS (m/z): 269.9 $[M - H]^-$.

2.4. Bioactivity

Compounds **3** and **14** demonstrated significant inhibition effects with MICs of 4 μg/mL and 8 μg/mL on *Candida albicans* ATCC 1023, respectively. Compounds **2**, **15**, and **20** displayed inhibitory activity against *Escherichia coli* ATCC 25922 with MICs of 8–32 μg/mL. Compounds **2**, **3**, **9**, and **15** also demonstrated antimicrobial activity against *Staphylococcus aureus* ATCC 25923, with MIC values of 16–32 μg/mL. In addition, compounds **3** and **14** had moderate antiproliferative effects on MCF-7, HepG2, and A549 cell lines, with IC_{50} values in the range of 19.88 ± 1.65 μM to 35.82 ± 2.70 μM (see Table 1 below).

Table 1. In vitro antimicrobial and antitumor effects of compounds **1–20**.

Compound	MIC Value (μg/mL)			IC_{50} Value (μM)		
	S. aureus ATCC 25923	*E. coli* ATCC 25922	*C. albicans* ATCC 10231	A549	MCF-7	HepG2
1	>64	>64	>64	>100	>100	>100
2	16	8	>64	>100	>100	>100
3	16	>64	4	21.52 ± 4.36	19.88 ± 1.65	35.82 ± 2.70
4	>64	>64	>64	>100	>100	>100
5	>64	>64	32	>100	>100	>100
6	>64	>64	>64	>100	>100	>100
7	>64	>64	>64	>100	>100	>100
8	>64	>64	>64	>100	>100	>100
9	32	>64	>64	>100	>100	>100
10	>64	>64	>64	>100	>100	>100
11	>64	>64	>64	>100	>100	>100
12	>64	>64	>64	>100	>100	>100
13	>64	>64	>64	>100	>100	>100
14	>64	>64	8	27.79 ± 6.70	35.57 ± 2.84	23.71 ± 2.89
15	32	32	>64	>100	>100	>100
16	>64	>64	>64	>100	>100	>100
17	>64	>64	>64	>100	>100	>100
18	>64	>64	>64	>100	>100	>100
19	>64	>64	>64	>100	>100	>100
20	>64	16	16	90.37 ± 2.46	>100	>100
Positive Control	0.25	1	0.25	14.86 ± 0.00	12.34 ± 0.01	15.30 ± 0.01
Negative Control	-	-	-	-	-	-

3. Discussion

In the OSMAC strategy, the best three media were selected, of which medium #2, with the addition of Cu^{2+}, produced salt–ion stress effects, affecting the regulation of metabolic pathways, balancing the water and ionic environments, and inducing antioxidant mechanisms [33,34]. The addition of 5-azacytidine to #7 medium may activate the expression of metabolite-related genes by inhibiting DNA methyltransferases and demethylating DNA, leading to changes in secondary metabolite production through an increase or decrease in the corresponding amino acids. Similarly, the addition of enriched nutrients promotes metabolic pathways, regulates metabolic homeostasis, and affects the production of secondary metabolites by *Streptomyces* [35,36]. To ensure the validity of the results, it is imperative to assess the efficacy of these mechanisms through gene expression analysis and metabolic analysis under varied metabolic conditions.

Compound **2**, 1-phenazinecarboxylic acid (PCA), induces the generation of reactive oxygen species (ROS) and regulates the apoptotic protein pathway, which drives bacterial cell lysis and death [37–39]. Compound **14,** staurosporine, demonstrates significant antitumor effects by means of multi-targeting and multi-pathway processes. It has been shown to significantly inhibit a variety of protein kinases. In addition, it was found to upregulate pro-apoptotic proteins (Bax) and downregulate anti-apoptotic proteins (Bcl-2). The resulting effect of these actions is a delay in the apoptosis of tumor cells, which in turn prevents tumor cell proliferation and metastasis [40–43].

4. Materials and Methods

4.1. General Experimental Procedures

The measurement of nuclear magnetic resonance (NMR) spectra was determined by a Bruker Avance III-600 MHz NMR instrument (Bruker, Fällande, Switzerland). ^{1}H NMR spectra were collected at 600 MHz; ^{13}C NMR spectra were obtained at 150 MHz. Electrospray ionization mass spectrometry (ESI-MS) analysis was performed on a SCIEX X500 B QTOF mass spectrometer (Framingham, MA, USA). Column chromatography (CC) was conducted on ODS reverse phase silica gel (YMC Co., Ltd., Kyoto, Japan), silica gel (Qingdao Marine Chemical Inc., Qingdao, China), and Sephadex LH-20 (GE Healthcare, Danderyd, Sweden). High-performance liquid chromatography (HPLC) was carried out on an Essentia LC-20AT apparatus (Shimadzu Co., Ltd., Shanghai, China) equipped with analytical columns (Phenomenex Synergi Hydro-RP, Torrance, CA, USA, 250 × 4.6 mm, 4 µm, and Phenomenex Luna C18, 250 × 4.6 mm, 5 µm). All solvents were of analytical grade except for the chromatographic grade used for HPLC.

4.2. Biological Materials

An off-white bacterial strain MNP-1 was isolated from an Arctic ore sample (No. BT08-1, Figure S67), contributed by Mr. Yanhui Dong, Second Institute of Oceanography, Ministry of Natural Resources of China. The classification of the strain as *Streptomyces* is substantiated by both morphological characteristics and the analysis of an internal transcribed spacer (ITS) sequence-based phylogeny tree. The strains of *S. aureus* ATCC 25923, *E. coli* ATCC 25922, and *C. albicans* ATCC 10231 were obtained from Nanjing Medical University, Nanjing, China; lung (A549), breast (MCF-7), and liver (HepG2) cancer cell lines were obtained from the American Type Culture Collection, Manassas, VA, USA.

4.3. Fermentation and Extraction

The strain MNP-1 was activated by using the PDA medium plate, and after being incubated at a constant temperature at 30 °C for 3 days, the appropriate amount of colonies was picked and inoculated into PDB medium and incubated at 28 °C and 180 rpm in shaking flasks for 3 days to make the MNP-1 strain fermentation seed cultures. Each flask (2 L) contained 1 L of culture medium, was autoclaved for 20 min at 121 °C, and inoculated with 5% seed cultures. Each flask (2 L) contained 1 L of medium, was sterilized by autoclaving at 121 °C for 20 min, and seeded with 5% inoculum.

Two fermentation modes of solid fermentation and liquid fermentation are adopted, selecting three types of culture media, the salt-stressed rice solid medium, Gauze's Synthetic Medium No. 1 with two epigenetic modifiers, and the nutrient-enriched liquid medium [44–47]. The fermentation solutions was separately diluted with equivalent volumes of EtOAC, facilitated by an ultrasonic device lasting 20 min. Subsequently, the combined organic phases were centrifuged (5000 rpm, 10 min) and the supernatants were concentrated to obtain the fermentation crude extracts I (25.5 g), II (4.3 g), and III (34.5 g).

4.4. Isolation and Purification

Crude extract I (25.5 g) obtained under medium #2 was subjected to fractionation and separated into 7 fractions (Fr.1–Fr.7, Figure S1) using a 200–300 mesh silica gel CC (CH_2Cl_2-CH_3OH, 100:0-0:100, *v*/*v*) in gradient elution. The Fr.1–Fr.7 segments were subjected to HPLC analysis, UPLC-MS/MS molecular network prediction, and an evaluation of inhibitory activity to further identify the fractions for further isolation and purification. Among them, Fr.2 (0.2546 g) was separated by ODS reverse-phase CC (CH_3OH-H_2O, 50:0–100:0, *v*/*v*) and HPLC to obtain compounds **1** (4.2 mg, t_R = 11.5 min) and **2** (4.6 mg, t_R = 10.1 min). Fr.3 (0.2840 g) was separated into 6 fractions (Fr.3.1–Fr.3.6), of which Fr.3.2 was further purified by Sephadex LH-20 and HPLC and afforded compounds **3** (1.3 mg, t_R = 18.8 min), **4** (1.6 mg, t_R = 5.0 min), and **5** (1.6 mg, t_R = 14.6 min). Fr.3.4 was isolated to obtain compound **6** (1.8 mg, tR = 9.0 min) by HPLC (Phenomenex Synergi, 4 μm, 250 × 4.6 mm; CH_3CN-0.1% formic acid, 11:89–15:85, *v*/*v*). Fr.4 was further isolated by ODS reverse-phase CC (CH_3OH-H_2O, 20:80–100:0, *v*/*v*) and HPLC to obtain compounds **7** (0.9 mg, t_R = 18.0 min), **8** (63.4 mg, t_R = 5.6 min), **9** (1.8 mg, t_R = 8.8 min), **10** (1.2 mg, t_R = 15.2 min), and **11** (1.2 mg, t_R = 4.7 min). Fr.5.7 was subjected to HPLC (Phenomenex Luna, 5 μm, 250 × 4.6 mm; 1.0 mL/min; CH_3CN-H_2O, 29:71, *v*/*v*) to provide compound **12** (2.0 mg, t_R = 12.7 min).

Crude extract II (4.3 g), separated under medium #7, was initially fragmented into 9 fractions (Fr.1–Fr.9, Figure S1) by MCI resin CC (CH_3OH-H_2O, 20:80–100:0, *v*/*v*). Based on the results of HPLC analysis, UPLC-MS/MS molecular network prediction and the evaluation of the inhibitory activity of Fr.1–Fr.9 segments, Fr.4 was further purified and separated using analytical CC (Agilent ZORBAX NH_2, Santa Clara, CA, USA, 5 μm, 250 × 4.6 mm; 1.0 mL/min; 218/302 nm; 20 min; CH_3CN-H_2O, 95:5–92:8, *v*/*v*) to elute compound **13** (4.6 mg, t_R = 10.6 min). Fr.7 was subjected to Sephadex LH-20 and HPLC (Phenomenex Luna, 5 μm, 250 × 4.6 mm; 1.0 mL/min; 210/254 nm; CH_3CN-0.1% formic acid, 65:35, *v*/*v*) to isolate compound **14** (2.1 mg, t_R = 11.5 min). The separation of Fr.8 was accomplished using analytical HPLC (Phenomenex Luna, 5 μm, 250 × 4.6 mm; 1.0 mL/min; 190/235 nm; 26 min; CH_3CN-H_2O, 78:22, *v*/*v*) to yield compounds **15** (26.7 mg, t_R = 14.7 min), **16** (15.4 mg, t_R = 21.3 min), and **17** (12.2 mg, t_R = 24.2 min).

Crude extract III (34.5 g), extracted under medium #16, was originally split into 6 fractions (Fr.1–Fr.6, Figure S1) by CC (Phenomenex Gemini Axia NX-C18, 10 μm, 50 × 21.2 mm; 210/254 nm; CH_3CN-H_2O, 10.0 mL/min). The Fr.1–Fr.6 segments were further separated and guided similarly to crude extract I. Based on the results of the antimicrobial activity assay, Fr.4 was separated using the analytical column (Phenomenex Luna, 5 μm, 250 × 4.6 mm; 1.0 mL/min; CH_3CN-H_2O, 29:71; 210/254 nm; 20 min) to obtain compounds **18** (58.3 mg, t_R = 9.3 min), **19** (6.2 mg, t_R = 11.1 min), and **20** (11.4 mg, t_R = 20.5 min), in that order (see Figure 4 below).

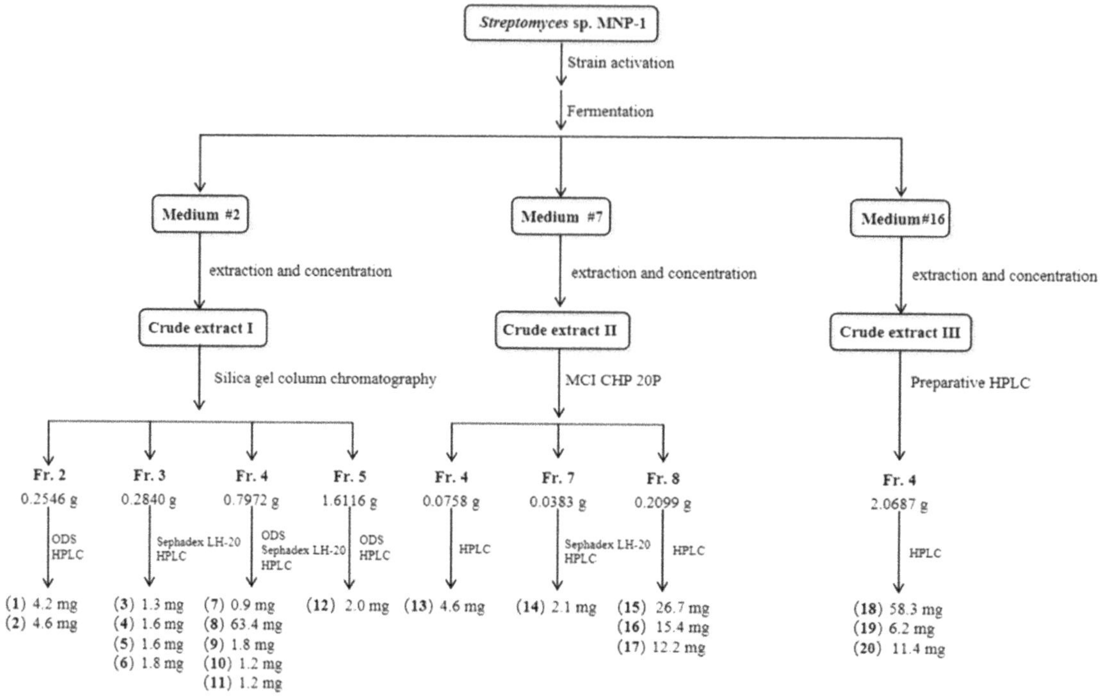

Figure 4. Flowchart of isolation and separation of compounds **1–20** from strain MNP-1.

4.5. Antimicrobial Assay

Antimicrobial activity was determined by the gradient dilution method [48]. Indicator strains were selected using *S. aureus* ATCC 25923, *E. coli* ATCC 25922, and *C. albicans* ATCC 10231 [49]. A certain concentration of the sample and antibiotic positive control solution (bacteria: ampicillin sodium; fungi: amphotericin B) was prepared using DMSO as the solvent, and an equivalent volume of CH_3OH was employed as the negative control. The seed solution of pathogenic bacteria was diluted using a blank medium so that its absorbance was 0.08 (UV detection wavelength 600 nm). Under the condition of aseptic operation, the concentration of the samples in each well was diluted in a gradient.

4.6. Cytotoxicity Assay

Three human tumor cell lines, A549, MCF-7, and HepG2, were selected to assess the antiproliferative effects of the purified chemicals using the MTT assay and were seeded into 96-well plates with a density of 5×10^3 cells/well and maintained in a 5% CO_2 tissue plate incubator at 37 °C for a total of 24 h [50]. Subsequently, 100 μL of 0.5 mg/mL MTT solution was placed into each well and incubated for 4 h. Cellular inhibition was calculated by enzyme labeling at 450 nm absorbance. Each experiment was performed in triplicate and the IC_{50} values of the samples were calculated by the GraphPad Prism 8.0 software. The determination of statistical significance was conducted using a one-way analysis of variance (ANOVA). (* $p < 0.05$, ** $p < 0.01$, *** $p < 0.001$). Doxorubicin was selected as the positive control while the negative control comprised the solvent DMSO.

5. Conclusions

In this research, 20 compounds (**1–20**) were isolated and characterized from three fermented crude extracts of *Streptomyces* sp. MNP-1. Of note, phenazine analog **3** and astrosporin derivative **14** displayed moderately inhibitory effects on *S. aureus* ATCC 25923 and *C. albicans* ATCC 10231, as well as on the tumor cell lines A549, MCF-7, and HepG2. To our knowledge, the inhibitory activity of the phenazine analog **3** was first reported against both Gram-positive bacteria and various human cancer cell lines.

Supplementary Materials: The following supporting information can be downloaded at: https://www.mdpi.com/article/10.3390/molecules30081657/s1, Table S1: Sequences producing significant alignments of strain MNP-1; Table S2: Media for fermentation of strain MNP-1; Table S3: Quality of crude extract of rice solid medium; Table S4: Quality of crude extract of Gauze's synthetic medium No. 1 (200 mL); Table S5: Quality of crude extract of nutrient-rich liquid medium (200 mL); Table S6: Antimicrobial activity results of rice solid medium test; Table S7: Antimicrobial activity results of Gauze's synthetic medium No. 1 test; Table S8: Antimicrobial activity of Gauze's synthetic medium No. 1 fermentation time test; Table S9: ^{1}H-NMR spectrum of compound **11;** Figure S1: ^{1}H-NMR of rice solid medium test; Figure S2: HPLC peak shape of rice solid medium test; Figure S3: ^{1}H-NMR of Gauze's synthetic medium No. 1 test; Figure S4: HPLC peak shape of Gauze's synthetic medium No. 1 test; Figure S5: Fermentation morphology of Gauze's synthetic medium No. 1; Figure S6: ^{1}H-NMR of Gauze's synthetic medium No. 1 fermentation time test; Figure S7: Antimicrobial activity of Gauze's synthetic medium No. 1 fermentation time; Figure S8: 1H-NMR of nutrient-rich liquid medium test; Figure S9: HPLC peak shape of nutrient-rich liquid medium test; Figure S10: Molecular network of nutrient-rich liquid medium test; Figure S11: ESI-MS spectrum of compound **1**; Figure S12: ^{1}H-NMR spectrum of compound **1;** Figure S13: ^{13}C-NMR spectrum of compound **1**; Figure S14: ESI-MS spectrum of compound **2**; Figure S15: ^{1}H-NMR spectrum of compound **2**; Figure S16: ^{13}C-NMR spectrum of compound **2**; Figure S17: HRESI-MS spectrum of compound **3**; Figure S18: ^{1}H-NMR spectrum of compound **3**; Figure S19: HRESI-MS spectrum of compound **4**; Figure S20: ^{1}H-NMR spectrum of compound **4**; Figure S21: ^{13}C-NMR spectrum of compound **4**; Figure S22: HRESI-MS spectrum of compound **5**; Figure S23: ^{1}H-NMR spectrum of compound **5**; Figure S24: HRESI-MS spectrum of compound **6**; Figure S25: ^{1}H-NMR spectrum of compound **6**; Figure S26: ^{13}C-NMR spectrum of compound **6**; Figure S27: HRESI-MS spectrum of compound **7**; Figure S28: ^{1}H-NMR spectrum of compound **7**; Figure S29: HRESI-MS spectrum of compound **8**; Figure S30: ^{1}H-NMR spectrum of compound **8**; Figure S31: ^{13}C-NMR spectrum of compound **8**; Figure S32: HRESI-MS spectrum of compound **9**; Figure S33: ^{1}H-NMR spectrum of compound **9**; Figure S34: ^{13}C-NMR spectrum of compound **9**; Figure S35: HRESI-MS spectrum of compound **10**; Figure S36: ^{1}H-NMR spectrum of compound **10**; Figure S37: ^{13}C-NMR spectrum of compound **10**; Figure S38: HRESI-MS spectrum of compound **11**; Figure S39: ^{1}H-NMR spectrum of compound **11**; Figure S40: ESI-MS spectrum of compound **12**; Figure S41: ^{1}H-NMR spectrum of compound **12**; Figure S42: ^{13}C-NMR spectrum of compound **12**; Figure S43: HRESI-MS spectrum of compound **13**; Figure S44: ^{1}H-NMR spectrum of compound **13**; Figure S45: ^{13}C-NMR spectrum of compound **13;** Figure S46: ESI-MS spectrum of compound **14**; Figure S47: ^{1}H-NMR spectrum of compound **14**; Figure S48: ^{13}C-NMR spectrum of compound **14**; Figure S49: HRESI-MS spectrum of compound **15**; Figure S50: ^{1}H-NMR spectrum of compound **15**; Figure S51: ^{13}C-NMR spectrum of compound **15**; Figure S52: HRESI-MS spectrum of compound **16**; Figure S53: ^{1}H-NMR spectrum of compound **16**; Figure S54: ^{13}C-NMR spectrum of compound **16**; Figure S55: HRESI-MS spectrum of compound **17**; Figure S56: ^{1}H-NMR spectrum of compound **17**; Figure S57: ^{13}C-NMR spectrum of compound **17**; Figure S58: ESI-MS spectrum of compound **18**; Figure S59: ^{1}H-NMR spectrum of compound **18**; Figure S60: ^{13}C-NMR spectrum of compound **18**; Figure S61: ESI-MS spectrum of compound **19**; Figure S62: ^{1}H-NMR spectrum of compound **19**; Figure S63: ^{13}C-NMR spectrum of compound **19**; Figure S64: ESI-MS spectrum of compound **20**; Figure S65: ^{1}H-NMR spectrum of compound **20**; Figure S66: ^{13}C-NMR spectrum of compound **20**, Figure S67: Morphology of the ore sample of Arctic origin.

Author Contributions: H.Z., M.W. and Z.L. conceived the idea and designed the experiment; M.W., Z.L., J.W. and W.H. performed the experiment; M.W., Z.L. and J.W. analyzed the data; H.Z. contributed to the material; H.Z. and M.W. wrote the paper. All authors have read and agreed to the published version of the manuscript.

Funding: This work was financially supported by the National Key Research and Development Program of China (2022YFC2804203).

Institutional Review Board Statement: Not applicable.

Informed Consent Statement: Not applicable.

Data Availability Statement: Data are contained within the article and Supplementary Materials.

Acknowledgments: The author would like to thank the above funding for the support of this study.

Conflicts of Interest: The authors declare no conflicts of interest.

Abbreviations

DMSO	Dimethyl sulfoxide
ESI-MS	Electrospray ionization mass spectrometry
^{1}H NMR	Proton nuclear magnetic resonance
HPLC	High-performance liquid chromatography
IC_{50}	Half maximal inhibitory concentration
MIC	Minimal inhibitory concentration
MTT	Methylthiazolyldiphenyl-tetrazolium bromide
ODS	Octadecylsilyl
OSMAC	One strain many compounds
ESI-MS	Electrospray ionization mass spectrometry

References

1. Rampelotto, P.H. Extremophiles and extreme environments. *Life* **2013**, *3*, 482–485. [CrossRef]
2. Wilson, Z.E.; Brimble, M.A. Molecules derived from the extremes of life: A decade later. *Nat. Prod. Rep.* **2021**, *38*, 24–82. [CrossRef]
3. Mast, Y.; Stegmann, E. Actinomycetes: The antibiotics producers. *Antibiotics* **2019**, *8*, 105. [CrossRef]
4. Lyudmila, P.T.; Gul, B.B.; Baiken, B.B.; Assya, S.B.; Saule, T.D.; Elmira, R.F.; John, A.B. Beyond traditional screening: Unveiling antibiotic potentials of actinomycetes in extreme environments. *Heliyon* **2024**, *10*, e40371. [CrossRef]
5. Magot, F.; Van Soen, G.; Buedenbender, L.; Li, F.; Soltwedel, T.; Grauso, L.; Mangoni, A.; Blümel, M.; Tasdemir, D. Bioactivity and metabolome mining of deep-sea sediment-derived microorganisms reveal new hybrid PKS-NRPS macrolactone from *Aspergillus versicolor* PS108-62. *Mar. Drugs* **2023**, *21*, 95. [CrossRef]
6. Martín-Aragón, V.R.; Millán, F.R.; Cuadrado, C.; Daranas, A.H.; Medarde, A.F.; López, J.M.S. Induction of new aromaticpolyketides from the marine actinobacterium *Streptomyces griseorubiginosus* through an OSMAC approach. *Mar. Drugs* **2023**, *21*, 526. [CrossRef]
7. Yu, H.-B.; Ning, Z.; Hu, B.; Zhu, Y.-P.; Lu, X.-L.; He, Y.; Jiao, B.-H.; Liu, X.-Y. Cytosporin derivatives from Arctic-derived fungus *Eutypella* sp. D-1 via the OSMAC approach. *Mar. Drugs* **2023**, *21*, 382. [CrossRef]
8. Hussain, A.; Rather, M.A.; Dar, M.S.; Aga, M.A.; Ahmad, N.; Manzoor, A.; Qayum, A.; Shah, A.; Mushtaq, S.; Ahmad, Z.; et al. Novel bioactive molecules from *Lentzea Violacea* strain AS 08 using one strain many compounds (OSMAC) approach. *Bioorg. Med. Chem. Lett.* **2017**, *27*, 2579–2582. [CrossRef]
9. Pan, R.; Bai, X.L.; Chen, J.W.; Zhang, H.W.; Wang, H. Exploring structural diversity of microbe secondary metabolites using OSMAC strategy: A literature review. *Front. Microbiol.* **2019**, *10*, 294. [CrossRef]
10. Pinedo-Rivilla, C.; Aleu, J.; Durán-Patrón, R. Cryptic metabolites from marine-derived microorganisms using OSMAC and epigenetic approaches. *Mar. Drugs* **2022**, *20*, 84. [CrossRef]
11. Rateb, M.E.; Houssen, W.E.; Harrison, W.T.A.; Deng, H.; Okoro, C.K.; Asenjo, J.A.; Andrews, B.A.; Bull, A.T.; Goodfellow, M.; Ebel, R.; et al. Diverse metabolic profiles of a *Streptomyces* strain isolated from a hyper-arid environment. *J. Nat. Prod.* **2011**, *74*, 1965–1971. [CrossRef] [PubMed]
12. Yao, F.H.; Liang, X.; Qi, S.H. Eight new cyclopentenone and cyclohexenone derivatives from the marine-derived fungus *Aspergillus* sp. SCSIO 41501 by OSMAC strategy. *Nat. Prod. Res.* **2020**, *35*, 3810–3819. [CrossRef] [PubMed]
13. Nakanishi, J.; Tatamidani, H.; Fukumoto, Y.; Chatani, N. A new synthesis of aldehydes by the palladium-catalyzed reaction of 2-pyridinyl esters with hydrosilanes. *Synlett* **2006**, *37*, 869–872. [CrossRef]
14. Mehnaz, S.; Saleem, R.S.; Yameen, B.; Pianet, I.; Schnakenburg, G.; Pietraszkiewicz, H.; Valeriote, F.; Josten, M.; Sahl, H.G.; Franzblau, S.G.; et al. Lahorenoic acids A-C, ortho-dialkyl-substituted aromatic acids from the biocontrol strain Pseudomonas aurantiaca PB-St2. *J. Nat. Prod.* **2013**, *76*, 135–141. [CrossRef]

15. Roemer, A.; Scholl, H.; Budzikiewicz, H.; Korth, H.; Pulverer, G. Bacterial constituents. part II. phenazines from pseudomonads. *Org. Chem.* **1981**, *36B*, 1037–1046.
16. Chen, C.; Pan, Y.; Zhao, H.; Xu, X.; Luo, Z.; Cao, L.; Xi, S.; Li, H.; Xu, L. Ruthenium(II)-catalyzed regioselective C-8 hydroxylation of 1,2,3,4-tetrahydroquinolines. *Org. Lett.* **2018**, *20*, 6799–6803. [CrossRef]
17. Li, C.; Wang, M.; Lu, X.; Zhang, L.; Jiang, J.; Zhang, L. Reusable brønsted acidic ionic liquid efficiently catalyzed N-formylation and *N*-acylation of amines. *ACS Sustain. Chem. Eng.* **2020**, *8*, 4353–4361.
18. Xu, H.J.; Liang, Y.F.; Cai, Z.Y.; Qi, H.X.; Yang, C.Y.; Feng, Y.S. CuI-nanoparticles-catalyzed selective synthesis of phenols, anilines, and thiophenols from aryl halides in aqueous solution. *J. Org. Chem.* **2001**, *76*, 2296–2300. [CrossRef]
19. Hund, H.K.; de Beyer, A.; Lingens, F. Microbial metabolism of quinoline and related compounds. VI. degradation of quinaldine by *Arthrobacter* sp. *Biol. Chem. Hoppe. Seyler.* **1990**, *371*, 1005–1008. [CrossRef]
20. Suzuki, H.; Ohnishi, Y.; Furusho, Y.; Sakuda, S.; Horinouchi, S. Novel benzene ring biosynthesis from C(3) and C(4) primary metabolites by two enzymes. *J. Biol. Chem.* **2006**, *281*, 36944–36951. [CrossRef]
21. Toyobo Co., Ltd. 3-Amino-4-hydroxybenzoic acid biosynthetic genes from Streptomyces griseus, and use in production of aminohydroxy aromatic carboxylic acid. JP2004283163. 14 October 2004.
22. Galm, U.; Dessoy, M.A.; Schmidt, J.; Wessjohann, L.A.; Heide, L. In vitro and in vivo production of new aminocoumarins by a combined biochemical, genetic, and synthetic approach. *Chem. Biol.* **2004**, *11*, 173–183. [CrossRef] [PubMed]
23. Higuchi, S.; Yasui, K. Interaction of N-6-substituted-9-methyladenines to poly-5-bromouridylic acid. *Nucleic Acids Symp. Ser.* **1983**, *12*, 185–188.
24. Seim, K.L.; Obermeyer, A.C.; Francis, M.B. Oxidative modification of native protein residues using cerium(IV) ammonium nitrate. *J. Am. Chem. Soc.* **2001**, *133*, 16970–16976. [CrossRef]
25. Jiao, J.; Zhang, X.R.; Chang, N.H.; Wang, J.; Wei, J.F.; Shi, X.Y.; Chen, Z.G. A facile and practical copper powder-catalyzed, organic solvent-and ligand-free ullmann amination of aryl halides. *J. Org. Chem.* **2011**, *76*, 1180–1183. [CrossRef]
26. Senior, M.M.; Williamson, R.T.; Martin, G.E. Using HMBC and adequate NMR data to define and differentiate long-range coupling pathways: Is the crews rule obsolete? *J. Nat. Prod.* **2013**, *76*, 2088–2093. [CrossRef]
27. Choi, J.Y.; Choi, E.H.; Jung, H.W.; Oh, J.S.; Lee, W.H.; Lee, J.G.; Lee, S.H. Melanogenesis inhibitory compounds from saussureae radix. *Arch. Pharm. Res.* **2008**, *31*, 294–299. [CrossRef] [PubMed]
28. Mori, K.; Tashiro, T.; Akasaka, K.; Ohrui, H.; Fattorusso, E. Determination of the absolute configuration at the two cyclopropane moieties of plakoside A, an immunosuppressive marine galactosphingolipid. *Tetrahedron Lett.* **2002**, *43*, 3719–3722. [CrossRef]
29. Huang, X.A.; Yang, R.Z. A new hydroquinone diglucoside from Lysimachia fordiana. *Chem. Nat. Compd.* **2004**, *40*, 457–459. [CrossRef]
30. Gutierrez-Lugo, M.T.; Woldemichael, G.M.; Singh, M.P.; Suarez, P.A.; Maiese, W.M.; Montenegro, G.; Timmermann, B.N. Isolation of three new naturally occurring compounds from the culture of *Micromonospora* sp. P1068. *Nat. Prod. Res.* **2005**, *19*, 645–652. [CrossRef]
31. He, J.; Fan, P.; Feng, S.; Shao, P.; Sun, P. Isolation and purification of two isoflavones from hericium erinaceum mycelium by high-speed counter-current chromatography. *Molecules* **2018**, *23*, 560. [CrossRef]
32. Selepe, M.A.; Drewes, S.E.; Van Heerden, F.R. Total synthesis of the pyranoisoflavone kraussianone 1 and related isoflavones. *J. Nat. Prod.* **2010**, *73*, 1680–1685. [CrossRef] [PubMed]
33. Kol, S.; Merlo, M.E.; Scheltema, R.A.; de Vries, M.; Vonk, R.J.; Kikkert, N.A.; Dijkhuizen, L.; Breitling, R.; Takano, E. Metabolomic characterization of the salt atress response in *Streptomyces coelicolor*. *Appl. Environ. Microbiol.* **2010**, *76*, 2574–2581.
34. Zhao, F.; Qin, Y.H.; Zheng, X.; Zhao, H.W.; Chai, D.Y.; Li, W.; Pu, M.X.; Zuo, X.S.; Qian, W.; Ni, P. Biogeography and adaptive evolution of *Streptomyces* strains from saline environments. *Sci. Rep.* **2016**, *6*, 32718. [CrossRef]
35. Novella, I.S.; Sánchez, J. Effects of 5-Azacytidine on physiological differentiation of *Streptomyces* antibioticus. *Res. Microbiol.* **1995**, *146*, 721–728. [CrossRef]
36. Kumar, J.; Sharma, V.K.; Singh, D.K.; Mishra, A.; Gond, S.K.; Verma, S.K.; Kumar, A.; Kharwar, R.N. Epigenetic activation of antibacterial property of an endophytic *Streptomyces coelicolor* strain AZRA 37 and identification of the induced protein using MALDI TOF MS/MS. *PLoS ONE* **2016**, *11*, e0147876. [CrossRef]
37. Karuppiah, V.; Alagappan, K.; Sivakumar, K.; Kannan, L. Phenazine-1-Carboxylic Acid-Induced Programmed Cell Death in Human Prostate Cancer Cells Is Mediated by Reactive Oxygen Species Generation and Mitochondrial-Related Apoptotic Pathway. *J. Appl. Biomed.* **2016**, *14*, 199–209. [CrossRef]
38. Cimmino, A.; Bahmani, Z.; Castaldi, S.; Masi, M.; Isticato, R.; Abdollahzadeh, J.; Amini, J.; Evidente, A. Phenazine-1-Carboxylic Acid (PCA), produced for the first time as an antifungal metabolite by truncatella angustata, a causal agent of Grapevine Trunk Diseases (GTDs) in iran. *J. Agric. Food Chem.* **2021**, *69*, 12143–12147. [CrossRef]
39. Zhang, L.L.; Tian, X.Y.; Kuang, S.; Liu, G.; Zhang, C.S.; Sun, C.M. Antagonistic activity and mode of action of phenazine-1-carboxylic acid, produced by marine bacterium Pseudomonas aeruginosa PA31x, against Vibrio anguillarum in vitro and in a zebrafish in vivo model. *Front. Microbiol.* **2017**, *8*, 289. [CrossRef]

40. Nishida, A.; Oda, A.; Takeuchi, A.; Lee, T.; Awano, H.; Hashimoto, N.; Takeshima, Y.; Matsuo, M. Staurosporine allows dystrophin expression by skipping of nonsense-encoding exon. *Brain Dev.* **2016**, *38*, 738–745. [CrossRef]
41. Yadav, S.S.; Prasad, C.B.; Prasad, S.B.; Pandey, L.K.; Singh, S.; Pradhan, S.; Narayan, G. Anti-Tumor Activity of Staurosporine in the Tumor Microenvironment of Cervical Cancer: An in Vitro Study. *Life Sci.* **2015**, *133*, 21–28. [CrossRef]
42. Zambrano, J.N.; Williams, C.J.; Williams, C.B.; Hedgepeth, L.; Burger, P.; Dilday, T.; Eblen, S.T.; Armeson, K.; Hill, E.G.; Yeh, E.S. Staurosporine, an inhibitor of hormonally up-regulated neu-associated kinase. *Oncotarget* **2018**, *9*, 35962–35973. [CrossRef] [PubMed]
43. Zhang, X.D.; Gillespie, S.K.; Hersey, P. Staurosporine Induces Apoptosis of Melanoma by Both Caspase-Dependent and -Independent Apoptotic Pathways. *Mol. Cancer Ther.* **2004**, *3*, 187–197. [CrossRef] [PubMed]
44. Rajeswari, P.; Jose, P.A.; Amiya, R.; Jebakumar, S.R.D. Characterization of saltern based *Streptomyces* sp. and statistical media optimization for its improved antibacterial activity. *Front. Microbiol.* **2015**, *5*, 2014. [CrossRef] [PubMed]
45. Mao, X.; Chen, S.; Shen, Y.; Wei, D.; Deng, Z. Effect of copper sulfate on biosynthesis of FR-008/candicidin complex production in *Streptomyces* sp. *World J. Microbiol. Biotechnol.* **2011**, *27*, 2033–2039. [CrossRef]
46. Bind, S.; Bind, S.; Sharma, A.K.; Chaturvedi, P. Epigenetic modification: A key tool for secondary metabolite production in microorganisms. *Front. Microbiol.* **2022**, *13*, 784109. [CrossRef]
47. Zdouc, M.M.; Iorio, M.; Maffioli, S.I.; Crüsemann, M.; Donadio, S.; Sosio, M. Planomonospora: A metabolomics perspective on an underexplored Actinobacteria genus. *J. Nat. Prod.* **2021**, *84*, 204–219. [CrossRef] [PubMed]
48. Li, Y.; Chen, L.; Han, G.; Zhou, J.; Zhao, Y. Determination of antibacterial activity of aucubigenin and aucubin. *Asian J. Chem.* **2014**, *26*, 559–561. [CrossRef]
49. Erkuş, B.; Özcan, H.; Zaim, Ö. Synthesis, antimicrobial activity, and ion transportation investigation of four new [1 + 1] condensed furan and thiophene-based cycloheterophane amides. *J. Heterocycl. Chem.* **2020**, *57*, 1956–1962. [CrossRef]
50. Shehabeldine, A.M.; Doghish, A.S.; El-Dakroury, W.A.; Hassanin, M.M.H.; Al-Askar, A.A.; AbdElgawad, H.; Hashem, A.H. Antimicrobial, antibiofilm, and anticancer activities of syzygium aromaticum essential oil nanoemulsion. *Molecules* **2023**, *28*, 5812. [CrossRef]

molecules

MDPI

Article

Secondary Metabolites from Australian Lichens *Ramalina celastri* and *Stereocaulon ramulosum* Affect Growth and Metabolism of Photobiont *Asterochloris erici* through Allelopathy

Martin Bačkor [1,2,*], Dajana Kecsey [2], Blažena Drábová [1], Dana Urminská [1], Martina Šemeláková [3] and Michal Goga [2]

[1] Department of Biochemistry and Biotechnology, Institute of Biotechnology, Faculty of Biotechnology and Food Sciences, Slovak University of Agriculture, Tr. A. Hlinku 2, 949 76 Nitra, Slovakia; blazena.drabova@uniag.sk (B.D.); dana.urminska@uniag.sk (D.U.)
[2] Department of Botany, Institute of Biology and Ecology, Faculty of Science, Pavol Jozef Šafárik University in Košice, Mánesova 23, 041 67 Košice, Slovakia; dajana.rucova@upjs.sk (D.K.); michal.goga@upjs.sk (M.G.)
[3] Department of Medical Biology, Faculty of Medicine, Pavol Jozef Šafárik University in Košice, Trieda SNP 1, 040 11 Košice, Slovakia; martina.semelakova@upjs.sk
* Correspondence: martin.backor@uniag.sk; Tel.: +421-905-428-396

Abstract: In the present work, the phytotoxic effects of secondary metabolites extracted from lichen *Ramalina celastri* (usnic acid) and lichen *Stereocaulon ramulosum* (a naturally occurring mixture of atranorin and perlatolic acid, approx. 3:1) on cultures of the aposymbiotically grown lichen photobiont *Asterochloris erici* were evaluated. Algae were cultivated on the surface of glass microfiber disks with applied crystals of lichen extracts for 14 days. The toxicity of each extract was tested at the two selected doses in quantities of 0.01 mg/disk and 0.1 mg/disk. Cytotoxicity of lichen extracts was assessed using selected physiological parameters, such as growth (biomass production) of photobiont cultures, content of soluble proteins, chlorophyll *a* fluorescence, chlorophyll *a* integrity, contents of chlorophylls and total carotenoids, hydrogen peroxide, superoxide anion, TBARS, ascorbic acid (AsA), reduced (GSH) and oxidized (GSSG) glutathione, and composition of selected organic acids of the Krebs cycle. The application of both tested metabolic extracts decreased the growth of photobiont cells in a dose-dependent manner; however, a mixture of atranorin and perlatolic acid was more effective when compared to usnic acid at the same dose tested. A higher degree of cytotoxicity of extracts from lichen *S. ramulosum* when compared to identical doses of extracts from lichen *R. celastri* was also confirmed by a more pronounced decrease in chlorophyll *a* fluorescence and chlorophyll *a* integrity, decreased content of chlorophylls and total carotenoids, increased production of hydrogen peroxide and superoxide anion, peroxidation of membrane lipids (assessed as TBARS), and a strong decrease in non-enzymatic antioxidants such as AsA, GSH, and GSSG. The cytotoxicity of lichen compounds was confirmed by a strong alteration in the composition of selected organic acids included in the Krebs cycle. The increased ratio between pyruvic acid and citric acid was a very sensitive parameter of phytotoxicity of lichen secondary metabolites to the algal partner of symbiosis. Secondary metabolites of lichens are potent allelochemicals and play significant roles in maintaining the balance between mycobionts and photobionts, forming lichen thallus.

Keywords: allelopathy; algae; antioxidants; citric acid; glutathione; lichens; Krebs cycle; organic acids; phenolic compounds; secondary metabolites

1. Introduction

Lichens are self-sustaining ecosystems forming a vegetative body (thallus) composed of a taxonomically defined fungal partner (mycobiont) and an extracellular arrangement of minimally one photosynthetic partner (photobiont), which is typically represented by algae or/and cyanobacteria. Within the "lichen" ecosystem, an indeterminate number of other microscopic organisms is typically present, representing the remaining components of the thalli, e.g., cystobasidiomycete yeast. However, it seems that the presence of cystobasidiomycete yeasts in lichen thalli is much less mycobiont-specific than the photobionts [1]. Lichens are, for this reason, considered the symbiotic phenotypes of lichen-forming fungi [2], and lichen names are determined by the presence of the fungal partner. The resulting "organism" does not resemble parental organisms when cultivated as a single symbiont in the laboratory [3], and the same mycobiont may produce a different phenotype when associated with a different photobiont [4].

Lichen plays an important role in the nutrient cycle. They can also be an important part of food chains as many herbivores feed on them, including gastropods, nematodes, mites, or even some mammalian herbivores, e.g., reindeer. They can grow on diverse surfaces and occur from sea level to high alpine elevation, from the polar regions to the equator. They have an important function in boreal forest ecology as they can affect the metabolism of plants that form biocenoses with them [3]. Lichens grow in diverse environments and, due to their symbiotic nature, can survive in extreme environments as extremophiles. Certain species can grow and develop in habitats affected by excessive cold, heat, drought, UV radiation, or metal pollution, where individual organisms forming lichen thallus can only barely survive [3].

A very important aspect of lichen-forming symbiosis is the production of more than one thousand substances that are unique to this group of organisms and are called secondary metabolites [5]. They usually constitute from 0.1 to 5.0% (*w*/*w*) of thallus dry weight and include dibenzofuran derivatives, depsides, and depsidones. Shortly, these metabolites are predominantly of mycobiont origin, extracellular phenols localized on the surfaces of mycobiont hyphae, and have many biological and ecological functions, e.g., antiviral and antimicrobial activities, allelopathy, antiherbivory, UV and excessive light screening of photobionts, and chelating of heavy metals [5,6].

Most of the lichen secondary metabolites, including types of depsides, depsidones, depsones, and dibenzofurans, are derived from the polyketide pathway, previously known as the acetyl-polymalonyl pathway [7]. Usnic acid (UA) is a yellowish pigment and dibenzofuran derivative, one of the most common cortical secondary metabolites of lichens, especially abundant in members of the genera *Alectoria*, *Usnea*, *Cladonia*, and *Xanthoparmelia*. Atranorin (ATR) is a depside and another widespread cortical secondary metabolite abundant in members of the lichen genera *Parmotrema*, *Cladonia*, or *Stereocaulon*. Perlatolic acid (PA) also belongs to lichen depsides; however, when compared to UA and ATR, localized in the medullar part of lichen thalli, PA is typically present in lichens in considerably lower quantities when compared to cortical compounds and present, for instance, in the lichen genera *Cladonia* and *Stereocaulon* [8,9].

So far, UA, ATR, and PA have previously been tested for their antibiotic, antiviral, antimycotic, antiprotozoal, antiherbivore, antiproliferative, anti-inflammatory, analgesic, antipyretic, allelopathic, and UV-protecting effects [5,10]. UA and ATR play significant photoprotective roles for symbiotic photobionts in lichen thalli; however, compound-specific screening mechanisms are still not sufficiently understood [11,12].

The mechanisms of the allelopathic effects of lichen secondary metabolites on plants in biocenoses containing lichens are still not sufficiently researched. Since the minimum requirement for a functional lichen thallus is the presence of mycobiont hyphae, which frequently produce typical lichen phenolic substances and photobiont cells, the involvement of lichen secondary metabolites in potential regulation of photobiont biology and ecology through allelopathy is another fascinating task for modern experimental plant biology.

As lichen photobionts are in direct contact with cortical and medullar extracellular metabolites produced by mycobiont, the main aim of this study was to evaluate their potential allelopathic effects on aposymbiotically grown lichen photobiont cultures of *A. erici* during 2-week prolonged cultivation on the surface of glass fiber disks coated with crystals of lichen secondary metabolites. Secondary metabolites extracted from lichen

Ramalina celastri (usnic acid) and lichen *Stereocaulon ramulosum* (the naturally occurring mixture of atranorin and perlatolic acid, approx. 3:1) were used at two selected doses, in quantities of 0.01 mg/disk and 0.1 mg/disk. Allelopathic effects of lichen extracts were assessed using selected physiological parameters, including growth (biomass production) of photobiont cultures, content of soluble proteins, chlorophyll *a* fluorescence, chlorophyll *a* integrity, contents of chlorophylls and total carotenoids, hydrogen peroxide, superoxide anion, TBARS production, content of ascorbic acid (AsA), reduced (GSH), and oxidized (GSSG) glutathione, or composition of selected organic acids included in the Krebs cycle.

2. Results and Discussion

2.1. Photobiont Growth

Biomass production (mg dw/disk) of two-week-old *A. erici* cultures decreased by the presence of extracts from *R. celastri* and. *S. ramulosum* in a quantitatively dependent manner (Figure 1A). While both tested quantities of *S. ramulosom* extracts significantly decreased photobiont growth, the lower tested amount of extract from *R. celastri* (0.01 mg/disk) was not strong enough to be significant.

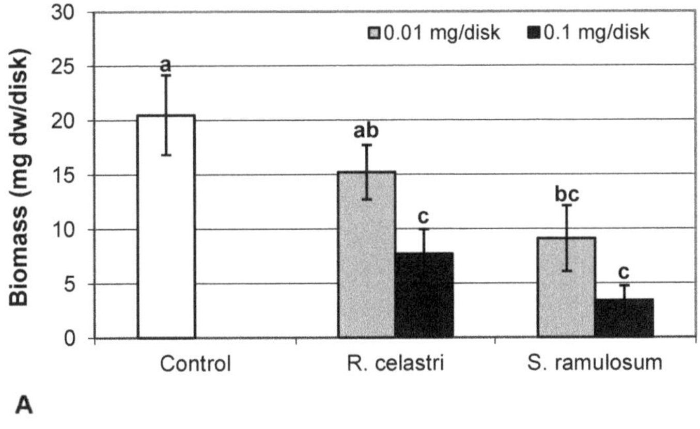

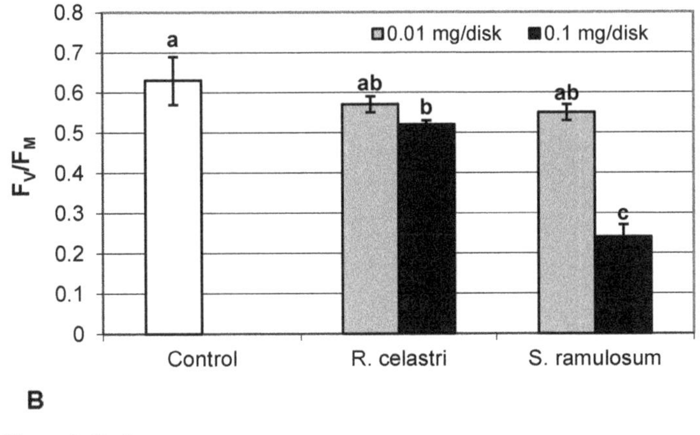

Figure 1. *Cont.*

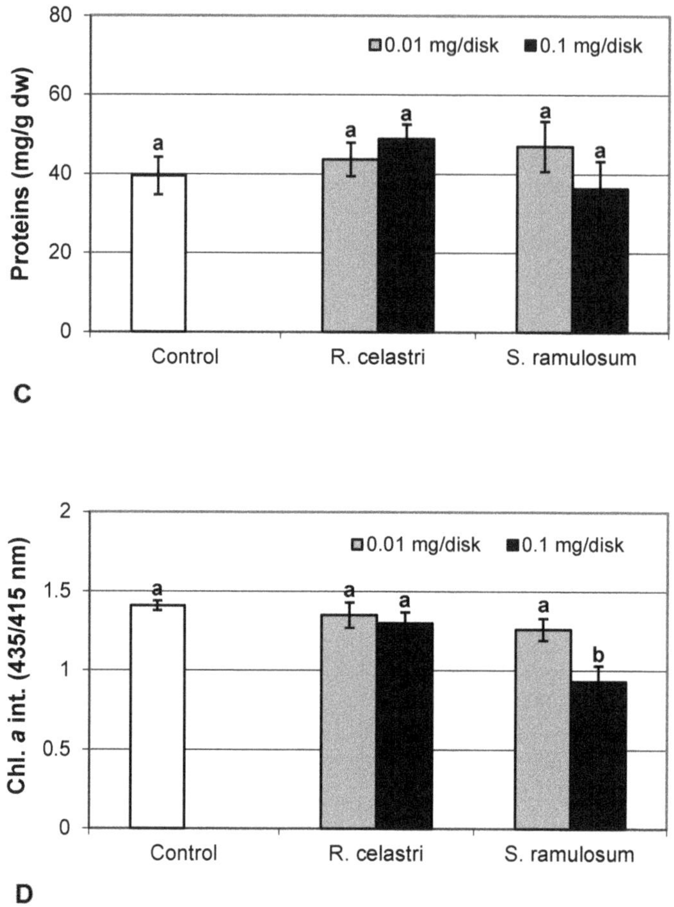

Figure 1. Biomass production ((**A**); mg dw/disk), chlorophyll *a* fluorescence ((**B**); F_V/F_M), content of soluble proteins ((**C**); mg/g dw), and chlorophyll *a* integrity ((**D**); A 435 nm/415 nm) of 2 week-old photobiont *Asterochloris erici* cultures cultivated on disks with addition of secondary metabolites extracts from lichens *Ramalina celastri* and *Stereocaulon ramulosum* (0.01 and 0.1 mg/disk). Values in vertical columns followed by the same letter(s) are not significantly different according to Tukey's test ($p < 0.05$), $n = 3$.

Secondary metabolites of lichens play many important biological and ecological roles. Mechanisms of their action in cases of antiproliferative or antibiotic effects are now relatively well documented for some common lichen cortical compounds. It has been previously found that both cortical lichen compounds used in the present study are potent inhibitors of the growth and development of some cancer cell lines. The suppression of viability and cell proliferation caused by the application of UA and ATR correlated strongly with an increased number of floating cells and a higher apoptotic index. The analysis of cell cycle distribution also revealed an accumulation of cells in S-phase [13].

Mechanisms of allelopathic role of lichen secondary metabolites are, on the other hand, still far from their full understanding. Allelopathy in lichens is probably a very complex phenomenon, and it appears that secondary metabolites of lichens from different chemical groups differ in the degree of their phytotoxicity. Firstly, lichen secondary metabolites on a global scale may influence boreal forest ecology through allelopathy [14] and can directly affect plant metabolism, e.g., confirmed in the case of seedlings of Norway spruce and

Scots pine seedlings, or indirectly through allelopathic effects on soil mycorrhizal fungi [14]. UA also has phytotoxic acid on cultures of lower plants, e.g., mosses, which typically form phytocenoses with lichens [15]. Secondly, the lichen thallus itself is formed as an ecosystem, and in addition to its own photobiont cells growing within its thallus, many other microscopic organisms, including cystobasidiomycete yeasts, may also be present [2]. It has been found in our previous studies that some lichen metabolites can directly decrease the growth of algal cultures, including lichen photobionts [16,17]. It can be the result of the effect of some lichen compounds (e.g., UA) on the spindle apparatus during the processes of mitosis [18,19]. However, the efficiency of lichen compounds in control of photobiont cell division is different, and some cortical metabolites can positively affect photobiont cells through their photoprotective roles, screening UV radiation, and excessive light-causing photoinhibition in photobiont cells [11,12].

2.2. *Activity of Photosystem II, Soluble Proteins, and Content of Assimilation Pigments in Photobiont*

Chlorophyll *a* fluorescence (expressed as F_V/F_M ratio) of two-week-old *A. erici* cultures decreased by the presence of the highest tested (0.1 mg/disk) amount of extracts from *R. celastri* and *S. ramulosum* (Figure 1B). The inhibition effect was more pronounced after exposure of algal cells to *S. ramulosum* extracts. The content of the soluble proteins (mg/g dw) in *A. erici* was a stable parameter and did not change regardless of the lichen source of tested extracts and extract quantities (Figure 1C).

Chlorophyll *a* integrity (expressed as a ratio of absorbance of pigment extracts at 435 nm/415 nm) of two-week-old *A. erici* cultures decreased only as a result of the highest tested (0.1 mg/disk) amount of extract from *S. ramulosum* (Figure 1D). Similarly, chlorophyll *a* content (Figure 2A), chlorophyll *b* content (Figure 2B), chlorophyll *a/b* ratio (Figure 2C), chlorophyll *a+b* content (Figure 2D), and content of total carotenoids (Figure 3A) of two-week-old *A. erici* cultures decreased only as a result of the highest tested (0.1 mg/disk) amount of extracts from *S. ramulosum*. Contents of chlorophylls and total carotenoids are expressed as mg/g dw of algal biomass. The ratio of total carotenoids/total chlorophyll in *A. erici* was a stable parameter and did not change regardless of the lichen source of tested extracts and extract quantities (Figure 3B).

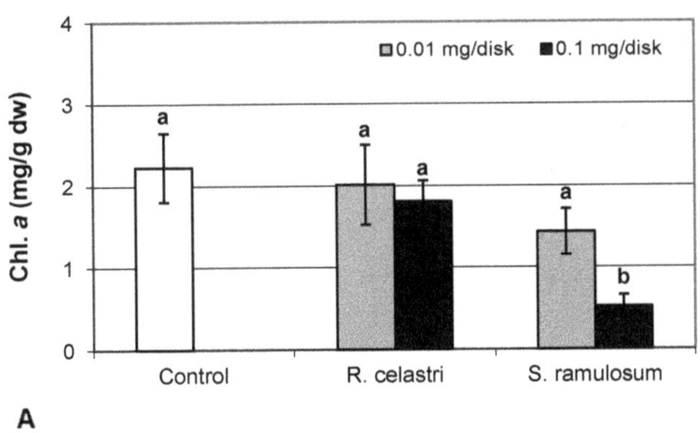

Figure 2. *Cont.*

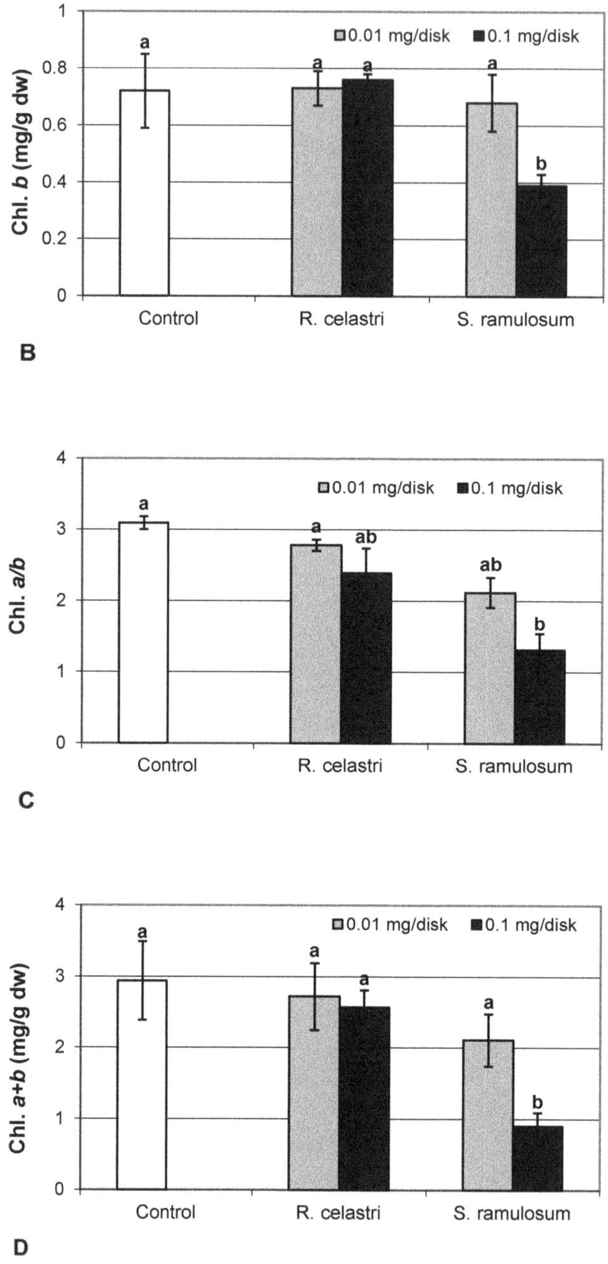

Figure 2. Chlorophyll *a* content ((**A**); mg/g dw), chlorophyll *b* content ((**B**); mg/g dw), chlorophyll *a*/*b* ((**C**), and chlorophyll *a*+*b* content ((**D**); mg/g dw) of 2-week-old photobiont *Asterochloris erici* cultures cultivated on disks with the addition of secondary metabolites extracts from lichens *Ramalina celastri* and *Stereocaulon ramulosum* (0.01 and 0.1 mg/disk). Values in vertical columns followed by the same letter(s) are not significantly different according to Tukey's test ($p < 0.05$), $n = 3$.

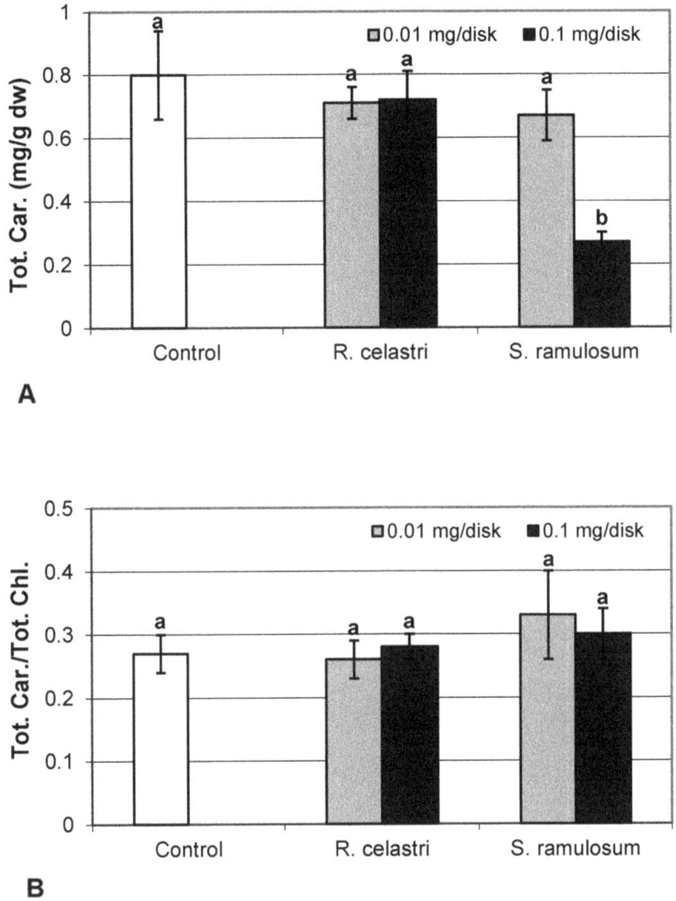

Figure 3. Content of total carotenoids ((**A**); mg/g dw) and total carotenoids/total chlorophyll (**B**) of 2-week-old photobiont *Asterochloris erici* cultures cultivated on disks with the addition of secondary metabolites extracts from lichens *Ramalina celastri* and *Stereocaulon ramulosum* (0.01 and 0.1 mg/disk). Values in vertical columns followed by the same letter(s) are not significantly different according to Tukey's test ($p < 0.05$), $n = 3$.

UA and a mixture of ATR+PA used in this study demonstrated strong inhibition effects on chlorophyll *a* fluorescence. The influence of lichen secondary metabolites on Photosystem II, including UA, was demonstrated previously, and the determination of chlorophyll *a* fluorescence in photobiont cells was found to be a useful marker for the assessment of their physiological status [20,21]. It seems that the utilization of other chlorophyll fluorescence methods may help explain the sensitivity or tolerance of complex photosynthetic processes in lichen algae [22,23].

The phytotoxic effect of secondary metabolites can also involve the alteration of assimilation pigment composition. In the previous study, Rapsch and Ascaso [24] studied the effect of medullary lichen secondary metabolite evernic acid on the chloroplast structure of spinach (*Spinacia oleracea*). Decreased total chlorophyll content and chlorophyll *a* content, as well as decreased chlorophyll *a* to chlorophyll *b* ratio due to the presence of evernic acid, was demonstrated also in free-living algae and lichen photobionts [25]. Bačkor et al. [16] found that the assimilation pigments (chlorophyll *a*, chlorophyll *b*, and content of total carotenoids) in the free-living alga *Scenedesmus quadricauda* were strongly decreased due to the presence of UA. However, the sensitivity of photobiont *A. erici* (syn. *Trebouxia erici* in the former study) cells was significantly lower. On the other hand, the assimilation pigments of *A. erici* were sensitive to the presence of a mixture of ATR+PA used in this study. This was also confirmed by decreased chlorophyll *a* integrity, a parameter very frequently employed in studies focused on lichen photobiont stress physiology [3].

2.3. Oxidative Status and Membrane Lipid Peroxidation in Photobiont

Hydrogen peroxide (μmol/g dw) content (Figure 4A) and content (μg/g dw) of superoxide anion (Figure 4B) of two-week-old *A. erici* cultures increased by the presence of the highest tested (0.1 mg/disk) amount of extracts from *R. celastri* and. *S. ramulosum*. Production of TBARS (nmol/g dw) in two-week-old *A. erici* cultures increased by the presence of extracts from *R. celastri* as well as *S. ramulosum* in a quantitatively dependent manner (Figure 4C). While both tested quantities of *S. ramulosom* extracts significantly increased TBARS production, the increase in TBARS caused by the lower tested amount of extract from *R. celastri* (0.01 mg/disk) was not strong enough to be significant (Figure 4C).

Oxidative stress may damage many biological molecules, including DNA and cell membrane lipids, which are significant targets of cellular injury (free radical attack). Plant polyphenols are well recognized for their antioxidant activities. Levels of hydrogen peroxide, superoxide, and TBARS significantly increased in photobiont cultures due to the highest tested concentrations of UA and a mixture of ATR + PA. Han et al. [26] found that usnic acid disrupted electron transport in mitochondria and induced oxidative stress in the hepatocytes. Caviglia et al. [27], on the other hand, demonstrated that UA extracted from lichen *Parmelia soredians* may act as an antioxidant and detoxify ROS produced after the application of the herbicide Paraquat.

Both tested compounds in the present study, UA and ATR, were previously found to be relatively effective anti-cancer compounds [28]. UA and ATR were capable of inducing a massive loss in the mitochondrial membrane potential, along with caspase-3 activation and phosphatidylserine externalization in tested cancer cell lines. The induction of both ROS and RNS was, at least in part, responsible for the cytotoxic effects of these cortical lichen secondary metabolites. Based on the detection of protein expression (PARP, p53, Bcl-2/Bcl-xL, Bax, p38, pp38), it has been found that UA and ATR may be activators of programmed cell death in cancer cell lines, probably through the mitochondrial pathway.

We know that UA increased levels of hydrogen peroxide and superoxide in the cells of lichen photobiont *Trebouxia erici* [16]. However, studies focused on the production of ROS or degradation of membrane lipids in lichens and their symbionts are still insufficiently intensively examined and require further study.

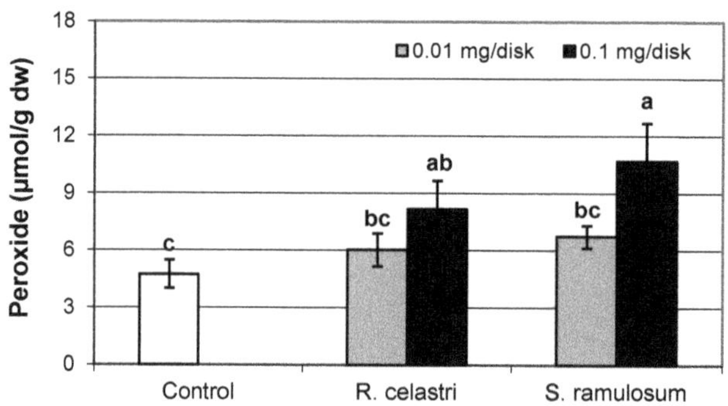

A

Figure 4. *Cont.*

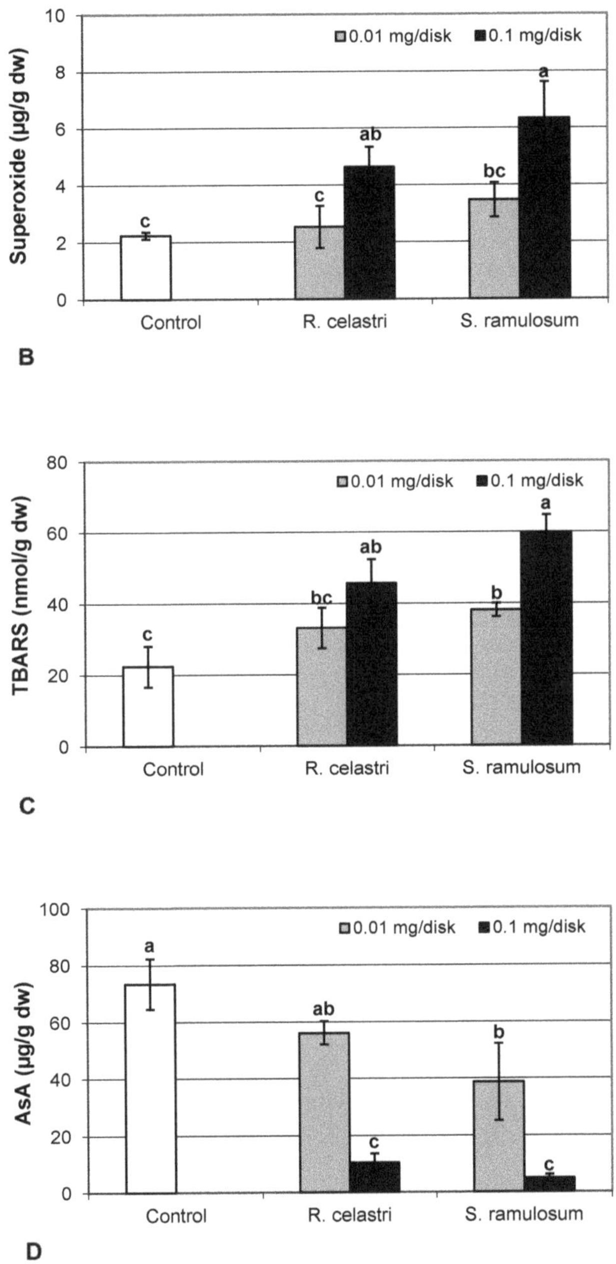

Figure 4. Content of hydrogen peroxide ((**A**); µmol/g dw), superoxide anion ((**B**); µg/g dw), TBARS ((**C**); nmol/g dw), and ascorbic acid (AsA, (**D**); µg/g dw) of 2-week-old photobiont *Asterochloris erici* cultures cultivated on disks with the addition of secondary metabolites extracts from lichens *Ramalina celastri* and *Stereocaulon ramulosum* (0.01 and 0.1 mg/disk). Values in vertical columns followed by the same letter(s) are not significantly different according to Tukey's test ($p < 0.05$), $n = 3$.

2.4. Production of Ascorbic Acid (AsA), Reduced (GSH) and Oxidized (GSSG) Glutathione

The content of AsA (μg/g dw) in two-week-old *A. erici* cultures decreased by the presence of a higher tested (0.1 mg/disk) amount of extract from *R. celastri* and both tested amounts of extracts from *S. ramulosum* (Figure 4D).

The content of GSH and GSSG (μg/g dw) in two-week-old *A. erici* cultures decreased by the presence of lichen extracts on the disk surface (Figure 5A,B). GSH and GSSG content decreased in a dose-dependent manner in both amounts of lichen extracts tested. While a significant decrease in GSH content was observed only after application of the highest tested dose (0.1 mg/disk) of extract from *R. celastri*, both tested amounts of extracts from *S. ramulosum* had a strong inhibition effect on GSH content. GSSG content in algae was inhibited only after higher (0.1 mg/disk) tested doses of extracts from both lichens (Figure 5B). Due to this reason, we observed a significant decrease in the GSH/GSSG ratio in photobiont cells after the application of both tested amounts of extracts from *S. ramulosum* (Figure 5C).

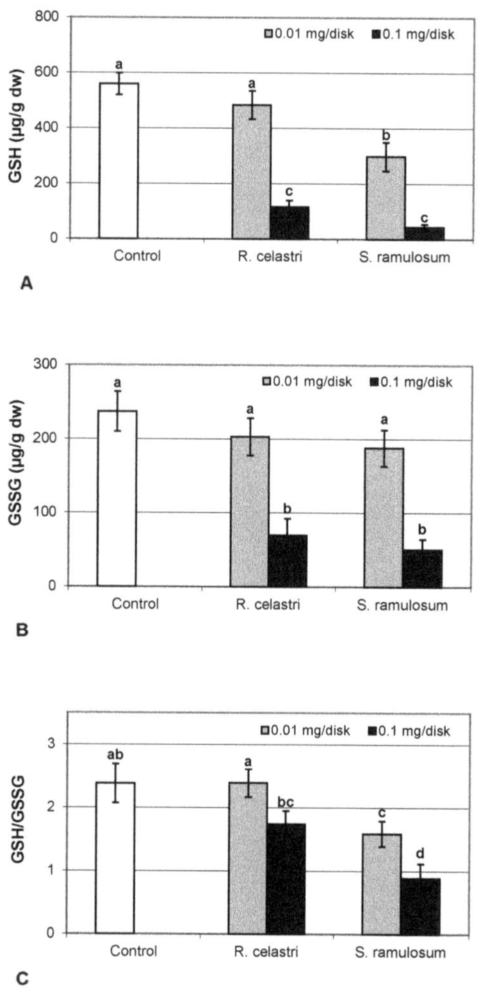

Figure 5. *Cont.*

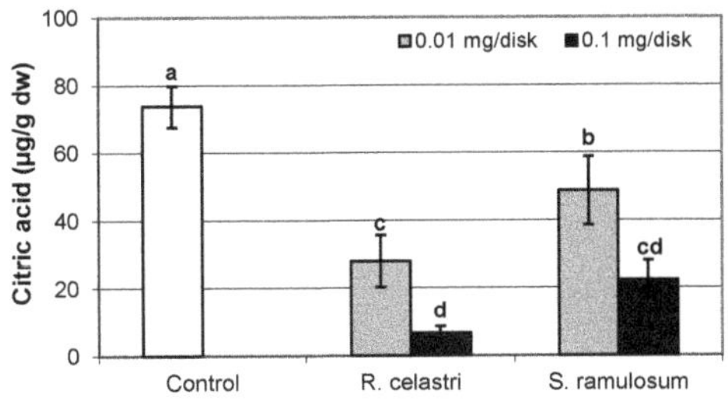

D

Figure 5. Content of reduced glutathione (GSH, (**A**); µg/g dw), oxidized glutathione (GSSG, (**B**); µg/g dw), GSH/GSSG (**C**), and citric acid ((**D**); µg/g dw) of 2-week-old photobiont *Asterochloris erici* cultures cultivated on disks with the addition of secondary metabolites extracts from lichens *Ramalina celastri* and *Stereocaulon ramulosum* (0.01 and 0.1 mg/disk). Values in vertical columns followed by the same letter(s) are not significantly different according to Tukey's test ($p < 0.05$), $n = 3$.

The response of lichen algae to the presence of allelochemicals, which are produced by their symbiotic partners, mycobiont, is still enigmatic. In general, when plants are exposed to biotic and/or abiotic stresses, they increase ROS production, which leads to changes in the balance between ROS and cellular antioxidants [29]. The Glutathione-Ascorbate (GSH-AsA) cycle is stimulated to detoxify ROS and ROS-generated toxic products of plant metabolism. The AsA and GSH are considered an antioxidant buffer to control ROS levels [30]. AsA can also play an important role in the regulation of xanthophyll pigment activity. Glutathione is found in all eukaryotes and is one of the most dominant cellular thiol compounds, which is important in defense against ROS. In the lichen, photobiont cells were found to play a significant role in metal defense [3]. Pools of GSH, GSSG, and AsA in photobiont *A. erici* decreased as a response to secondary metabolites presence on cultivation disks.

2.5. Photobiont Organic Acids

In the present study, we also analyzed the content of selected organic acids (µg/g dw) related to the Krebs cycle, or the so-called "citric acid cycle". Citric acid content (Figure 5D) in two-week-old *A. erici* cultures decreased in a dose-dependent manner in the case of both tested lichen extracts; however, extracts from *R. celastri* were more effective when compared to identical amounts of extracts from *S. ramulosum*.

Fumaric acid content (Figure 6A) and lactic acid content (Figure 6D) were stable parameters in photobiont cells and did not change regardless of the lichen extract tested and extract quantities applied. Contents of glutamic acid (Figure 6B) in *A. erici* decreased in a dose-dependent manner for both extracts tested, whereas only the higher doses tested were effective for ketoglutaric acid (Figure 6C).

The content of malic acid (Figure 7A) and succinic acid (Figure 7D) in algae decreased as a response to the application of both tested lichen extracts, reflecting the increase in their quantity. For the lower dose of *R. celastri* extracts tested, the decrease in succinic acid was not strong enough to be significant. In contrast, pyruvic acid content in photobiont cells increased with increased tested amounts of extracts (Figure 7B). Contents of quinic acid (Figure 7C) and tartaric acid (Figure 8A) in *A. erici* were stable at all tested amounts of both lichen extracts. The ratio of pyruvic acid/citric acid contents (Figure 8B) in photobionts increased in response to the application of lichen secondary metabolite extracts, more pronounced at extracts from *R. celastri*.

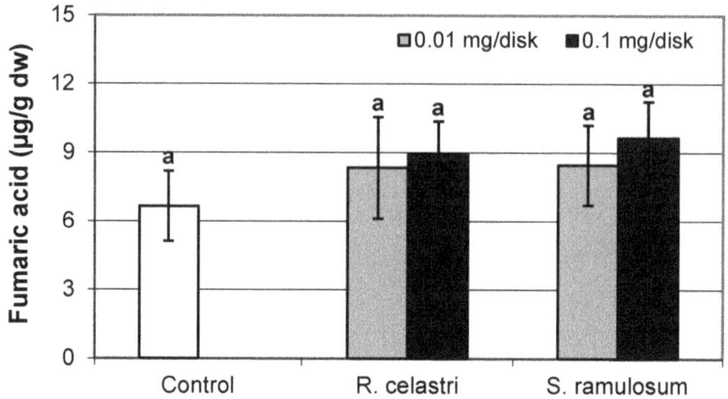

A

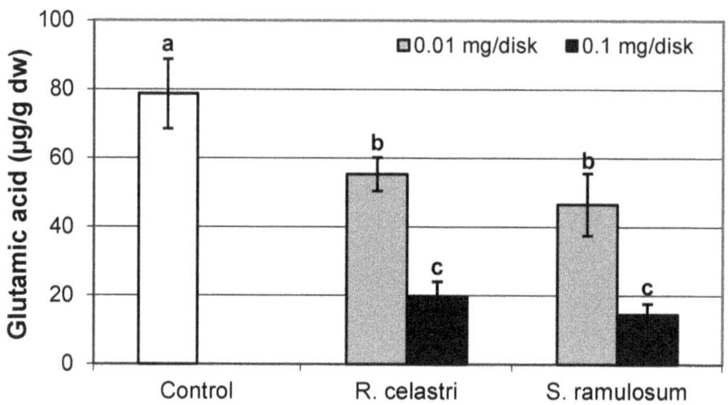

B

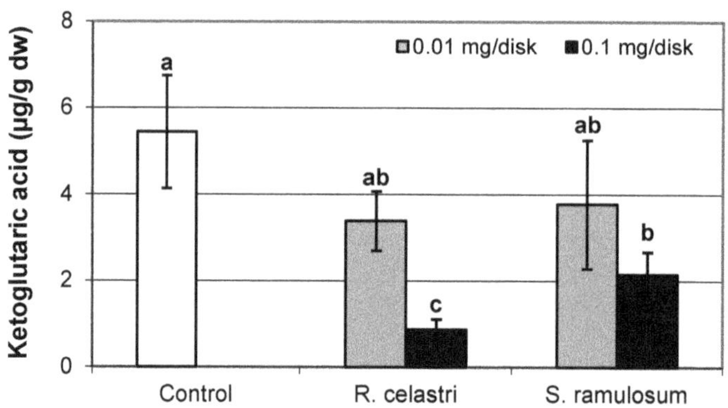

C

Figure 6. *Cont.*

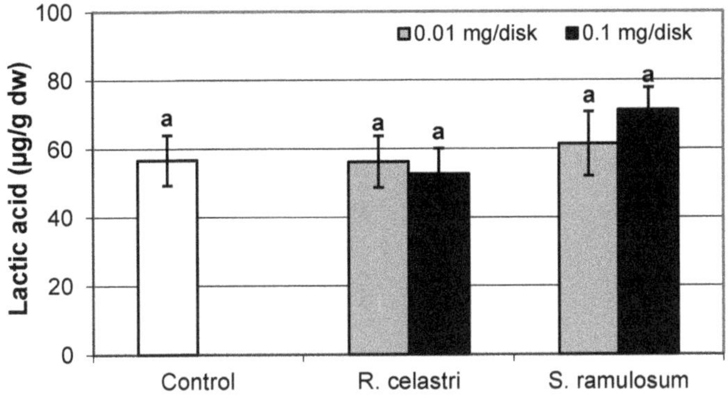

D

Figure 6. Content of fumaric acid ((**A**); µg/g dw), glutamic acid ((**B**); µg/g dw), ketoglutaric acid ((**C**); µg/g dw), and lactic acid ((**D**); µg/g dw) of 2-week-old photobiont *Asterochloris erici* cultures cultivated on disks with the addition of secondary metabolites extracts from lichens *Ramalina celastri* and *Stereocaulon ramulosum* (0.01 and 0.1 mg/disk). Values in vertical columns followed by the same letter(s) are not significantly different according to Tukey's test ($p < 0.05$), $n = 3$.

Pyruvic acid ensures the energy to the cells by the citric acid cycle, and it is a key product between catabolism and anabolism of carbohydrates, fats, and proteins. The increase in pyruvate in response to allelochemicals responds similarly to that in response to other xenobiotics (e.g., heavy metals) and is likely directed toward the production of ATP and NADH for other metabolic processes [25]. However, we know almost virtually nothing about the role of organic acids from the Krebs cycle in lichens and their symbionts.

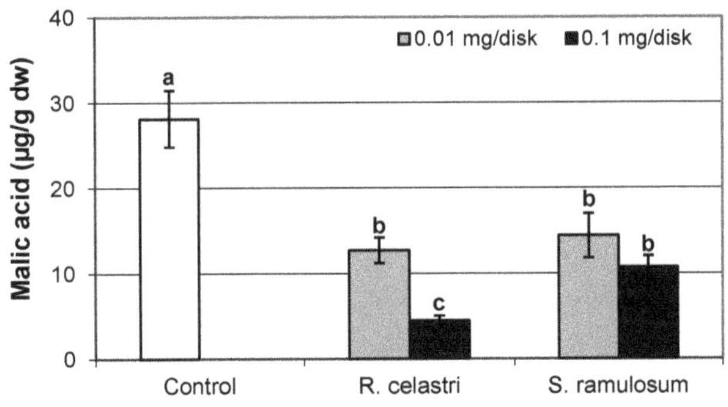

A

Figure 7. *Cont.*

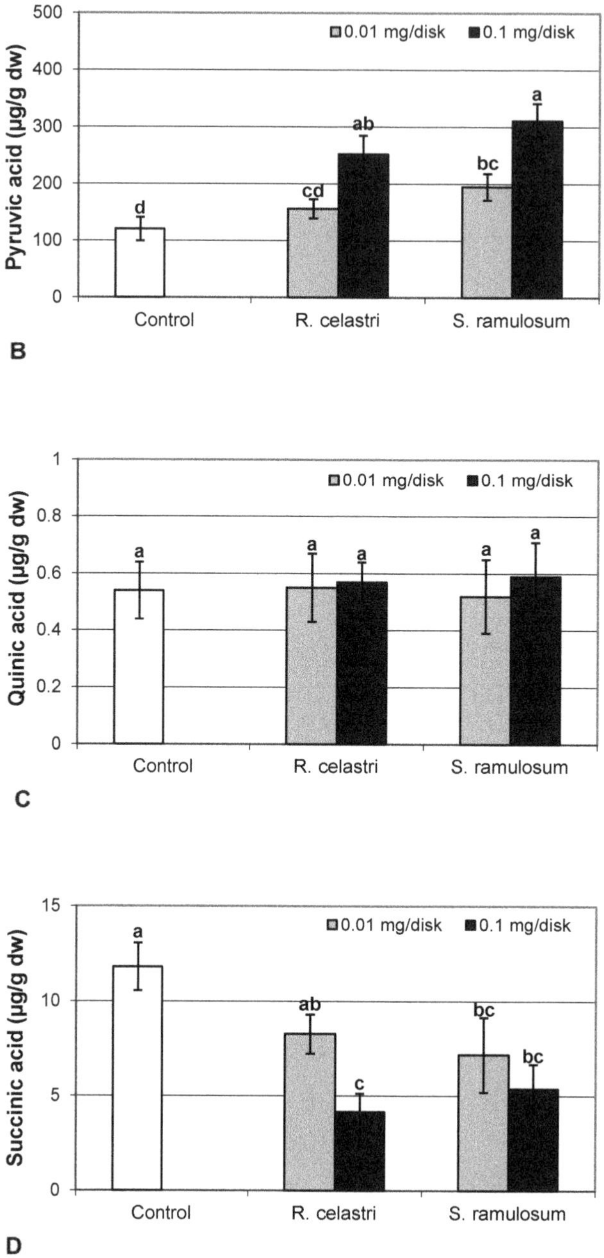

Figure 7. Content of malic acid ((**A**); µg/g dw), pyruvic acid ((**B**); µg/g dw), quinic acid ((**C**); µg/g dw), and succinic acid ((**D**); µg/g dw) of 2-week-old photobiont *Asterochloris erici* cultures cultivated on disks with the addition of secondary metabolites extracts from lichens *Ramalina celastri* and *Stereocaulon ramulosum* (0.01 and 0.1 mg/disk). Values in vertical columns followed by the same letter(s) are not significantly different according to Tukey's test ($p < 0.05$), $n = 3$.

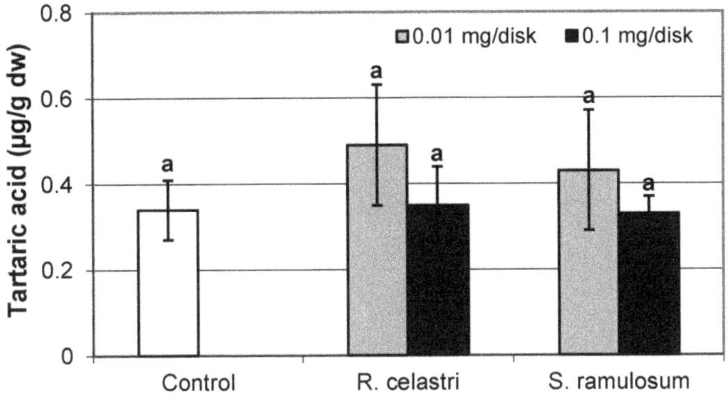

A

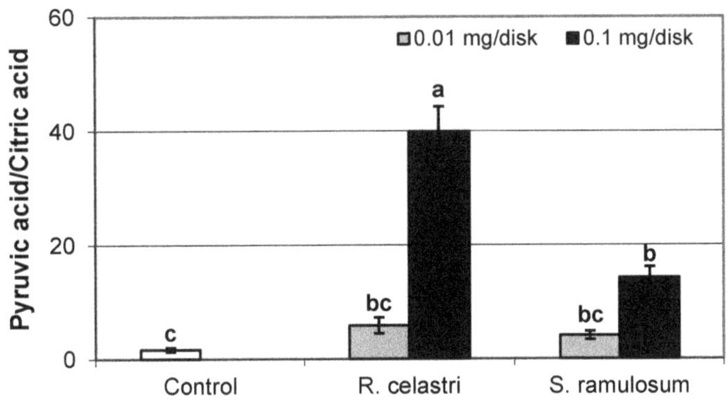

B

Figure 8. Content of tartaric acid ((**A**); µg/g dw) and pyruvic acid/citric acid (**B**) of 2-week-old photobiont *Asterochloris erici* cultures cultivated on disks with the addition of secondary metabolites extracts from lichens *Ramalina celastri* and *Stereocaulon ramulosum* (0.01 and 0.1 mg/disk). Values in vertical columns followed by the same letter(s) are not significantly different according to Tukey's test ($p < 0.05$), $n = 3$.

3. Materials and Methods

3.1. Organism, Culture Conditions, and Lichen Extracts

The lichen photobiont *Asterochloris erici* (Ahmadjian) Skaloud et Peksa (syn. *Trebouxia erici* Ahmadjian, UTEX 911), isolated from the lichen *Cladonia cristatella* Tuck., was used in this study. This lichen contains typical secondary metabolites, e.g., usnic acid, barbatic acid, and didimic acid [8]. Photobiont stock cultures were cultivated on agar medium as described in our previous studies [25] and maintained at 22 °C under a 16-h photoperiod and 30 $\mu mol \cdot m^{-2} \cdot s^{-1}$ artificial irradiance ("cool white" tubes).

For the assessment of the role of lichen secondary metabolites on the growth and metabolism of photobiont cells, acetone extracts from two lichen species were prepared. Usnic acid was isolated from the lichen *Ramalina celastri* (Spreng.) Krog. & Swinscow collected from the stems of Norfolk Island pine trees (*Araucaria heterophylla*), seacoast area near Kiama city, NSW, Australia, on 12 September 2022. The naturally occurring mixture of atranorin and perlatolic acid (approximately 3:1) was obtained from the terrestrial lichen *Stereocaulon ramulosum* (Sw.) Raeusch. also collected at the seacoast near Kiama

City, NSW, Australia, on 13 September 2022. Lichens were collected and identified by Prof. Martin Bačkor. Voucher specimens of both lichen species were stored in our laboratory for future reference. The purity of compounds was assessed by High-Performance Liquid Chromatography (HPLC) and Thin Layer Chromatography as described previously [25]. The mean contents of secondary metabolites in lichen *R. celastri* was 0.2% for usnic acid and in *S. ramulosum* was 0.32% for atranorin and 0.11% for perlatolic acid. Identification of secondary metabolites was carried out using the standard of (+)-usnic acid (CAS No.: 7562-61-0, Sigma-Aldrich, St. Louis, MO, USA). Atranorin (CAS No.: 479-20-9) and perlatolic acid (CAS No.: 529-47-5) were isolated according to previous studies performed in our laboratories using a protocol described by Elečko et al. [31].

3.2. Allelopathic Assay

Photobionts were cultivated on the surface of glass fiber filter disks (Whatman CF/C filters), 25 mm in diameter, as described in previous studies [25].

For quantitative photobiont cultivation, sterilized 25 mm (in diameter) glass fiber disks were subjected to five different pretreatments. Crystals of extracts from lichens *R. celestri* and *S. ramulosum* (0.1 and 0.01 mg/disk) were dissolved in acetone (volume 30 μL) and applied by automatic pipette on the surface of disks while the same volume of acetone was used for control disks. After 4 h of acetone evaporation, individual disks from all treatments were transferred to the surface of solid *Trebouxia* medium in a separate Petri dish, 6 cm in diameter, and 20 μL of algal suspensions were inoculated into the center of each disk. Disk pores allow supplemental nutrient media to pass through the disk and permit growth to be easily determined from changes in biomass. The total mass of cultures was calculated by subtracting the mean fresh weight (fw) of a glass fiber disk saturated by an identical medium from the fw of a disk supporting algal cultures after 14 days of cultivation [25]. Each treatment was replicated twenty times.

3.3. Assimilation Pigments and Chlorophyll a Integrity

Weighed disks with grown photobiont cells were directly extracted in the dark for 1 h at 65 °C in 5 mL of dimethylsulfoxide (DMSO). To maximize chlorophyll extraction, cell aggregates were homogenized using mortar; glass fibers of disks facilitated the disruption of the cell walls of algae. After cooling to ambient temperature, the absorbance of the extracts, as a reflection of turbidity, was determined at 750 nm with a spectrophotometer to be certain that it was always less than 0.01. The absorbance of extracts was then read at 665, 649, 435, and 415 nm to assess chlorophyll content and the possibility of chlorophyll *a* degradation [32]. To determine the content of "total" carotenoids, absorbance was read at 480 nm. Chlorophyll *a*, chlorophyll *b*, chlorophyll *a* + *b*, and total carotenoids were calculated using equations derived from specific absorption coefficients for pure chlorophyll *a* and chlorophyll *b* in DMSO. Chlorophyll *a*/*b* was used to assess the physiological competence of algal cells. The ratios of absorbances at 435 and 415 nm, termed the phaeophytinization quotient, were calculated as a reflection of the ratio of chlorophyll *a* to phaeophytin *a* and provide an indication of the integrity of photobiont chlorophyll *a* [33]. Each treatment was replicated three times.

3.4. Activity of Photosystem II

Chlorophyll *a* fluorescence was measured in algae grown on glass fiber disks on the surface of *Trebouxia* agar media. While still on the surface of media in Petri dishes to minimize desiccation, photobiont cultures were dark-adapted for 30 min before the measurement. The potential quantum yield of photosystem II (PSII) was measured using a FluorCam MF-800 fluorescence-imaging camera (Photon Systems Instruments Ltd., Brno, Czech Republic) by applying a saturating flash of light 2000 μmol s^{-1} m^{-2} for 1 s. The maximum efficiency of the PSII was assessed by the F_V/F_M ratio. $F_V/F_M = (F_M - F_0)/F_M$, where F_0 represents ground fluorescence in the dark-adapted state and F_M represents maximum fluorescence at a saturating radiation pulse in the dark-adapted state. Chlorophyll fluorescence parameters were taken from three separate positions on each disk, and the mean value was used as the overall observation [25]. Each treatment was replicated three times.

3.5. Content of Soluble Proteins

Photobiont cultures grown on Whatman CF/C disks were homogenized directly with the disks in an ice-cold mortar in phosphate buffer (50 mM). After centrifugation at 10,000× *g* at 4 °C for 20 min, the water-soluble protein content of supernatants was measured using the methods of Bradford [34]. Supernatants (100 µL) were pipetted into 900 µL of Bradford assay kit (Bio-Rad, Hercules, CA, USA) in a spectrophotometric cuvette and mixed. After 10 min, the absorbance of samples was spectrophotometrically measured at 595 nm. Bovine serum albumin was used as a calibration standard [33]. Each treatment was replicated three times.

3.6. Oxidative Status

Hydrogen peroxide and superoxide were measured in homogenates with potassium phosphate buffer prepared for assay of determination of soluble proteins. Hydrogen peroxide was quantified by the $TiCl_4$ method, and superoxide was estimated by monitoring the formation of nitrite from hydroxylamine at 530 nm [16]. The reaction mixture contained 0.27 mL of potassium phosphate buffer, 0.03 mL of 10 mM hydroxylamine, 0.3 mL of supernatant, 0.3 mL of 17 mM sulfanilamide, 0.3 mL of 7 mM α-naphthylamine, and 0.3 mL of diethyl ether. Each treatment was replicated three times.

3.7. Membrane Lipid Peroxidation

The membrane lipid peroxidation state in photobiont cultures was estimated using the thiobarbituric acid reactive substances assay (TBARS) as described in [33]. Algae grown on Whatman CF/C disks were homogenized with disks in a mortar using ice-cold 10% (*w*/*v*) trichloroacetic acid (TCA). The homogenate (2 mL final volume) was centrifuged at 10,000× *g* for 10 min. The supernatant (1 mL) was added to 1 mL of 0.6% thiobarbituric acid (TBA) in 10% TCA. After the treatment of samples at 98 °C for 25 min and immediate cooling in an ice bath, the mixture was again centrifuged at 10,000× *g* for 10 min. The absorbance of the supernatant was measured at 532 nm (extinction coefficient for MDA-TBA complex 155 mM^{-1} cm^{-1}) and corrected for non-specific absorption at 600 nm. Three replicates were used for each variant of the experiment.

3.8. Determination of Ascorbic Acid (AsA), Glutathione (GSH, GSSG) and Organic Acids

Reduced (GSH), oxidized glutathione (GSSG), and AsA were extracted from photobiont samples with 0.1 M HCl (0.1 g fw/1 mL) and quantified using LC–MS/MS (Agilent 1200 Series Rapid Resolution LC system coupled on-line to an MS detector Agilent 6460 Triple Quadrupole with Agilent Jet Stream Technologies, Santa Clara, CA, USA) at *m*/*z* values 308/76, 613/231, and 177/95 in positive MRM mode, respectively. Separation was conducted using column Zorbax EC-C18 100 × 4.6 mm, 2.7 µm particle size, and a mobile phase consisting of 0.2% acetic acid and methanol (95:5). The flow rate was 0.6 mL/min, and the column temperature was set at 25 °C. Freshly prepared standards were used for calibration and quantification [35].

Liquid chromatography with tandem mass spectrometry using a triple-quadrupole MS detector was used to analyze organic acids in the algae. A volume of 1 mL of 80% methanol and glass beads 0.5 mm in diameter were added to samples of algal cells. Samples were homogenized at 6800× *g*, centrifuged at 16,000× *g*, and then filtered through Whatman Mini-UniPrep syringeless filters at 0.45 µm before analysis by liquid chromatography with mass detection. The samples were analyzed using an Agilent 1200 Rapid Resolution LC system. The system was connected to an Agilent Technologies 6460 triple-quadrupole MS detector with an Agilent Jet Stream, all from Agilent Technologies, Waldbronn, Germany. This system is described in more detail in previous publications [17,25,35].

3.9. Statistical Analysis

One-way analysis of variance and Tukey's pairwise comparisons (MINITAB Release 17, 2017, Minitab Inc., State College, PA, USA) were used to determine the significance ($p < 0.05$) of differences in all measured parameters.

4. Conclusions

As we noted previously, lichens are self-sustaining ecosystems forming thallus, which is composed of a fungal partner and an extracellular arrangement of minimally one photosynthetic partner, alga, or/and cyanobacteria. There may also be an indeterminate number of other microscopic organisms within the thallus that are of greater or lesser importance to the functioning of the thallus as a "single" unit. Lichen photobionts are in permanent contact with their mycobionts, and lichen symbiosis is a long-term association (some lichens may survive for hundreds of years); we assume that the production of typical secondary metabolites by mycobionts is a key element for maintaining homeostasis in lichen thallus, preventing overgrowth of biomass of photobionts resulting in an imbalance between bionts.

Although the biological and ecological roles of lichen secondary metabolites are diverse, due to a high number of so far discovered chemical compounds and their origin, some of them possess allelopathic effects. The phytotoxicity of lichen compounds to algal partners of symbiosis has only been investigated in a few studies so far. Toxicity symptoms involve growth (biomass production), inhibition of photobionts, alteration of the composition of assimilation pigments, a decrease in chlorophyll *a* fluorescence, and chlorophyll *a* integrity. Phytotoxicity of lichen secondary metabolites is manifested with increased production of ROS and peroxidation of membrane lipids. Regulation of the antioxidant defense system in the *A. erici* photobiont is followed by decreased content of ascorbic acid in the cells and decreased contents of reduced and oxidized glutathione. Contents of selected organic acids involved in the Krebs cycle were altered in photobiont cells exposed to crystals of lichen substances. One of the most prominent markers of phytotoxicity of lichen substances in lichen algae was the increased pyruvic acid/citric acid ratio employed in the present study.

Author Contributions: Conceptualization, M.B.; Methodology, M.B., D.K., B.D., D.U., M.Š. and M.G.; Software, B.D.; Validation, M.B. and B.D.; Formal analysis, M.B., B.D. and D.U.; Investigation, M.B., D.K., B.D., D.U., M.Š. and M.G.; Resources, M.B. and D.U.; Data curation, M.B. and M.G.; Writing—original draft, M.B. and M.Š.; Writing—review & editing, M.B.; Visualization, D.K., B.D. and M.G.; Supervision, M.B.; Project administration, D.K. and D.U.; Funding acquisition, M.B. and M.Š. All authors have read and agreed to the published version of the manuscript.

Funding: This paper is dedicated to the memory of our dear co-worker Prof. Bořivoj Klejdus from Mendel University in Brno, Czech Republic. This work was financially supported by the Slovak Research and Development Agency under contract No. APVV-21-0289, Slovak Grant Agency KEGA under contracts No. 008SPU-4/2023, 024UPJŠ-4/2024, and 009UPJŠ-4/2023, and Slovak Grant Agency VEGA (VEGA 1/0252/24).

Institutional Review Board Statement: Not applicable.

Informed Consent Statement: Not applicable.

Data Availability Statement: The data will be provided upon request.

Conflicts of Interest: The authors reported no potential conflicts of interest.

Abbreviations

AsA—ascorbic acid; ATR—atranorin; dw—dry weight; fw—fresh weight; GSH—reduced glutathione; GSSG—oxidized glutathione; PA—perlotolic acid; RNS—reactive nitrogen species; ROS—reactive oxygen species; TBARS—thiobarbituric acid reactive substances; UA—usnic acid.

References

1. Mark, K.; Laanisto, L.; Bueno, C.G.; Niinemets, U.; Keller, C.; Scheidegger, C. Contrasting co-occurrence patterns of photobiont and cystobasidiomycete yeast associated with common epiphytic lichen species. *New Phytol.* **2020**, *227*, 1362–1375. [CrossRef] [PubMed]
2. Hawksworth, D.L.; Grube, M. Lichens redefined as complex ecosystems. *New Phytol.* **2020**, *227*, 1281–1283. [CrossRef] [PubMed]
3. Bačkor, M.; Loppi, S. Interactions of lichens with heavy metals. *Biol. Plant.* **2009**, *53*, 214–222. [CrossRef]
4. Lücking, R.; Leavitt, S.D.; Hawksworth, D.L. Species in lichen-forming fungi: Balancing between conceptual and practical considerations, and between phenotype and phylogenomics. *Fungal Divers.* **2021**, *109*, 99–154. [CrossRef]
5. Goga, M.; Elečko, J.; Marcinčinová, M.; Ručová, D.; Bačkorová, M.; Bačkor, M. Lichen metabolites: An overview of some secondary metabolites and their biological potential. In *Co-Evolution of Secondary Metabolites*; Mérillon, J.M., Ramawat, K., Eds.; Springer: Cham, Switzerland, 2020; pp. 175–209.
6. Xu, M.; Heiðmarsson, S.; Olafsdottir, E.S.; Buonfglio, R.; Kogej, T.; Omarsdottir, S. Secondary metabolites from cetrarioid lichens: Chemotaxonomy, biological activities and pharmaceutical potential. *Phytomedicine* **2016**, *23*, 441–459. [CrossRef]
7. Elshobary, M.E.; Becker, M.G.; Kalichuk, J.L.; Chan, A.C.; Belmonte, M.F.; Piercey-Normore, M.D. Tissue-specific localization of polyketide synthase and other associated genes in the lichen, *Cladonia rangiferina*, using laser microdissection. *Phytochemistry* **2018**, *156*, 142–150. [CrossRef]
8. Brodo, I.M.; Sharnoff, S.D.; Sharnoff, S. *Lichens of North America*; Yale University Press: New Haven, CT, USA, 2001.
9. Smeds, A.I.; Kytöviita, M.M. Determination of usnic and perlatolic acids and identification of olivetoric acids in Northern reindeer lichen (*Cladonia stellaris*) extracts. *Lichenologist* **2010**, *42*, 739–749. [CrossRef]
10. Cocchietto, M.; Skert, N.; Nimis, P.L.; Sava, G. A review on usnic acid, an interesting natural compound. *Naturwissenschaften* **2002**, *89*, 137–146. [CrossRef]
11. Beckett, R.P.; Minibayeva, F.; Solhaug, K.A.; Roach, T. Photoprotection in lichens: Adaptations of photobionts to high light. *Lichenologist* **2021**, *53*, 21–33. [CrossRef]
12. Solhaug, K.E.; Eiterjord, G.; Løken, M.H.; Gauslaa, Y. Non-photochemical quenching may contribute to the dominance of the pale mat-forming lichen *Cladonia stellaris* over the sympatric melanic *Cetraria islandica*. *Oecologia* **2024**, *204*, 187–198. [CrossRef]
13. Bačkorová, M.; Bačkor, M.; Mikeš, J.; Jendželovský, R.; Fedoročko, P. Variable responses of different human cancer cells to the lichen compounds parietin, atranorin, usnic acid and gyrophoric acid. *Toxicol. Vitro* **2011**, *25*, 37–44. [CrossRef]
14. Pizňak, M.; Kolarčik, V.; Goga, M.; Bačkor, M. Allelopathic effects of lichen metabolite usnic acid on growth and physiological responses of Norway spruce and Scots pine seedlings. *S. Afr. J. Bot.* **2019**, *124*, 14–19. [CrossRef]
15. Goga, M.; Antreich, S.; Bačkor, M.; Weckwerth, W.; Lang, I. Lichen secondary metabolites affect growth of *Physcomitrella patens* by allelopathy. *Protoplasma* **2017**, *254*, 1307–1315. [CrossRef] [PubMed]
16. Bačkor, M.; Klemová, K.; Bačkorová, M.; Ivanova, V. Comparison of the phytotoxic effects of usnic acid on cultures of free-living alga *Scenedesmus quadricauda* and aposymbiotically grown lichen photobiont *Trebouxia erici*. *J. Chem. Ecol.* **2010**, *36*, 405–411. [CrossRef] [PubMed]
17. Kolackova, M.; Chaloupsky, P.; Cernei, N.; Klejdus, B.; Huska, D.; Adam, V. Lycorine and UV-C stimulate phenolic secondary metabolites production and miRNA expression in *Chlamydomonas reinhardtii*. *J. Hazard. Mater.* **2020**, *391*, 12208839. [CrossRef]
18. Al-Bekairi, A.M.; Qureshi, S.; Chaudhry, M.A.; Krishna, D.R.; Shah, A.H. Mitodepressive, clastogenic and biochemical effects of (+)-usnic acid in mice. *J. Ethnopharmacol.* **1991**, *33*, 217–220. [CrossRef]
19. Cardarelli, M.; Serino, G.; Campanella, L.; Ercole, P.; De Cicco Nardone, F.; Alesiani, O.; Rosiello, F. Antimitotic effects of usnic acid on different biological systems. *Cell. Mol. Life Sci.* **1997**, *53*, 667–672. [CrossRef]
20. Endo, T.; Takahagi, T.; Kinoshita, Y.; Yamamoto, Y.; Sato, F. Inhibition of photosystem II of spinach by lichen-derived depsides. *Biosci. Biotechnol. Biochem.* **1998**, *62*, 2023–2027. [CrossRef]
21. Takahagi, T.; Endo, T.; Yamamoto, Y.; Sato, F. Lichen photobionts show tolerance against lichen acids produced by lichen mycobionts. *Biosci. Biotech. Bioch.* **2008**, *72*, 3122–3127. [CrossRef]
22. Váczi, P.; Gauslaa, Y.; Solhaug, K.A. Efficient fungal UV-screening provides a remarkably high UV-B tolerance of photosystem II in lichen photobionts. *Plant Physiol. Biochem.* **2018**, *132*, 89–94. [CrossRef]
23. Barták, M.; Hájek, J.; Halici, M.G.; Bednaříková, M.; Casanova-Katny, A.; Váczi, P.; Puhovkin, A.; Mishra, K.B.; Giordano, D. Resistance of primary photosynthesis to photoinhibition in Antarctic lichen *Xanthoria elegans*: Photoprotective mechanisms activated during a short period of high light stress. *Plants* **2023**, *12*, 2259. [CrossRef]
24. Rapsch, S.; Ascaso, C. Effect of evernic acid on structure of spinach chloroplasts. *Ann. Bot-London* **1985**, *56*, 467–473. [CrossRef]
25. Bačkor, M.; Goga, M.; Ručová, D.; Urminská, D.; Bačkorová, M.; Klejdus, B. Allelopathic effects of three lichen secondary metabolites on cultures of aposymbiotically grown lichen photobionts and free-living alga *Scenedesmus quadricauda*. *S. Afr. J. Bot.* **2023**, *162*, 688–693. [CrossRef]
26. Han, D.; Matsumaru, K.; Rettori, D.; Kaplowitz, N. Usnic acid-induced necrosis of cultured mouse hepatocytes: Inhibition of mitochondrial function and oxidative stress. *Biochem. Pharmacol.* **2004**, *67*, 439–451. [CrossRef]
27. Caviglia, A.M.; Nicora, P.; Giordani, P.; Brunialti, G.; Modenesi, P. Oxidative stress and usnic acid content in *Parmelia caperata* and *Parmelia soredians* (Lichenes). *Farmaco* **2001**, *56*, 379–382. [CrossRef] [PubMed]
28. Bačkorová, M.; Jendželovský, R.; Kello, M.; Bačkor, M.; Mikeš, J.; Fedoročko, P. Lichen secondary metabolites are responsible for induction of apoptosis in HT-29 and A2780 human cancer cell lines. *Toxicol. Vitro* **2012**, *26*, 462–468. [CrossRef] [PubMed]

29. Sun, X.M.; Lu, Z.Y.; Liu, B.Y.; Zhou, Q.H.; Zhang, Y.Y.; Wu, Z.B. Allelopathic effects of pyrogallic acid secreted by submerged macrophytes on *Microcystis aeruginosa*: Role of ROS generation. *Allelopath. J.* **2014**, *33*, 121–130.
30. Li, H.; Zhou, X.; Huang, Y.; Liao, B.; Cheng, L.; Ren, B. Reactive oxygen species in pathogen clearance: The killing mechanisms, the adaption response, and the side effects. *Front. Microbiol.* **2020**, *11*, 622534.
31. Elečko, J.; Vilková, M.; Frenák, R.; Routray, D.; Ručová, D.; Bačkor, M.; Goga, M. A Comparative Study of Isolated Secondary Metabolites from Lichens and Their Antioxidative Properties. *Plants* **2022**, *11*, 1077. [CrossRef]
32. Wellburn, A.R. The spectral determination of chlorophylls a and b, as well as total carotenoids, using various solvents with spectrophotometers of different resolutions. *J. Plant Physiol.* **1994**, *144*, 307–313. [CrossRef]
33. Lokajová, V.; Bačkorová, M.; Bačkor, M. Allelopathic effects of lichen secondary metabolites and their naturally occurring mixtures on cultures of aposymbiotically grown lichen photobiont *Trebouxia erici* (Chlorophyta). *S. Afr. J. Bot.* **2014**, *93*, 86–91. [CrossRef]
34. Bradford, M.M. Rapid and sensitive method for quantitation of microgram quantities of protein utilizing principle of protein-dye binding. *Anal. Biochem.* **1976**, *72*, 248–254. [CrossRef] [PubMed]
35. Bačová, R.; Klejdus, B.; Ryant, P.; Cernei, N.; Adam, V.; Húska, D. The effects of 5-azacytidine and cadmium on global 5-methylcytosine content and secondary metabolites in the freshwater microalgae *Chlamydomonas reinhardtii* and *Scenedesmus quadricauda*. *J. Phycol.* **2019**, *55*, 329–342. [CrossRef] [PubMed]

molecules

MDPI

Article

Screening of Antioxidant Maillard Reaction Products Using HPLC-HRMS and Study of Reaction Conditions for Their Production as Food Preservatives

Sara Bolchini [1], Roberto Larcher [2,*], Ksenia Morozova [1], Matteo Scampicchio [1] and Tiziana Nardin [2]

[1] Faculty of Agricultural, Environmental and Food Science, Free University of Bolzano, 39100 Bolzano, Italy; sbolchini@unibz.it (S.B.); ksenia.morozova@unibz.it (K.M.); matteo.scampicchio@unibz.it (M.S.)

[2] Centro di Trasferimento Tecnologico, Fondazione Edmund Mach, 38010 San Michele all'Adige, Italy; tiziana.nardin@fmach.it

* Correspondence: roberto.larcher@fmach.it

Abstract: The Maillard reaction (MR) involves interactions between reducing sugars and amino acids or proteins during heating, producing Maillard reaction products (MRPs) that influence food flavour, aroma, and colour. Some MRPs exhibit antioxidant properties, prompting interest in their potential as natural food preservatives. This study aimed to develop a method for detecting and identifying antioxidant MRPs using high-pressure liquid chromatography (HPLC) coupled with high-resolution mass spectrometry (HRMS). By improving chromatographic conditions, the separation of antioxidant MRPs was optimised using known antioxidant MRPs as reference signals. This work also examined the effects of pH, reaction time, and different sugar–amino acid combinations on the production and composition of antioxidant MRPs. Results indicated that neutral to basic pH facilitated faster reactions, with pH 7 selected as optimal. A library of 50 *m/z* signals for potential antioxidant MRPs was created, and the best combinations of amino acids and sugars for their production were identified. These findings pave the way for more precise analyses of antioxidant MRPs, with future research focusing on isolating and characterising specific MRPs to understand their structures and mechanisms, ultimately contributing to the development of functional foods with natural antioxidant properties.

Keywords: Maillard reaction; antioxidants; food preservatives; high-resolution mass spectrometry

1. Introduction

The Maillard reaction (MR) is a complex series of chemical reactions that occur between reducing sugars and amino acids or proteins during heating. This reaction leads to the formation of Maillard reaction products (MRPs) that are responsible for flavours, aromas, and colour formation in food [1–3]. Furthermore, some MRPs have been reported to possess antioxidant activity, offering potential health and food maintenance benefits [4–6]. Indeed, MR can be considered a potential way to obtain natural antioxidant compounds from widely common molecules, especially reducing sugars and amino acids, present in food and food waste. In the food chain, antioxidant compounds represent an important class of food preservatives because they prevent oxidative deterioration of food, thereby extending its shelf life and maintaining quality.

Moreover, natural food preservatives have gained attention from the scientific community due to increasing concerns about the health impacts of synthetic additives, the rise of the clean label trend, and the growing demand for sustainable food practices. As consumers seek safer and more wholesome options, there is a need for alternative preservation methods derived from natural sources [7].

Although the antioxidant activity developed with MR in both model systems and real food has been widely studied using mainly antioxidant assays like the spectrophotometric assay based on 2,2-diphenyl-1-picrylhydrazyl (DPPH) change of colour, ferric reducing power assay (FRAP), or oxygen radical absorbance capacity assay (ORAC) [3,5,8], limited information is available regarding their structure and composition. Moreover, although the influence of important variables like pH, type of reducing sugar, amino compounds, or time of reaction on MR has been extensively studied in previous literature [9], their influence on the production and composition of antioxidant MRPs is not well understood.

Accordingly, the general aim of this work was to define the optimal MR conditions to obtain, on an industrial scale, antioxidant MRPs from solutions of amino acids and sugars, with the objective of producing natural food antioxidants. Initially, an analytical method was developed to enable the detection and, when possible, identification of potential antioxidant MRPs. This was feasible using High-Pressure Liquid Chromatography (HPLC) coupled with High-Resolution Mass Spectrometry (HRMS). After proper optimisation of the chromatographic method, it allowed the separation and detection of a large selection of molecules produced during MR. To develop and optimise the analytical method, HRMS signals corresponding to 10 known antioxidant MRPs (KAMs) were used as references [10,11].

Additionally, the influence of pH, time of reaction, and different sugars and amino acids on the production, composition, and structure of KAMs was studied using the previously set signals as an index to evaluate which conditions produced the maximum yield of KAMs.

Since MR is very complex and KAMs may not represent all the antioxidants produced during the reaction, it became essential to increase the number of signals studied to evaluate the influence of variables (pH, time, and reagent composition) on the production of additional potential antioxidant MRPs (PAMs). Therefore, through untargeted HPLC-HRMS analyses of a mix of sugars and amino acids at pH 7, a pool of ions corresponding to PAMs was selected and monitored during the reaction with a radical initiator, which caused a decrease in the corresponding signals. Once the signals corresponding to PAMs were selected, their production was monitored in MR samples with different pH, time of reaction, and combinations of common sugars and amino acids in food, allowing for a more comprehensive evaluation of each variable's influence on the production of PAMs.

2. Results and Discussion

2.1. Optimisation of Chromatographic Method

The optimisation of the chromatographic method was done using as a reference the signals corresponding to KAMs reported in Table 1. Some compounds were tentatively identified based on *m/z* values and fragmentation profiles (MS^2) compared with the literature, while for others (indicated in Table 1 with *), the identification was confirmed with standard reagents. For 1-methyl-2-pyrrole-carboxaldehyde and 2-acetyl-1-methylpyrrole, three different peaks were detected. In both cases, all had the same *m/z* and fragmentation patterns, suggesting the isomeric nature of the molecules responsible for the peaks. Since no standard compounds were available, all three peaks were considered in the following integration of areas. This choice was supported by the fact that in the presence of a radical initiator, all three peaks decreased, indicating that the molecules represented by the peaks are potential antioxidants and corroborating the hypothesis that these molecules are isomers.

Table 1. Antioxidant compounds produced by MR and reported in the literature, with exact mass, theoretical and measured *m/z*, ionisation, and literature reference. * Compounds verified with analytical standards. DDMP: dihydro-dihydroxymethylpyrone.

Compound	Theoretical *m/z*	Measured *m/z*	Δppm	Retention Time	Ionisation	Reference
Maltol *	127.03897	127.03859	0.00038	13.27	[M + H]+	[10]
Maltol isomer	127.03897	127.03905	0.00008	10.98	[M + H]+	[10]
2-acetylpyrrole *	110.06004	110.06005	0.00001	18.79	[M + H]+	[10]
Sotolon *	129.05462	129.04928	0.00534	14.56	[M + H]+	[10]
Norfuraneol *	115.03897	115.03938	0.00041	9.25	[M + H]+	[10]
Furaneol	129.05462	129.06004	0.00542	12.62	[M + H]+	[10]
DDMP	145.04954	145.05029	0.00075	10.50	[M + H]+	[10]
2-pyrrolecarboxaldehyde	96.04439	96.04450	0.00011	4.00	[M + H]+	[11]
1-methyl-2-pyrrole carboxaldehyde	110.06004	110.06099	0.00095	6.00	[M + H]+	[11]
2-acetyl-1-methylpyrrole	124.07569	124.07662	0.00093	9.30	[M + H]+	[11]

In Figure 1, the extracted ion chromatograms (EIC) referring to the *m/z* of the best separation obtained for the KAMs are reported. The same figures obtained with different columns are available in the supporting information (Figure S1).

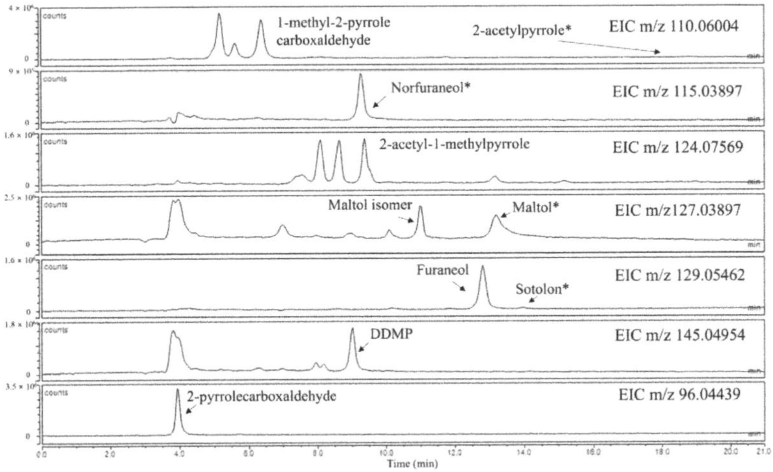

Figure 1. EICs corresponding to the *m/z* of the KAMs considered. (*) indicates the confirmed identifications with standard compounds.

The final conditions selected after the optimisation of the method and applied in all the subsequent experiments are reported here: the Dionex IonPac NS2 column (4 × 150 mm, 5 μm particle size Thermo Fisher Scientific, USA) was used for the analysis. The column temperature was kept constant at 30 °C. The mobile phases were Milli-Q water with 0.5% formic acid (*v/v*) (A) and acetonitrile with 0.5% formic acid (*v/v*) (B). Before injection, the samples were diluted 1:10 with Milli-Q water, and the injection volume was set to 2 μL. The analysis was conducted at a constant flow rate of 0.35 mL·min^{-1} using the following gradient for optimal chromatographic separation: from 0 to 1 min, 5% of eluent B; from 1 to 10 min, 15% eluent B; from 10 to 15 min, 35% eluent B; from 15 to 21 min, 5% eluent B.

2.2. Influence of Initial pH on Antioxidant Production in MR

The influence of pH at different reaction times was evaluated. In literature it is reported that basic pH favours the MR, while at acidic pH the reaction is slowed down [12]. The production of 10 KAMs and 5-hydroxymethylfurfural (HMF) was monitored every 20 min in amino acid and reducing sugar solutions at pH 6, 7, and 8 using HRMS. HMF was evaluated because it is a known MRP considered to be a health hazard [13], and its production is considered negative. The samples were compared by analysing the areas of the integrated peaks corresponding to the analytes of interest. As reported in Figure 2, at pH 6, the lowest production of antioxidant compounds but the highest production of HMF was detected. This result agrees with previous literature where HMF production is reported to be induced by acidic pH [14]. The low development of antioxidant compounds at pH 6 increased at pH 7 and 8. In particular, most of the analysed antioxidant compounds show similar kinetics of production in the first part of the reaction (40 min) at pH 7 and 8. This result suggests that their mechanism of formation is favoured by neutral and basic pH. Although, some of them, like DDMP (dihydro-dihydroxymethylpyrone), norfuraneol, and maltol isomer, seem to decrease in concentration after the first part of the reaction, particularly at basic pH. This could indicate that these compounds are MR intermediates and further react and degrade, especially at basic pH, where the reaction is favoured. The development of some other KAMs, like sotolon and maltol, seems to be more induced by neutral pH than basic.

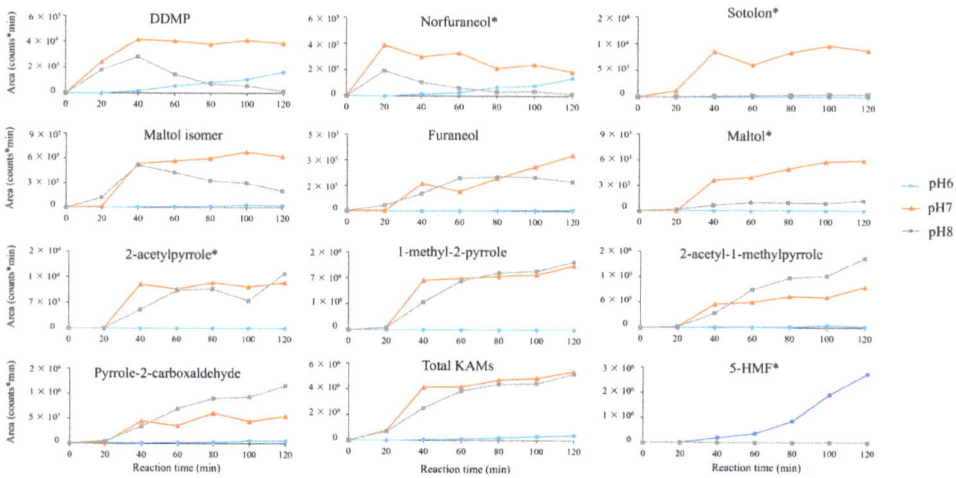

Figure 2. Kinetics of production of 10 KAMs at pH 6, 7, and 8. The sum of the signals of the known antioxidant MRPs (KAMs) and the kinetics of production of 5-hydroxymethylfurfural (HMF) are also reported. Measurements were performed in duplicate with a relative standard deviation (RSD) lower than 10%. (*) indicates the confirmed identifications with standard compounds.

2.3. *Influence of Amino Acids and Sugars on Antioxidant Production in MR*

The influence of amino acids and sugars on the production of KAMs was evaluated keeping the pH constant at 7 and testing all the 20 amino acids and 6 sugars in binary combinations. All the samples obtained were analysed using HPLC-HRMS, and the peaks corresponding to the *m/z* signals of the KAMs and HMF were integrated to obtain areas for comparison.

Figure 3 displays the normalised values for each antioxidant compound, where the highest recorded area for each compound is set to 100, and the areas for all other samples are normalised relative to this maximum. The values are color-coded based on their numerical value. The experiments were performed in duplicate, and the relative standard deviation (RSD) measured was lower than 20%.

Here are some observations on the analysis of this heatmap. Firstly, it is evident that HMF is mainly produced in the presence of acidic reagents such as glutamic and aspartic acid. This result agrees with previous findings [12]. Secondly, the combination of amino acids and sugars that allowed the highest total production of antioxidants was threonine combined with disaccharides (maltose and lactose). In addition, maltol, as well as the maltol isomer, are mainly produced with disaccharides, while their production is low (maltol isomer) or absent (maltol) with monosaccharides. This result confirms the findings available in the published literature [15].

Moreover, DDMP was produced with all combinations of amino acids and sugars, except for arabinose, the only pentose sugar tested. This suggests that DDMP is an antioxidant MRP produced only with hexose monosaccharides and disaccharides. On the contrary, norfuraneol seems to be mainly produced in the presence of arabinose. Furaneol originates in the presence of tryptophan, while the production of 2-acetyl-1-methylpyrrole is high in MR with threonine and asparagine but low with cysteine, aspartic acid, proline, and tyrosine. 2-pyrrole carboxaldehyde and 1-methyl-2-pyrrole are produced in all combinations, but not with proline (both) and with tyrosine and valine (1-methyl-2-pyrrole). Sotolon shows a peak of production with lysine as the amino acid, even if it is produced with almost all other amino acids. Finally, 2-acetylpyrrole is highly produced in the reaction with methionine but is almost not present with all other amino acids.

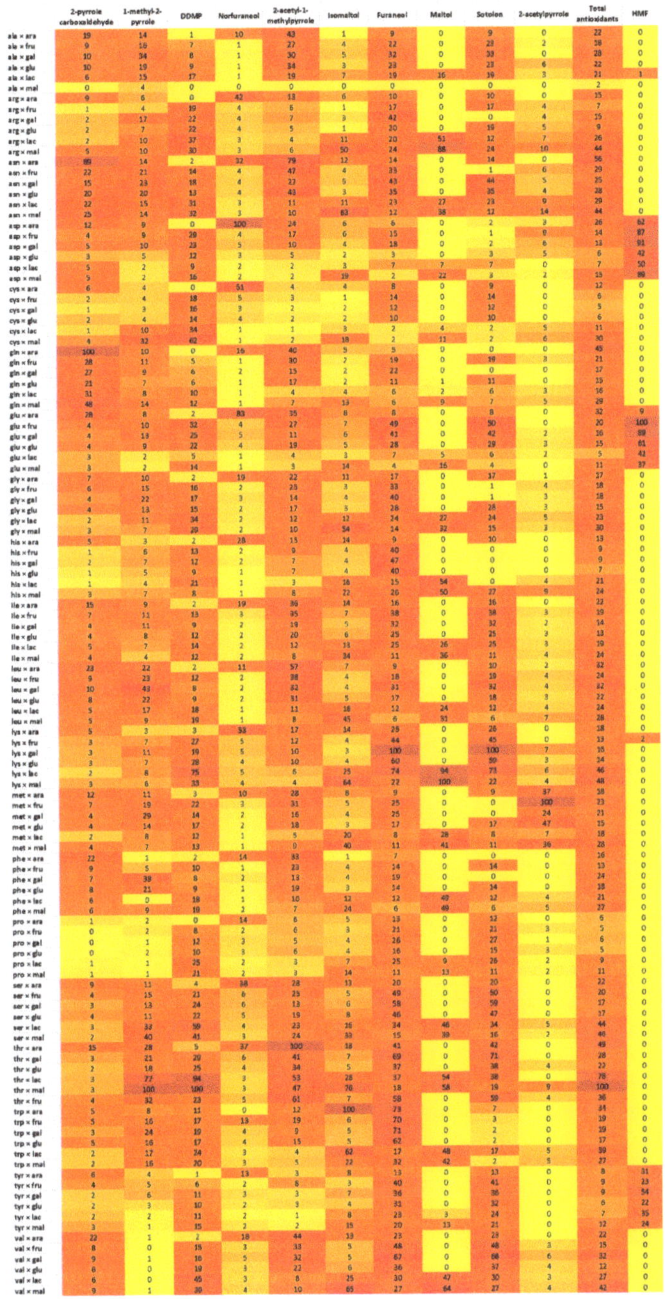

Figure 3. Heatmap representing normalised values of peak areas obtained from HRMS analyses corresponding to *m/z* signals of KAMs for all binary combinations of amino acids and sugars studied. Colours range from red (high value) to yellow (low value). Values are normalised based on the highest for each KAM. Experiments were performed in duplicate; RSD lower than 20%.

In Figure 4, the Principal Component Analyses (PCA) of the dataset obtained from the analysis of 120 binary combinations of amino acids and sugars, composed of 11 variables (areas of HRMS peaks corresponding to KAMs and HMF) and 120 observations, is shown. The main differentiation that was obtained regarding the sugar composition, where on the left, samples with monosaccharides (arabinose, fructose, galactose, glucose) are separated from samples containing disaccharides (maltose and lactose), on the right. Moreover, it is possible to understand the variables responsible for this separation, in particular it is evident that maltol, maltol isomer, DDMP, and 2-acetylpyrrole are mainly present in MR samples with disaccharides as reagents, while all the other molecules are characteristic of MR between amino acids andmonosaccharides.

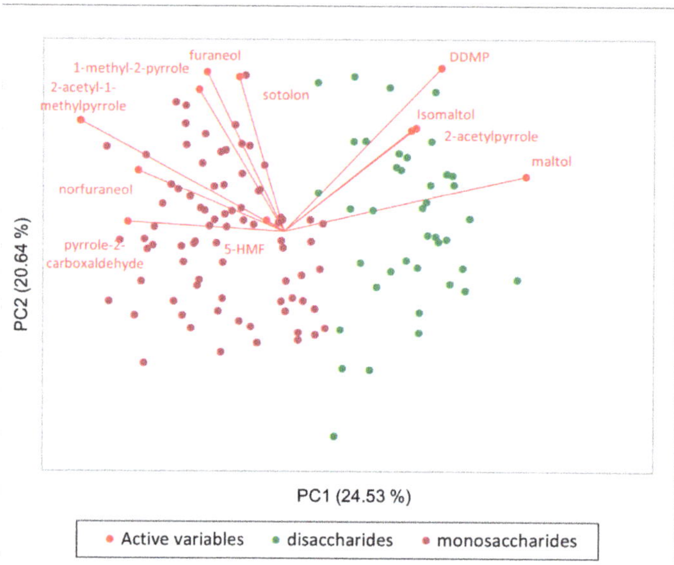

Figure 4. Principal Component Analyses (PCA) of 120 samples with 10 HRMS peak areas corresponding to KAMs and HMF.

2.4. *Consumption of Reagents in MR*

Sugars and amino acids consumption in a mix sample containing all 20 amino acids and sugars was evaluated to have a more complete understanding of which are the most reactive compounds, and which are less. This information is fundamental to develop further applications of MR as a source of antioxidant compounds.

In Figure 5, the percentage of consumption of amino acids and reducing sugars before 0 min, after 60 min, and 120 min of MR is reported. As evidenced, the most consumed, and therefore reactive amino acids are arginine, asparagine, lysine, and glutamine. This result was expected since it is known that the amino groups of amino acids are the reactive moieties in MR [16], and the most consumed amino acids present amino groups in the side chain. The most consumed sugars, on the other hand, are arabinose, maltose, and lactose. Interesting result is that the disaccharides (lactose and maltose) were consumed more than monosaccharides like glucose and galactose. This aspect can be explained by the thermal degradation of disaccharides at 140 °C, which produces glucose and galactose as products (monosaccharides present in maltose and lactose), thereby increasing their concentration even as they are consumed by MR.

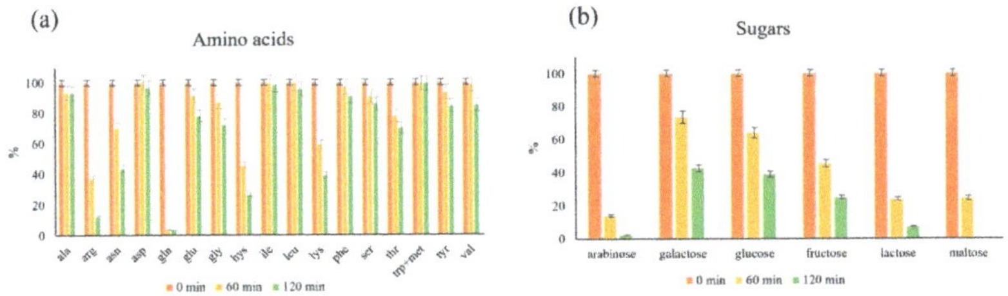

Figure 5. Percentage of consumption of amino acids (**a**) and reducing sugars (**b**) at pH 7 before the MR (0 min) and after 60 and 120 min of MR. Measurements were performed in duplicates, and RSD was lower than 5%.

2.5. Untargeted Approach to Create a Library of Potential Antioxidant MRPs

To have a more exhaustive idea of the real potential antioxidant molecules produced through MR, an untargeted approach using HRMS was used. The MR sample obtained from a mix of 20 amino acids and sugars was incubated with a radical initiator (AAPH) to detect, using HRMS, the *m/z* that were produced by MR that also decreased after incubation with the radical initiator. In fact, if a signal decreases in the presence of the radical initiator, it means that the corresponding molecule was oxidised by the radical initiator, suggesting its potential antioxidant properties. Compound Discoverer software was used to analyse the data corresponding to the amino acid and sugar solution before the heating process, after the heating process (and MR), and after 1 and 2 h of incubation with the radical initiator. Figure 6 shows the trend of peak areas during the experiment corresponding to *m/z* of interest. Peaks were selected when their signals were low before MR, increased after MR, and then decreased after incubation with the radical initiator.

During the creation of the library, the 10 KAMs also resulted showing the same boxplot reported in Figure 6, as expected, and in Table 2 the KAMs are reported, ordered based on the percentage of degradation in the presence of the radical initiator: the higher the percentage of degradation, the quicker the oxidation reaction, and the stronger the antioxidant. Moreover, the best combination of amino acids and sugars to obtain each specific compound is reported, highlighting the possible condition to be followed to produce specific antioxidant MRPs.

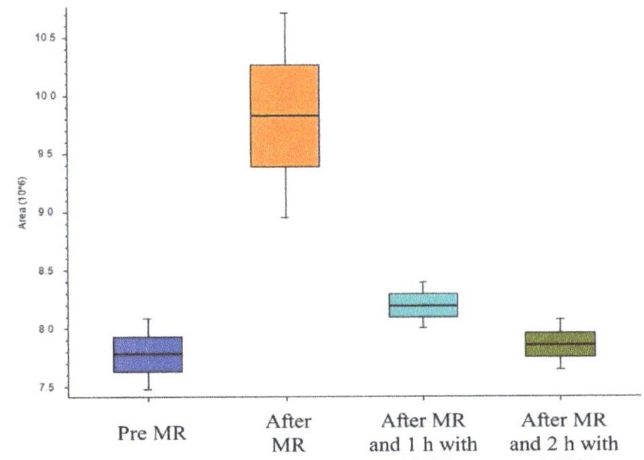

Figure 6. Boxplot used to select the *m/z* signals to create the library of potential antioxidants.

Table 2. KAMs, ordered based on the percentage of degradation after 1 h of incubation with the radical initiator. *m/z*, retention time, and the best amino acid × sugar combination to obtain each specific compound are also reported.

Compound	% of Degradation after 1 h with Radical Initiator	Theoretical *m/z*	Retention Time (min)	Best AA × Sugar
2-pyrrolecarboxaldehyde	88%	96.0444	4	Glutamine × arabinose
Norfuraneol	70%	115.039	9.25	Aspartic acid × arabinose
Furaneol	69%	129.055	12.62	Lysine × galactose
1-methyl-2-pyrrole carboxaldehyde	60%	110.06	6	Threonine × maltose
DDMP	42%	145.05	10.5	Threonine × maltose
2-acetyl-1-methylpyrrole	38%	124.076	9.3	Threonine × arabinose
Maltol	35%	127.039	13.27	Lysine × maltose
Maltol isomer	31%	127.039	10.98	Tryptophan × arabinose
2-acetylpyrrole	29%	110.06	18.79	Methionine × fructose
Sotolon	13%	129.055	14.56	Lysine × galactose

Using this approach, the first library obtained, based only on *m/z* signals and not MS2 fragmentation patterns, included 170 *m/z* signals of potential antioxidants that underwent further selection based on the MS2 fragmentation patterns, obtaining 77 interesting signals. At last, the intensity of signals (areas that were lower than 10^6 counts * min after MR were excluded) was used to create the final library, which includes 50 *m/z* signals and is available in the supporting information (Table S1).

Study of Variable Influence on Antioxidant MRPs Production Using the Library Created

The HRMS spectra obtained from previous experiments (Sections 2.2 and 2.3) were re-processed with Compound Discoverer software specifically looking for the signals included in the library created.

In Figure 7, the total areas of the HRMS peaks corresponding to the 50 *m/z* signals present in the library are reported for each sample at different pHs. It is interesting to note that at pH 7, the highest amount of potential antioxidants is produced, as previously evidenced with the KAMs, but an increase in signals is also observed at pH 6. This result may mean that, even if at pH 6 the KAMs do not develop, probably some other potential antioxidants are produced.

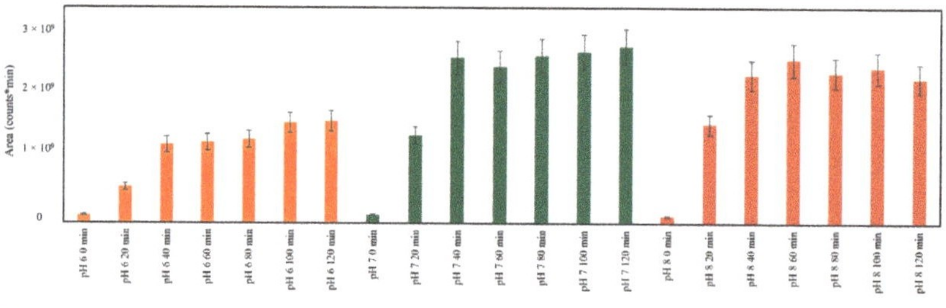

Figure 7. Sum of areas of HRMS peaks corresponding to the 50 *m/z* signals present in the library for each combination of pH and time of MR.

The spectra corresponding to the binary combinations of amino acids and sugars were also reprocessed, leading to Figure 8a, where S values are shown, divided into ranges (range 1: from 0 to 25 percentile; range 2: from 25 to 50 percentile; range 3: from 50 to 75 percentile; range 4: from 75 to 100 percentile). The table containing all the S values calculated for each sample is available in the Supplementary Material (Table S2). They were obtained by summing the areas of HRMS peaks corresponding to the 50 *m/z* signals present in the library for each combination of amino acids and sugars, each corrected as described by Equation (1). This approach allows the analysis of the combinations of amino acids and sugars, not only considering which pair produces the highest amount of antioxidant molecules but also taking into consideration the tendency of degradation of these PAMs in the presence of a radical initiator. The formula used to process the data is reported below:

$$S = \sum \left(\frac{An}{\%Dn} \right) \quad (1)$$

where S is the final value corresponding to each sample plotted in the graph, An is the area of each peak recorded with HRMS, and $\%Dn$ is the percentage of degradation of the peak, calculated as the difference of peak areas * 100 after 1 h of incubation with a radical initiator.

(a)

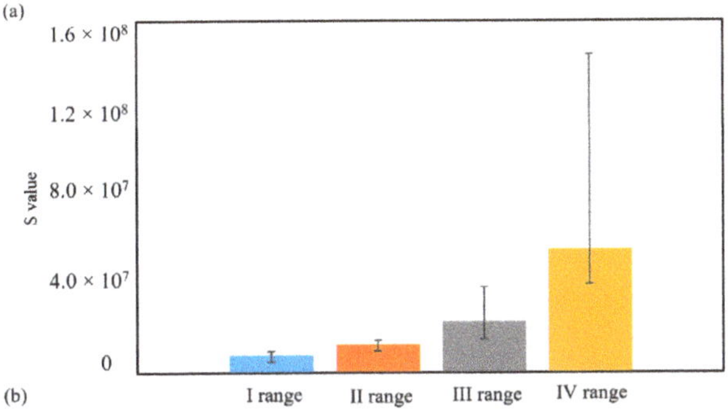

(b)

Range I 0–25 percentile		Range II 25–50 percentile		Range III 50–75 percentile		Range IV 75–100 percentile	
tyr × arab	pro × lact	gln × glu	gln × arab	gln × gal	val × fruct	trp × lact	his × glu
asp × glu	lys × fruct	lys × malt	ala × arab	asn × gal	isoleu × arab	trp × malt	arg × gal
pro × arab	glut × malt	asp × gal	asn × fruct	isoleu × fruct	ser × lact	thr × arab	his × fruc
tyr × glu	lys × glu	glut × arab	ala × lact	leu × malt	leu × arab	trp × glu	trp × arab
lys × arab	asp × arab	tyr × gal	ala × glu	leu × fruct	isoleu × gal	val × gal	arg × arab
tyr × fruct	asp × malt	met × malt	asn × arab	leu × glu	leu × lact	his × malt	phe × malt
cys × fruct	met × lact	glut × gal	isoleu × glu	ser × arab	thr × glu	trp × gal	phe × fruct
pro × glu	glut × lact	gln × lact	isoleu × malt	ala × gal	val × arab	gly × glu	gly × arab
asp × fruct	cys × gal	ala × malt	met × gal	ser × fruct	thr × gal	his × gal	phe × arab
cys × lact	glut × glu	cys × malt	asn × glu	val × malt	leu × gal	gly × gal	phe × glu
tyr × malt	lys × gal	ala × fruct	gln × malt	ser × gal	arg × lact	trp × fruct	arg × glu
pro × fruct	pro × gal	lys × lact	ser × glu	ser × malt	thr × malt	his × arab	phe × lact
pro × malt	met × arab	asn × malt		val × glu	thr × lact	arg × malt	phe × gal
cys × glu	met × glu	asn × lact		isoleu × lact	thrl × fruct	gly × fruct	arg × fruct
tyr × lact	aspx lact	met × fruct		val × lact		gly × lact	
cys × arab	glut × fruct	gln × fruct		his × lact		gly × malt	

Figure 8. (**a**) S values shown in ranges (range 1: from 0 to 25 percentile; range 2: from 25 to 50 percentile; range 3: from 50 to 75 percentile; range 4: from 75 to 100 percentile) obtained by summing the areas of HRMS peaks corresponding to the 50 *m/z* signals present in the library for each combination of amino acids and sugars, each corrected as described by equation (1). The error bars represent the maximum and minimum values belonging to that range. (**b**) The combinations of amino acids and sugars present in the samples, belonging to ranges I, II, III, and IV.

The PAMs present in samples belonging to range I are the ones degrading faster with a radical initiator, compared to the others. PAMs belonging to range IV, instead, are the ones degrading slower, and are, therefore, more stable in presence of a radical initiator.

As evidenced in Figure 8b, some interesting aspects are revealed. The first one regards the presence of PAMs derived from proline in range I. This means that most of the PAMs produced by proline are very reactive in scavenging radicals and, therefore, quickly degraded. This is interesting because, if compared to the sum of the intensity of the signals produced, not weighted for their % of degradation (available in Figure S2), proline samples are in the average of production, but not the highest producer of PAMs. On the contrary, tryptophane and arginine PAMs belong mostly to range IV, indicating that they are not so reactive with the radical initiator.

Unfortunately, no other patterns regarding the presence of amino acids or sugars can be drawn based on the production and reactivity of PAMs.

In Table 3, the PAMs included in the library are reported, ordered based on the % of their degradation after 1 h in presence of the radical initiator AAPH. Moreover, the best combination of amino acid and sugar to obtain each compound, based on the data previously presented, is reported. These results give an idea of the optimal conditions to be maintained if the production of one of the specific PAMs is the aim.

Table 3. PAMs, ordered based on the percentage of degradation after 1 h of incubation with radical initiator. *m/z*, retention time, and the best amino acid × sugar combination to obtain each specific compound are also reported.

Name	% Degradation after 1 h with AAPH	*m/z*	RT [min]	Best AA × Sugar
Unknown 1	96.2	230.1140	8.41	Arginine × fructose
Unknown 2	93.0	199.1077	10.36	Asparagine × fructose
Unknown 3	92.6	154.0497	4.39	Glycine × arabinose
Unknown 4	91.1	124.0757	10.12	Asparagine × glucose
Unknown 5	91.0	164.0818	5.07	Arginine × arabinose
Unknown 6	86.8	170.0811	4.59	Leucine × galactose
Unknown 7	86.0	135.0552	9.21	Histidine × maltose
Unknown 8	83.3	230.1140	6.76	Arginine × fructose
Unknown 9	82.5	212.1028	3.10	Histidine × glucose
Unknown 10	79.7	230.1140	4.13	Arginine × fructose
Unknown 11	79.6	127.0390	6.97	Tryptophan × arabinose
Unknown 12	79.2	151.123	6.65	Leucine × lactose
Unknown 13	77.1	124.0757	8.39	Threonine × fructose
Unknown 14	74.8	140.0706	4.61	Threonine × arabinose
Unknown 15	74.5	119.0350	4.73	Valine × galactose
Unknown 16	73.8	127.0389	3.93	Aspartic acid × lactose
Unknown 17	72.1	154.0498	4.72	Glycine × arabinose
Unknown 18	69.1	212.1034	8.03	Arginine × fructose
Unknown 19	65.0	184.1085	4.72	Arginine × fructose
Unknown 20	64.5	123.0914	3.48	Glycine × fructose
Unknown 21	63.7	135.0553	8.77	Histidine × maltose
Unknown 22	59.7	110.0600	5.15	threonine × maltose
Unknown 23	58.1	166.0861	10.08	Phenylalanine × lactose
Unknown 24	56.2	124.0756	10.56	Asparagine × arabinose
Unknown 25	53.8	127.0389	10.12	Tryptophan × maltose
Unknown 26	53.6	123.0915	3.71	Glycine × fructose
Unknown 27	53.1	110.0600	5.61	Threonine × arabinose
Unknown 28	51.7	230.1141	5.53	Arginine × galactose
Unknown 29	51.6	196.0224	7.01	Glutamine × lactose
Unknown 30	49.9	110.0600	6.38	Leucine × galactose
Unknown 31	47.3	164.0817	6.20	Histidine × fructose
Unknown 32	47.1	143.0349	3.78	Glutamic acid × lactose
Unknown 33	42.0	144.0807	15.71	Tryptophan × galactose
Unknown 34	39.5	124.0757	8.06	Threonine × arabinose
Unknown 35	39.3	124.0756	4.34	Alanine × maltose
Unknown 36	38.2	210.1127	8.29	Threonine × maltose
Unknown 37	37.6	124.0756	15.82	Alanine × fructose
Unknown 38	35.6	169.0972	3.27	Serine × galactose
Unknown 39	34.4	127.039	3.76	Serine × lactose
Unknown 40	31.3	124.0757	7.49	Glutamine × fructose
Unknown 41	30.0	212.1035	6.82	Arginine × fructose
Unknown 42	28.3	217.0971	17.69	Tryptophan × arabinose

Table 3. *Cont.*

Name	% Degradation after 1 h with AAPH	*m/z*	RT [min]	Best AA × Sugar
Unknown 43	25.5	144.0807	13.62	Tryptophan × galactose
Unknown 44	25.0	144.0807	17.69	Tryptophan × arabinose
Unknown 45	22.0	124.0757	8.63	Asparagine × arabinose
Unknown 46	21.0	143.0350	6.95	Tryptophan × arabinose
Unknown 47	20.1	230.1139	8.12	Arginine × fructose
Unknown 48	14.9	95.06033	3.11	Glycine × galactose
Unknown 49	12.2	151.1229	6.06	Valine × galactose
Unknown 50	10.0	143.035	5.25	Methionine × fructose

3. Materials and Methods

3.1. Analytical Methods

3.1.1. HPLC-HRMS Analyses

The samples were analysed using a Dionex UltiMate3000 HPLC system (Thermo Fisher Scientific, Waltham, MA, USA) equipped with two binary pumps and an autosampler with a temperature control system. The detector used was a Q-Exactive Orbitrap high-resolution mass spectrometer (HRMS, Thermo Fisher Scientific, Waltham, MA, USA). The heated electrospray source (HESI) was operated in positive and negative ionisation mode with a capillary voltage of 2.50 kV and a capillary temperature of 330 °C. The full MS scan was acquired from 50 to 750 *m/z* with a resolution of 70.000 full width at half maximum (FWHM) at 200 *m/z*, an automatic gain control (AGC) target of 3×10^6 and a maximum injection time of 100 ms. Data-dependent MS analyses (MS2) were conducted to obtain the fragmentation patterns of targeted and untargeted species. In this case, the AGC target was set at 1×10^5, the maximum injection time was 50 ms, and the resolution was 17,500 FWHM with an isolation window of 4.0 *m/z*. The spectrometer was calibrated prior to analyses with Pierce LTQ Velos ESI-positive and negative calibration solution (Thermo Fisher Scientific, Waltham, MA, USA). The data were collected and analysed using Chromeleon 7.3, Compound Discoverer 3.3.3.200, and Mass Frontier 8.3 software (ThermoFisher Scientific, Waltham, MA, USA).

3.1.2. Amino Acids Quantification

Amino acids were quantified using HPLC coupled with a fluorescence detector (FLD) after derivatisation with ortho-phthalaldehyde (OPA) as described by [17]. Briefly, the measurements were performed using an HPLC 1260 Infinity system (Agilent Technologies, Santa Clara, CA, USA) equipped with a fluorescence detector (Ex = 336 nm, Em = 445 nm). Separation was carried out with sodium acetate 0.05 M (pH 6.9; eluent A) and methanol (eluent B) using a Chromolith Performance RP-18e 100 × 4.6 mm column (Merck, Darmstadt, Germany) with a Guard Cartridge Chromolith RP-18e 10 × 4.6 mm (Merck, Darmstadt, Germany) at 40 °C. The flow rate was set at 2 $mL \cdot min^{-1}$. The analytical gradient for eluent B was as follows: 0% from 0–1 min, 20% from 1–11 min, 40% from 11–16 min, 100% from 16–25 min, 10% from 25–27 min, 0% from 27–30 min. The samples (10 µL), after proper dilution (1:50) with Milli-Q water, adjustment of pH to 7, and filtration with 0.2 nm polytetrafluoroethylene (PTFE) filters, were kept at 10 °C by the autosampler. The derivatisation was automatically carried out by the instrumentation, with the introduction of 10 µL of sample into the loop, addition of 10 µL derivatising solution, mixing for 1 min, and injection. The derivatising mix was 4.5 $g \cdot L^{-1}$ of OPA (Sigma-Aldrich, St. Louis, MO, USA) in sodium tetraborate 0.1 M, adjusted to pH 10.5, 10% methanol, and 2% 2-mercaptoethanol. Agilent OpenLab CDS 3.1 software was used for data acquisition and processing. Quantification was obtained with external amino acids standard calibration curves and internal standard (β-glutamic acid) as a control.

3.1.3. Sugars Quantification

The chromatographic separation and quantification of sugars were performed according to the method developed by [18]. The system used was an ICS 5000 ion chromatographer (Dionex, Thermo Fisher Scientific, Waltham, MA, USA) equipped with an eluent generator, an autosampler, a quaternary gradient pump, a column oven, and a pulsed amperometric detector (PAD) consisting of a gold working electrode and a palladium reference electrode. Separation of sugars (both monosaccharides and disaccharides) was performed by injecting 5 μL of sample onto a CarboPac PA200 3 × 250 mm analytical column, preceded by a CarboPac PA200 3 × 50 mm guard column (Dionex, Thermo Fisher Scientific, Waltham, MA, USA). The column stationary phase consisted of a hydrophobic polymeric pellicular resin bonded to quaternary ammonium as an anion-exchange resin functional group. Both columns were operated at a constant temperature of 30 °C. The flow rate was adjusted to 0.4 mL·min^{-1} using an eluent generator that allowed the automatic preparation of potassium hydroxide (KOH) eluent by controlling the electrical current applied for the electrolysis of deionised water. Isocratic KOH elution at 0.1 mM was carried out from 0 to 18 min, followed by gradient elution from 0.1 to 100 mM from 18 to 21.5 min, and held until 27.5 min. The KOH concentration was then reduced to 0.1 mM, allowing the column to equilibrate for 5 min. Deionised water was continuously purged with helium to avoid the formation of carbonates. Sugar detection was performed using PAD with the working pulse potential quaternary curve (Table 4) with respect to a palladium reference electrode.

Table 4. PAD potentials and duration with Ag/AgCl electrode as reference.

Time (s)	Potential (V vs. Ag/AgCl)	Integration
0.00	1.35	Off
0.20	1.35	On
0.40	1.35	Off
0.41	−1.15	Off
0.42	−1.15	Off
0.43	1.45	Off
0.44	1.15	Off
0.50	1.15	Off

3.2. Sample Preparation and Maillard Reaction

In all the MR experiments, 5 mL of each sample were placed in a 10 mL Pyrex flask sealed with a PTFE hermetic plug and heated in an oven set at 140 °C. This temperature was chosen because it is sufficient to guarantee the active form of the sugars, with an open chain, allowing a faster reaction [19]. The samples were then cooled in an ice bath and kept at −20 °C prior to analyses for a maximum of one week.

3.2.1. Optimisation of Chromatographic Method

The optimisation of the chromatographic method to analyse the antioxidant compounds produced by MR consisted mainly of testing different HPLC columns using as reference the HRMS signals of 10 known antioxidant molecules produced by MR on a mix of amino acids and sugars. The columns and eluents were selected based on the chemical characteristics of the known antioxidant molecules.

The columns tested were: Acclaim Trinity P1 3 μm 2.1 × 100 mm (Thermo Fisher Scientific, USA); Raptor Biphenyl 2.7 μm 3 × 150 mm (Restek Corporation, Bellefonte, PA, USA); Ionpac NS2 5 μm 4 × 150 mm (Thermo Fisher Scientific, Waltham, MA, USA); Eclipse XDB-C8 μm 4.6 × 150 mm (Agilent Technologies, Santa Clara, CA, USA); and Poroshell 120 HILIC-Z 2.7 μm 2.1 × 100 mm (Agilent Technologies, Santa Clara, CA, USA).

The samples analysed were obtained by heating for 90 min at 140 °C a water solution containing equimolar concentrations of 20 amino acids (1.25 mM), equimolar concentrations of 6 sugars (4 mM), and phosphate salts to obtain a phosphate buffer solution (PBS) 0.1 M, pH 7. The concentrations of reagents were chosen to achieve a comparable amount of reactive amino groups from amino acids and reactive hydroxyl groups from sugars. Since acid compounds are produced during the MR, the pH level usually decreases [12]; therefore, a buffer solution was used to keep the pH as constant as possible.

3.2.2. Influence of Initial pH on Antioxidants Production in MR

Once the chromatographic method to detect and separate the KAMs was optimised, the influence of initial pH on antioxidant production in MR was investigated. Three different initial pH were tested: pH 6, pH 7, and pH 8. PBS 0.1 M was used to maintain the pH as constant as possible. Experiments were performed in duplicate. The solutions were prepared in bulk (50 mL), and then six Pyrex flasks with hermetic plugs for each sample were filled with 5 mL. The samples were heated at 140 °C for 120 min, and every 20 min a sample was collected and analysed to evaluate antioxidant and 5-hydroxymethylfurfural (HMF) production. HMF was evaluated because it is a known MRP considered to be a health hazard [13], and its production is considered as negative.

3.2.3. Influence of Amino Acids and Sugar Composition on Antioxidant Production in MR

The influence of sugar and amino acid composition on KAMs production was evaluated. The amino acids and sugars tested, the most common in food, are reported in Table 5. Since only reducing sugars react in MR [20], non-reducing sugars like sucrose and trehalose were not tested, even though they are common in food samples.

Table 5. Amino acids and reducing sugars tested, with abbreviations.

Amino Acids			Reducing Sugars
Arginine (arg)	**Histidine (his)**	**Lysine (lys)**	**Glucose (glu)**
Aspartic acid (asp)	Glutamic acid (glu)	Serine (ser)	Galactose (gal)
Threonine (thr)	Asparagine (asn)	Glutamine (gln)	Fructose (fru)
Cysteine (cys)	Glycine (gly)	Proline (pro)	Arabinose (ara)
Alanine (ala)	Valine (val)	Isoleucine (Ile)	Maltose (mal)
Leucine (leu)	Methionine (met)	Phenylalanine (phe)	Lactose (lac)
Tyrosine (tyr)	Tryptophan (trp)		

All the amino acids and sugars were combined one-to-one, for a total of 120 different combinations. All the experiment were conducted in PBS 0.1 M at pH 7, with equimolar concentrations of amino acids and sugars (25 mM), except for tyrosine, which was tested at a concentration of 2 mM (always equimolar with sugars) due to its low solubility in water. The samples were kept at 140 °C for 90 min and then cooled and analysed with HPLC coupled with HRMS to detect antioxidant compounds and HMF.

3.2.4. Consumption of Reagents in MR

Amino acid and sugar consumption (with respect to the non-heated sample) was evaluated in MR samples obtained by heating in an oven at 140 °C a mix solution of all 20 amino acids and 6 sugars at pH 7 for 60 min and 120 min to evaluate which compounds are the most reactive. The mix solution contained equimolar concentrations of 20 amino acids (1.25 mM), equimolar concentrations of 6 sugars (4 mM), and phosphate salts (PBS 0.1 M, pH 7).

3.2.5. Untargeted Approach to Create a Library of Potential Antioxidant MRPs

The MR is a very complex reaction, and MRPs are a wide group of molecules with different bioactivity. Because of these, to obtain a more comprehensive idea of the influence of the variable studied on the production of PAMs, an untargeted approach was used to create a library of *m/z* that represents the main potential molecules produced by MR. To do so, the MR samples were incubated with a radical initiator (AAPH) that activated the oxidation of all potential antioxidants. In a 10 mL volumetric flask, 1 mL of MR sample was added to 100 mM of 2,2′-Azobis(2-methylpropionamidine) dihydrochloride (AAPH) radical initiator in water. The mixture was then separated into four 2 mL vials and incubated at 37 °C for 2 h, taking a sample after 1 h. At 37 °C, the radical initiator starts the oxidation process, and the potential antioxidant compounds are oxidised. This method is based on a previously published study [21] that demonstrated that the peak areas of compounds with potential antioxidant activity in HPLC chromatograms are significantly reduced or disappear after incubation with AAPH, which can release ROO• at 37 °C [21]. Once cooled in ice, the samples were analysed using HPLC coupled with HRMS, and all the spectra were processed using Compound discover software 3.3.3.200 (Thermo Fischer Scientific).

The library of *m/z* obtained was then used to reprocess, using Compound discover software, all the spectra and information obtained from previous experiments (Sections 3.2.2 and 3.2.3). This approach allowed to obtain a clearer, even if complex, idea of how reaction variables like pH and reagent composition influence antioxidant production in MR.

4. Conclusions

In conclusion, this study elucidated the significant influence of some key variables in MR, such as pH, time, and reagent composition, on antioxidant compound production through the development of a method for the detection and quantification of these compounds. The way for more precise and comprehensive analyses of antioxidant MRPs has been paved, and the next crucial step in this research would be to identify within the large group of new MRPs identified here (PAMs), the molecules with the highest antioxidant activity. This will involve isolating MRPs and characterising their structures and mechanisms of action. Such efforts will not only enhance our understanding of the Maillard reaction's contribution to food chemistry but also open new avenues for the development of functional foods that valorise these natural antioxidants.

Supplementary Materials: The following supporting information can be downloaded at: https://www.mdpi.com/article/10.3390/molecules29204820/s1. Figure S1: Extracted ion chromatograms (EIC) referring to *m/z* of known antioxidant MRPs are reported for all the tested and discarded columns; Table S1: *m/z* PAMs library with retention times, reference ion and main MS2 fragments when available; Table S2: S values obtained as described in the manuscript (eq. 1) corresponding to each sample and divided in ranges based on the percentile; Figure S2: Sum of areas of HRMS peaks corresponding to the 50 *m/z* signals present in the library for each combination of amino acids and sugars. In yellow, the samples with amino acids that contain a nitrogen atom in the side chain; in blue, all the others.

Author Contributions: Conceptualization, R.L., T.N. and S.B.; formal analysis, S.B.; investigation, S.B.; writing—original draft preparation, S.B.; writing—review and editing, T.N., R.L., K.M. and M.S.; supervision, T.N. and R.L. All authors have read and agreed to the published version of the manuscript.

Funding: This research received no external funding.

Data Availability Statement: The authors confirm that the original contributions presented in the study are included in the article/Supplementary Material ; further inquiries can be directed to the corresponding author/s.

Conflicts of Interest: The authors declare no conflicts of interest.

References

1. Nooshkam, M.; Varidi, M.; Bashash, M. The Maillard reaction products as food-born antioxidant and antibrowning agents in model and real food systems. *Food Chem.* **2019**, *275*, 644–660. [CrossRef] [PubMed]
2. Salter, L.J.; Mottram, D.S.; Whitfield, F.B. Volatile compounds produced in maiilard reactions involving glycine, ribose and phospholipid. *J. Sci. Food Agric.* **1989**, *46*, 227–242. [CrossRef]
3. Sun, A.; Wu, W.; Soladoye, O.P.; Aluko, R.E.; Bak, K.H.; Fu, Y.; Zhang, Y.H. Maillard reaction of food-derived peptides as a potential route to generate meat flavor compounds: A review. *Food Res. Int.* **2022**, *151*, 110823. [CrossRef] [PubMed]
4. Michalska, A.; Amigo-Benavent, M.; Zielinski, H.; del Castillo, M.D. Effect of bread making on formation of Maillard reaction products contributing to the overall antioxidant activity of rye bread. *J. Cereal Sci.* **2008**, *48*, 123–132. [CrossRef]

5. Nooshkam, M.; Varidi, M.; Verma, D.K. Functional and biological properties of Maillard conjugates and their potential application in medical and food: A review. *Food Res. Int.* **2020**, *131*, 109003. [CrossRef]
6. Wang, H.Y.; Qian, H.; Yao, W.R. Melanoidins produced by the Maillard reaction: Structure and biological activity. *Food Chem.* **2011**, *128*, 573–584. [CrossRef]
7. Novais, C.; Molina, A.K.; Abreu, R.M.V.; Santo-Buelga, C.; Ferreira, I.C.F.R.; Pereira, C.; Barros, L. Natural Food Colorants and Preservatives: A Review, a Demand, and a Challenge. *J. Agric. Food Chem.* **2022**, *70*, 2789–2805. [CrossRef] [PubMed]
8. Fu, Y.; Zhang, Y.; Soladoye, O.P.; Aluko, R.E. Maillard reaction products derived from food protein-derived peptides: Insights into flavor and bioactivity. *Crit. Rev. Food Sci. Nutr.* **2020**, *60*, 3429–3442. [CrossRef] [PubMed]
9. Liu, X.; Xia, B.; Hu, L.T.; Ni, Z.J.; Thakur, K.; Wei, Z.J. Maillard conjugates and their potential in food and nutritional industries: A review. *Food Front.* **2020**, *1*, 382–397. [CrossRef]
10. Kanzler, C.; Haase, P.T.; Schestkowa, H.; Kroh, L.W. Antioxidant Properties of Heterocyclic Intermediates of the Maillard Reaction and Structurally Related Compounds. *J. Agric. Food Chem.* **2016**, *64*, 7829–7837. [CrossRef] [PubMed]
11. Yanagimoto, K.; Lee, K.G.; Ochi, H.; Shibamoto, T. Antioxidative activity of heterocyclic compounds formed in Maillard reaction products. *Int. Congr. Ser.* **2002**, *1245*, 335–340. [CrossRef]
12. Martins, S.I.F.S.; Van Boekel, M.A.J.S. Kinetics of the glucose/glycine Maillard reaction pathways: Influences of pH and reactant initial concentrations. *Food Chem.* **2005**, *92*, 437–448. [CrossRef]
13. Zhang, L.; Kong, X.; Yang, X.; Zhang, Y.Y.; Sun, B.G.; Chen, H.T.; Sun, Y. Kinetics of 5-hydroxymethylfurfural formation in the sugar—Amino acid model of Maillard reaction. *J. Sci. Food Agric.* **2019**, *99*, 2340–2347. [CrossRef] [PubMed]
14. Delgado-andrade, C.; Seiquer, I.; Haro, A.; Castellano, R.; Navarro, M.P. Development of the Maillard reaction in foods cooked by different techniques. Intake of Maillard-derived compounds. *Food Chem.* **2010**, *122*, 145–153. [CrossRef]
15. Yaylayan, V.A.; Mandeville, S. Stereochemical Control of Maltol Formation in Maillard Reaction. *J. Agric. Food Chem.* **1994**, *42*, 771–775. [CrossRef]
16. Amaya-Farfan, J.; Rodriguez-Amaya, D.B. (Eds.) The Mailard Reactions. In *Chemical Changes During Processing and Storage of Foods*; Academic Press: Cambridge, MA, USA, 2021. [CrossRef]
17. Gallo, A.; Guzzon, R.; Ongaro, M.; Paolini, M.; Nardin, T.; Malacarne, M.; Roman, T.; Larcher, R. Biological acidification of 'Vino Santo di Gambellara' by mixed fermentation of L. thermotolerans and S. cerevisiae. Role of nitrogen in the evolution of fermentation and aroma profile. *Oeno. One* **2023**, *57*, 205–217. [CrossRef]
18. Di Lella, S.; Tognetti, R.; La Porta, N.; Lombardi, F.; Nardin, T.; Larcher, R. Characterization of Silver fir Wood Decay Classes Using Sugar Metabolites Detected with Ion Chromatography. *J. Wood Chem. Technol.* **2019**, *39*, 90–110. [CrossRef]
19. Van Boekel, M.A.J.S. Kinetic aspects of the Maillard reaction: A critical review. *Food/Nahrung* **2001**, *45*, 150–159. [CrossRef] [PubMed]
20. Hodge, J.E. Browning Reaction Theories Integrated in Review Chemistry of Browning Reactions in Model Systems. *Agric. Food Chem.* **1953**, *1*, 928–943. [CrossRef]
21. Zhuang, G.D.; Gu, W.T.; Xu, S.H.; Cao, D.M.; Deng, S.M.; Chen, Y.S.; Wang, S.M.; Tang, D. Rapid screening of antioxidant from natural products by AAPH-Incubating HPLC-DAD-HR MS/MS method: A case study of Gardenia jasminoides fruit. *Food Chem.* **2023**, *401*, 134091. [CrossRef] [PubMed]

molecules

MDPI

Article

Development of Delivery Systems with Prebiotic and Neuroprotective Potential of Industrial-Grade *Cannabis sativa* L.

Szymon Sip [1], Anna Stasiłowicz-Krzemień [1], Anna Sip [2], Piotr Szulc [3], Małgorzata Neumann [3], Aleksandra Kryszak [4] and Judyta Cielecka-Piontek [1,4,*]

1. Department of Pharmacognosy and Biomaterials, Faculty of Pharmacy, Poznań University of Medical Sciences, Rokietnicka 3, 60-806 Poznań, Poland; szymonsip@ump.edu.pl (S.S.); astasilowicz@ump.edu.pl (A.S.-K.)
2. Department of Biotechnology and Food Microbiology, Poznań University of Life Sciences, Wojska Polskiego 48, 60-627 Poznań, Poland; anna.sip@up.poznan.pl
3. Department of Agronomy, Poznań University of Life Sciences, Dojazd 11, 60-632 Poznań, Poland; piotr.szulc@up.poznan.pl (P.S.); malgorzata.neumann@up.poznan.pl (M.N.)
4. Department of Pharmacology and Phytochemistry, Institute of Natural Fibres and Medicinal Plants, Wojska Polskiego 71b, 60-630 Poznań, Poland; aleksandra.kryszak@iwnirz.pl

* Correspondence: jpiontek@ump.edu.pl

Abstract: This study delves into the transformative effects of supercritical carbon dioxide ($scCO_2$) cannabis extracts and prebiotic substances (dextran, inulin, trehalose) on gut bacteria, coupled with a focus on neuroprotection. Extracts derived from the Białobrzeska variety of Cannabis sativa, utilising supercritical fluid extraction (SFE), resulted in notable cannabinoid concentrations (cannabidiol (CBD): 6.675 ± 0.166; tetrahydrocannabinol (THC): 0.180 ± 0.006; cannabigerol (CBG): 0.434 ± 0.014; cannabichromene (CBC): 0.490 ± 0.017; cannabinol (CBN): 1.696 ± 0.047 mg/gD). The assessment encompassed antioxidant activity via four in vitro assays and neuroprotective effects against acetylcholinesterase (AChE) and butyrylcholinesterase (BChE). The extract boasting the highest cannabinoid content exhibited remarkable antioxidant potential and significant inhibitory activity against both enzymes. Further investigation into prebiotic deliveries revealed their proficiency in fostering the growth of beneficial gut bacteria while maintaining antioxidant and neuroprotective functionalities. This study sheds light on the active compounds present in the Białobrzeska variety, showcasing their therapeutic potential within prebiotic systems. Notably, the antioxidant, neuroprotective, and prebiotic properties observed underscore the promising therapeutic applications of these extracts. The results offer valuable insights for potential interventions in antioxidant, neuroprotective, and prebiotic domains. In addition, subsequent analyses of cannabinoid concentrations post-cultivation revealed nuanced changes, emphasising the need for further exploration into the dynamic interactions between cannabinoids and the gut microbiota.

Keywords: *Cannabis Sativa*; SFE; antioxidant activity; neuroprotective activity; prebiotic; microbiome; industrial hemp

1. Introduction

Alterations in gut microbiota are increasingly recognised as pivotal contributors to the pathophysiology of neurodegenerative diseases, specifically Alzheimer's and Parkinson's. In the context of Alzheimer's disease, studies indicate a shift in gut microbiota composition, characterised by a decrease in beneficial bacterial species and an increase in pro-inflammatory bacteria [1–3]. This dysbiosis is associated with the accumulation of amyloid-beta plaques in the brain, a hallmark of Alzheimer's pathology [4,5]. The gut–brain axis dysfunction in Alzheimer's is implicated in neuroinflammatory responses and oxidative stress, further exacerbating cognitive decline [6,7]. In Parkinson's disease, a distinct pattern of gut microbiota alterations emerges. The abundance of specific bacterial taxa, such as *Prevotellaceae* and *Lactobacillaceae*, appears to be reduced, while the presence of potentially pathogenic bacteria like *Enterobacteriaceae* increases. These microbial composition shifts correlate with alpha-synuclein aggregation, a key pathological feature of Parkinson's [8,9]. The gut–brain axis disruption in Parkinson's contributes to neuroinflammation, potentially facilitating the propagation of alpha-synuclein pathology from the gut to the brain [10]. The mechanistic links between gut microbiota changes and neurodegenerative diseases involve the production of neuroactive metabolites, modulation of the immune system, and the influence on the integrity of the gut epithelial barrier. Specific microbial metabolites, including short-chain fatty acids, neurotransmitters, and lipopolysaccharides, can enter systemic circulation and affect the central nervous system, contributing to neuroprotection and reduction in oxidative stress [11–13].

Understanding the dynamics of gut microbiota alterations in neurodegenerative diseases provides valuable insights into potential therapeutic targets. Strategies aimed at restoring a balanced and diverse gut microbiome, such as probiotics, prebiotics, and dietary interventions, hold promise for mitigating the progression of neurodegenerative disorders by modulating the gut–brain axis and addressing systemic inflammation [14,15].

Cannabis sativa boasts a rich history of medicinal use spanning centuries, harnessing a diverse array of bioactive compounds such as cannabinoids, terpenes, and flavonoids, each demonstrating notable health-promoting properties [16–18]. Cannabinoids, including the extensively studied non-psychoactive cannabidiol (CBD), interact with the human body's endocannabinoid system (ECS). The ECS, a regulatory network, influences various physiological processes, prominently governing gut function. While the precise impact of cannabinoids on the gut microbiome is an ongoing area of investigation, the dynamic interplay between the gut microbiota and the ECS is increasingly recognised. This reciprocal relationship implies that alterations in the gut microbiome can influence ECS signalling and vice versa, thereby influencing gut health [19,20].

In the context of neurodegenerative diseases, the therapeutic potential of modulating the endocannabinoid system using cannabinoids has garnered significant attention. CBD, renowned for its neuroprotective effects, emerges as an up-and-coming candidate [21]. Its ability to safeguard the nervous system from damage or degeneration positions it as a potential therapeutic agent for conditions characterised by neurodegeneration. The well-known psychoactive cannabinoid tetrahydrocannabinol (THC) also interacts with the endocannabinoid system, impacting mood, perception, and cognitive functions [22]. While THC's psychoactive nature is associated with the "high" from cannabis use, its potential therapeutic role in mitigating neurodegenerative processes warrants exploration [23]. Cannabinol (CBN), a by-product of THC degradation found in aged cannabis, presents intriguing possibilities for addressing sleep disorders, with some studies suggesting sedative effects [24,25]. Cannabichromene (CBC), a non-psychoactive cannabinoid, holds promise in anti-inflammatory, anti-depressant, and anti-cancer applications, expanding the scope of research in this area [26,27]. Cannabigerol (CBG), often called the "mother cannabinoid", is gaining recognition for potential anti-inflammatory, antibacterial, and neuroprotective effects [28,29]. The modulation of the endocannabinoid system through cannabinoids offers a potential avenue for therapeutic interventions in neurodegenerative diseases. Utilising cannabinoids like CBD to influence the ECS may mitigate neuroinflammation, oxidative stress, and other pathological processes associated with conditions such as Alzheimer's and Parkinson's [30]. The intricate interplay between cannabinoids, the gut microbiota, and the endocannabinoid system thus represents a multifaceted area of research with implications for understanding and treating various neurodegenerative conditions.

The robust gut microbiome supports the immune system, enhancing its ability to fight pathogens and maintain health [31]. The prebiotic effect of *Cannabis sativa* can also contribute to a more substantial gut barrier, potentially reducing the risk of gastrointestinal issues. A well-functioning gut barrier helps prevent the leakage of harmful substances into the bloodstream and maintains a healthy gut lining [32,33]. This, in turn, supports the overall health of the digestive system. The diverse compounds in *Cannabis sativa* and prebiotic properties can foster beneficial gut bacteria growth [34]. A well-functioning gut barrier ensures that the integrity of the gut lining is maintained, reducing the risk of gastrointestinal issues. The combination of *Cannabis sativa* and prebiotics creates a symbiotic relationship that supports the flourishing of beneficial gut bacteria, fortifies the immune system, and reinforces the protective barrier of the gut [35,36]. Developing prebiotic systems represents a potent strategy for delivering plant-derived active compounds endowed with multifaceted effects. This approach assumes particular significance in neurodegenerative diseases, where the imperative for comprehensive therapeutic interventions is paramount.

Ongoing research explores using *Cannabis sativa* varieties characterised by low cannabinoid content, aiming to unlock their potential applications. Remarkably, extracts from the leaves of industrial varieties like Białobrzeska, Henola, and Tygra, traditionally cultivated for fibres and seeds, exhibited significant antioxidant potential. The study, employing ultrasound-assisted extraction, demonstrated the robust antioxidant capacity of these varieties, hinting at potential applications in medicine or targeted processing for active compound extraction [37]. Furthermore, investigations on the inflorescences of these *Cannabis sativa* varieties and extracts obtained using supercritical carbon dioxide ($scCO_2$) extraction unveiled additional neuroprotective potential. The inhibition of acetylcholinesterase (AChE), butyrylcholinesterase (BChe), and tyrosinase indicated untapped avenues for obtaining active compounds. This sheds light on potential applications in various *Cannabis sativa* industry sectors and suggests novel development paths [38].

Prebiotics play a crucial role in modulating the gut microbiota, influencing various aspects of human health. Among the diverse compounds with potential prebiotic properties, dextran, inulin, and trehalose have garnered significant interest for their unique characteristics and potential health benefits.

A significant aspect of this study lies in its comprehensive approach, encompassing not only the analysis of cannabinoids derived from an industrially cultivated *Cannabis sativa* variety primarily intended for seed production but also the development of advanced delivery systems with prebiotic attributes. This innovative strategy aims to fully exploit the therapeutic potential of the *Cannabis sativa* variety by facilitating multifaceted actions. By integrating the investigation of cannabinoids from an industrially cultivated *Cannabis sativa* strain with the refinement of prebiotic delivery systems, this research introduces a holistic perspective to cannabis investigation. This approach transcends traditional boundaries, offering substantial potential in elucidating the reservoir of cannabinoids within *Cannabis sativa*, traditionally cultivated for seed production. Simultaneously, the study pioneers the development of sophisticated delivery systems with prebiotic potential, exemplifying a multifaceted approach to harnessing the complete therapeutic benefits of cannabinoids.

2. Results and Discussion

Our investigation commenced with the creation of a series of extracts derived from a specific raw material: the panicle of the *Cannabis sativa* variety *Białobrzeska*. This particular variety, classified as industrial *Cannabis sativa* and cultivated in Poland, is predominantly known for its fibre, which contributes to producing textiles, paper, and various industrial goods. Beyond its fibre content, the *Białobrzeska* variant yields seeds notably rich in protein and other essential nutrients, rendering them a sought-after ingredient in health foods and supplements. While recognised as an industrial *Cannabis sativa* variety, *Białobrzeska* typically maintains low levels of THC, the psychoactive cannabinoid associated with cannabis [39]. Nonetheless, it may contain other non-psychoactive cannabinoids, including CBD, CBG, and CBC. We meticulously delineated critical process conditions to procure the most potent extract, instrumental in the subsequent formulation of the prebiotic delivery system. These encompassed precise specifications for temperature, pressure, and the volume of CO_2 employed in the extraction process, as outlined in Table 1. In-depth exploration of these critical parameters was integral to optimising the extraction process, ensuring the concentration of bioactive compounds conducive to our study's objectives. As detailed in Table 1, the modifications applied aimed to enhance the extraction efficiency and enrich the resulting extract with targeted cannabinoids, laying the foundation for the subsequent phases of our investigation.

Table 1. Process conditions used in the SFE used for obtaining 9 extracts.

Extract	Temperature [°C]	Pressure [psi/bar]	CO_2 Volume [mL]
1	50	5000/344.738	100
2	70	4000/275.790	100
3	50	3000/206.843	300
4	30	5000/344.738	200
5	30	4000/275.790	300
6	70	3000/206.843	200
7	30	3000/206.843	100
8	50	4000/275.790	200
9	70	5000/344.738	300

We used High-Performance Liquid Chromatography (HPLC) to carefully analyse the content of five necessary cannabinoids: cannabidiol (CBD), cannabigerol (CBG), cannabinol (CBN), tetrahydrocannabinol (THC), and cannabichromene (CBC) in all our extracted samples. This method allowed us to measure these cannabinoids accurately and their total amount, as shown in Table 2. HPLC is a precise tool that ensures we can detect even tiny amounts of these compounds. This thorough analysis is crucial for understanding the chemical makeup of the *Białobrzeska Cannabis sativa* variety extracts. By including the concentrations of each cannabinoid and their total sum, we obtain a detailed picture of what is in the extract. The results in Table 2 show that our extraction process worked well in concentrating these cannabinoids. This detailed profile gives us essential information about potentially using the *Białobrzeska Cannabis sativa* extracts, especially in developing prebiotic delivery systems.

Table 2. The results of determining the content of active compounds using the HPLC method of obtained extracts 1–9.

	CBD	CBG	CBN	THC	CBC	Sum of Cannabinoids
Extract	[mg/gD *]	[mg/gD *]	[mg/gD *]	[mg/gD *]	[mg/gD *]	[mg/gD *]
1	3.078 ± 0.077	0.092 ± 0.003	0.500 ± 0.014	0.057 ± 0.002	0.145 ± 0.005	3.872 ± 0.107
2	1.157 ± 0.029	0.049 ± 0.002	0.275 ± 0.008	0.026 ± 0.001	0.082 ± 0.003	1.589 ± 0.044
3	5.097 ± 0.127	0.173 ± 0.005	0.863 ± 0.024	0.11 ± 0.004	0.323 ± 0.011	6.566 ± 0.181
4	5.300 ± 0.132	0.161 ± 0.005	0.529 ± 0.015	0.071 ± 0.002	0.342 ± 0.012	6.404 ± 0.176
5	6.071 ± 0.151	0.245 ± 0.008	1.034 ± 0.028	0.131 ± 0.005	0.350 ± 0.012	7.831 ± 0.216
6	2.574 ± 0.064	0.127 ± 0.004	0.556 ± 0.015	0.057 ± 0.002	0.169 ± 0.006	3.484 ± 0.096
7	1.335 ± 0.033	0.066 ± 0.002	0.282 ± 0.008	0.034 ± 0.001	0.093 ± 0.003	1.810 ± 0.050
8	5.065 ±0.126	0.145 ± 0.005	0.460 ± 0.013	0.056 ± 0.002	0.336 ± 0.011	6.062 ± 0.167
9	6.675 ±0.166	0.434 ± 0.014	1.696 ± 0.047	0.180 ± 0.006	0.490 ± 0.017	9.475 ± 0.261

gD *—dry mass of plant material.

Upon analysing the active compound content, extract 9 emerged as the most potent, displaying the highest concentrations, affirming the effectiveness of the defined critical parameters in the supercritical fluid extraction (SFE) process. Notably, the consistently low levels of THC across all extracts align with expectations for an industrial *Cannabis sativa* variety primarily cultivated for fibre production. Despite its industrial focus, the *Białobrzeska* variety showcases the presence of cannabinoids, indicating its potential not only for medicinal applications in varieties with high cannabinoid content but also for utilising by-products in food or textile production. This underscores industrial *Cannabis sativa*'s versatility, suggesting avenues for further processing to derive therapeutic compounds for medical applications. The detailed cannabinoid profiles presented in Table 2 provide a robust foundation for understanding the chemical composition of the extracts. The varying concentrations of CBD, CBG, CBN, THC, and CBC underscore the nuanced nature of *Białobrzeska* variety, presenting opportunities for tailored applications in different industries. These findings contribute to the broader discourse on harnessing the potential of industrial *Cannabis sativa* for diverse purposes, including medicine and sustainable material production.

We explored the entourage effect in plant raw materials, highlighting how different active compounds work together synergistically. This means that having a high concentration of one compound does not necessarily mean that it will have the most significant biological effect. Instead, it is the combination and balance of multiple compounds that matter. We conducted a detailed analysis of the extracts to understand their antioxidative activity using specific lab tests, as shown in Figure 1. This analysis aimed to uncover how the mixture of cannabinoids, terpenes, and other active compounds influences the overall antioxidative potential.

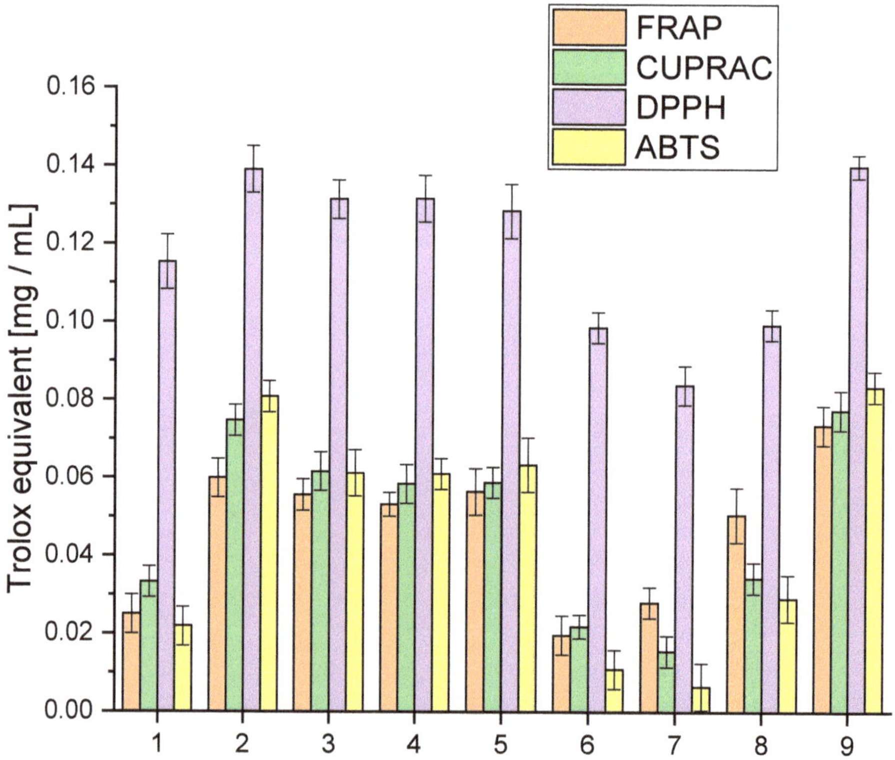

Figure 1. Analysis of the obtained extracts 1–9 antioxidant activity using the model FRAP, Cuprac, DPPH and ABTS expressed in Trolox equivalent.

Simultaneously, our study extended to evaluating acetyl- and butyrylcholinesterase inhibition (AChE and BChE), an essential facet in the quest for neuroprotective potential, as depicted in Figure 2. We sought to unravel the extracts' potential in safeguarding neural functions by strategically directing our focus on these enzymatic activities. The numerical data of the obtained results of antioxidant and biological activity are included in the Supplementary Materials (Tables S2 and S3).

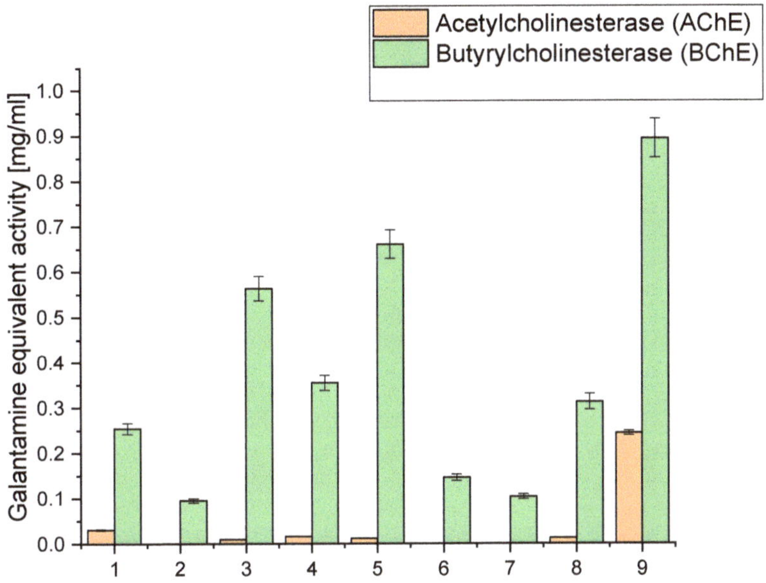

Figure 2. Analysis of acetyl- and butyrylcholinesterase inhibition of the obtained extracts 1–9 expressed as galanthamine equivalent.

The analysis of antioxidant activity indicated that extract 9 was the most active in all the in vitro models used. All tested extracts were characterised by high activity in the DPPH model, giving activity at the Trolox equivalent level of 0.0837–0.13995 mg/mL. The highest activity variability was observed for the ABTS model, which occurred at the Trolox equivalent level of 0.0065–0.0833 mg/mL. In the case of the applied Cuprac and FRAP models, differences were observed at the Trolox equivalent level of 0.0155–0.0774 mg/mL and 0.0196–0.0735 mg/mL, respectively. The observed variability in activity underscores the intricate interplay of compounds within each extract, revealing a spectrum of antioxidant capacities. These findings not only position extract 9 as a standout performer but also emphasise the dynamic and versatile nature of the entire spectrum of extracts, fostering a nuanced comprehension of their antioxidant capabilities.

However, the cannabinoid content alone does not fully explain the antioxidant activity of the extracts. Other compounds, such as terpenes and flavonoids, also contribute significantly to the antioxidant properties. Terpenes, which include compounds like β-caryophyllene, myrcene, and limonene, can enhance the antioxidant capacity through synergistic effects, known as the entourage effect. This effect refers to the phenomenon where different compounds within the extract work together to produce a more robust antioxidant response than any single compound alone [40,41].

For extracts 6–8, the lower antioxidant activity might also be due to a suboptimal combination of cannabinoids and terpenes, which could result in weaker synergistic interactions. The lower amounts of CBG, known for its potent antioxidant properties, especially in extracts 6 and 7, could significantly impact their overall antioxidant capacity. Furthermore, the balance and interaction between various cannabinoids, such as the ratio of CBD to THC or the presence of CBN and CBC, play crucial roles in determining antioxidant efficacy.

The assessment of antioxidant activity is a complex issue and requires different models reflecting various mechanisms of action [42]. In our study, we observed significant differences in the activity of the extracts in four different assays: FRAP, Cuprac, DPPH, and ABTS. Such differences can be attributed to the diversity of active compounds in the extracts and the different mechanisms underlying the specific assay. It is well known that a range of active compounds in plant extracts contribute to their antioxidant properties. These include phenolic acids, flavonoids, and terpenoids, which exhibit varying antioxidant activity and can act through different mechanisms [43,44]. For example, phenolic acids such as caffeic acid and ferulic acid act as electron and hydrogen atom donors, while flavonoids such as quercetin and rutin scavenge free radicals through single electron transfer reactions [45]. Terpenoids, such as β-carotene and lycopene, act as singlet oxygen quenchers [46]. The distinctive activity profiles observed in our study find their roots in the diverse mechanisms underpinning the assays employed. Each assay, a unique window into the extract's antioxidant potential, unveils specific facets of its functional dynamics. In the FRAP and Cuprac assays, the extracts' prowess to reduce Fe(III) and Cu(II) ions takes centre stage, respectively. These assays delve into the extract's capacity to donate electrons, offering insights into its electron transfer capabilities within distinct chemical contexts.

On the other hand, the DPPH and ABTS assays pivot on the extract's ability to scavenge free radicals, a pivotal attribute in neutralising oxidative stress. Here, the focus shifts to the extract's capacity to intercept and nullify these highly reactive species. The spectrum of active compounds within the extract exhibits varying degrees of efficacy, contingent upon the specific mechanism tested. The intricate interplay of cannabinoids, terpenes, and other bioactive compounds manifests diverse functional roles, shaping the extract's overall antioxidant activity. Our study underscores the imperative of employing a battery of assays. This multifaceted approach not only elucidates the diverse antioxidant capacities of the extracts but also unravels the underlying mechanisms steering their activity. By embracing this comprehensive strategy, we navigate the complex landscape of antioxidant potential, laying the foundation for a more profound understanding of the extract" multifaceted contributions to oxidative stress mitigation.

Inhibition of acetylcholinesterase (AChE) and butyrylcholinesterase (BChE) stands as a pivotal strategy in neuroprotection, particularly concerning neurodegenerative diseases like Alzheimer's disease (AD) [47]. Dysregulated cholinesterase activity, notably elevated AChE and BChE levels, is a hallmark of AD pathology, contributing to cognitive decline and memory impairment. The cholinergic hypothesis underscores the significance of cholinergic neurotransmission in AD pathogenesis, emphasising the role of AChE-mediated acetylcholine hydrolysis in cognitive deficits [48]. Therefore, inhibiting AChE activity represents a potential avenue for ameliorating symptoms and slowing disease progression. The neuroprotective effects of AChE and BChE inhibition stem from their ability to increase acetylcholine levels, thereby enhancing cholinergic neurotransmission and promoting neuronal survival. This mechanism is crucial for preserving cognitive function and attenuating neurodegeneration in AD. Clinical trials have demonstrated the efficacy of cholinesterase inhibitors such as donepezil, rivastigmine, and galantamine in improving cognitive function and delaying disease progression in AD patients. These drugs exert their effects by inhibiting AChE and BChE, leading to elevated acetylcholine levels in the brain [49–51].

Our study investigated the extract's capability to inhibit AChE and BChE, critical enzymes involved in neurotransmission processes. AChE and BChE are integral to maintaining cognitive functions, as they regulate the levels of neurotransmitters in the brain. Inhibiting these enzymes is a sought-after strategy in neuroprotection, as it can modulate neurotransmitter levels and potentially mitigate cognitive decline. Our findings shed light on the neuroprotective potential of the extracts by demonstrating their ability to inhibit AChE and BChE activities. This comprehensive evaluation underscores the multifaceted contributions of the extracts to neuroprotection, offering promising avenues for further research in cognitive health and neurological well-being (Figure 2).

When examining the inhibition against AChE and BChE, we observed that extract 9 exhibited the highest AChE inhibition, with a value of 0.2438, equivalent to galantamine. This extract had a relatively high CBD content (6.675 mg/g) compared to the other extracts. Similarly, it displayed the highest BChE inhibition, with a value of 0.894, equivalent to galantamine. Notably, extract 9 had the highest sum of cannabinoids among all the extracts (9.475 mg/g), indicating a rich composition of active compounds. Extracts 3, 4, 5, and 8 also showed moderate AChE and BChE inhibition. These extracts also had notable CBD content, ranging from 5.065 mg/g to 6.071 mg/g. High CBD levels in these extracts suggest a potential correlation between CBD content and cholinesterase inhibitory activity. On the other hand, extracts 1, 2, 6, and 7 exhibited lower inhibitory activity against both AChE and BChE. These extracts had relatively lower cannabinoid content than the others, which might contribute to their reduced inhibitory effects.

The significantly higher cholinesterase inhibitory activities observed in extract 9 compared to extracts 6–8 can be linked to its higher total concentration of cannabinoids, as well as the presence and higher concentration of specific cannabinoids that may contribute more effectively to cholinesterase inhibition. Extract 9 contains 6.675 mg/g of CBD, 0.434 mg/g of CBG, 1.696 mg/g of CBN, 0.180 mg/g of THC, and 0.490 mg/g of CBC, summing up to a total of 9.475 mg/g of cannabinoids. The specific roles of these cannabinoids in cholinesterase inhibition are significant; for instance, CBN has been reported to have notable effects in such biochemical activities. The relatively lower inhibitory activities in extracts 6 and 7 can be attributed to their lower concentrations of cannabinoids and other active compounds. Extract 6 has a total of 3.484 mg/g of cannabinoids, while extract 7 has only 1.810 mg/g. The lower concentrations of these bioactive compounds directly correlate with their diminished ability to inhibit AChE and BChE.

Other bioactive compounds, such as terpenes and flavonoids, are crucial in cholinesterase inhibition. Terpenes, like α-pinene, limonene, and β-caryophyllene, may enhance the inhibitory effects through synergistic interactions with cannabinoids, contributing to the overall efficacy. The entourage effect, where different compounds in the extract work together to produce a more potent inhibitory response, is essential in understanding the varying degrees of activity among different extracts [41,52]. Extract 8, although having a higher concentration of cannabinoids (6.062 mg/g) than extracts 6 and 7, still demonstrates lower inhibitory activities than extract 9. This suggests that the total concentration and the specific composition and ratio of cannabinoids and other compounds influence the inhibitory efficacy. The balance between different cannabinoids and their interaction with terpenes might be less optimal in extract 8, resulting in reduced cholinesterase inhibition compared to extract 9.

While our initial findings are promising, they represent just the beginning of our exploration. We need more thorough and detailed studies to understand the specific relationships and mechanisms behind these inhibitory effects. It is essential to recognise that many factors, including other plant chemicals and how they interact, contribute to the overall activity of the extracts. As we continue to delve into this complex area of bioactivity, these early findings urge us to dig deeper. By unravelling the layers of complexity, we can better understand the precise mechanisms behind the extracts' inhibitory abilities. This, in turn, will pave the way for more nuanced therapeutic applications and a deeper grasp of their potential for pharmacological use.

As observed in this study, the inhibitory activity of *Cannabis sativa* extracts against AChE and BChE aligns with previous research investigating the potential of *Cannabis sativa*-derived compounds in modulating these enzymes [53,54]. Various studies have explored the effects of *Cannabis sativa* extracts and individual cannabinoids on cholinesterase activity, providing valuable insights into their mechanisms of action.

Several cannabinoids found in *Cannabis sativa*, including delta-9-tetrahydrocannabinol (THC), cannabidiol (CBD), and cannabigerol (CBG), have been studied for their activities on AChE and BChE. THC has been shown to possess moderate inhibitory effects on these enzymes, suggesting its potential role in modulating cholinergic transmission [55]. However, THC is also known for its psychoactive properties, which may limit its widespread use for therapeutic purposes.

Conversely, CBD has been found to have minimal direct inhibitory effects on AChE and BChE [56]. Instead, CBD modulates these enzymes' activity indirectly by influencing other molecular targets within the endocannabinoid system and other signalling pathways. CBD's neuroprotective and anti-inflammatory properties have been proposed to contribute to its potential therapeutic effects on neurodegenerative diseases [57]. CBG, a minor cannabinoid found in *Cannabis sativa*, has shown promising inhibitory effects on AChE, suggesting its potential as a therapeutic agent for cholinergic dysfunction conditions. CBG has also demonstrated antioxidant and anti-inflammatory properties, highlighting its potential neuroprotective effects [58].

When these individual cannabinoids are combined in a supercritical *Cannabis sativa* extract, the entourage effect may come into play. The entourage effect suggests that the synergistic interactions between cannabinoids, terpenes, and other phytochemicals present in the extract can enhance their overall therapeutic potential [59,60]. While THC, CBD, and CBG may have their activities on AChE and BChE, their combined presence in a *Cannabis sativa* extract could lead to a more significant inhibitory effect on these enzymes than isolated cannabinoids alone. It is worth noting that the specific composition and ratios of cannabinoids and other bioactive compounds in a *Cannabis sativa* extract can vary depending on the extraction method and the strain of *Cannabis sativa* used [61,62]. Thus, the extent and nature of the entourage effect in inhibiting AChE and BChE may differ among different *Cannabis sativa* extracts.

However, it is essential to acknowledge the limitations of this study and the existing literature. The observed relationships between cannabinoid content and inhibitory activity are correlative and do not establish causation. Further mechanistic studies are required to elucidate the specific interactions between cannabinoids and cholinesterase enzymes. In conclusion, the findings from this study, combined with previous research, suggest that *Cannabis sativa* extracts and individual cannabinoids, particularly CBD, exhibit inhibitory activity against AChE and BChE. These results support the potential of *Cannabis sativa*-derived compounds as valuable candidates for developing cholinesterase inhibitors for various applications, including neurodegenerative diseases. Further research is warranted to understand these compounds' underlying mechanisms and therapeutic potential fully.

Moving forward with our investigation, we delved into prebiotics, carefully designing systems infused with prebiotic potential. Through a meticulous lyophilisation process, we created three distinct systems tailored to leverage the established prebiotic properties of dextran, inulin, and trehalose. These well-known substances in prebiotic research were chosen for their unique characteristics. After thoroughly examining their prebiotic properties, we integrated them into our systems to explore their interaction with the complex gut microbiome.

Dextran, a polysaccharide produced by the fermentation of sucrose, has shown promising prebiotic effects. It acts as a substrate for beneficial gut bacteria, such as Bifidobacterium and *Lactobacillus*, stimulating their growth and activity. Dextran also exhibits potential immunomodulatory properties and contributes to maintaining intestinal barrier function [63]. Inulin, a fructan-type carbohydrate, has been widely recognised as a prebiotic compound. It undergoes fermentation in the colon, selectively promoting the growth of beneficial bacteria, particularly *Bifidobacteria*. Inulin has been associated with improved bowel regularity, increased calcium absorption, and potentially positive effects on lipid metabolism [64]. Trehalose, a disaccharide composed of two glucose molecules, has shown prebiotic potential by selectively promoting the growth of beneficial gut bacteria. It has been reported to enhance the growth of *Bifidobacterium* and *Lactobacillus* strains while inhibiting the proliferation of pathogenic bacteria. Trehalose also exhibits antioxidative properties and has been investigated for its potential neuroprotective effects [65].

The prepared systems (1-dextran, 2-inulin, 3-trehalose) were further evaluated to assess the content of active compounds (Table 3).

Table 3. The results of determining the content of active compounds using the HPLC method in obtained systems 1–3.

	CBD	CBG	CBN	THC	CBC	Sum of Cannabinoids
System	[mg/gDs *]	[mg/gDs*]	[mg/gDs *]	[mg/gDs *]	[mg/gDs *]	[mg/gDs *]
1	6.665 ± 0.106	0.414 ± 0.011	1.691 ± 0.027	0.180 ± 0.004	0.478 ± 0.014	9.428 ± 0.231
2	6.666 ± 0.161	0.431 ± 0.010	1.690 ± 0.045	0.178 ± 0.003	0.488 ± 0.015	9.453 ± 0.222
3	6.668 ± 0.145	0.427 ± 0.017	1.688 ± 0.044	0.175 ± 0.005	0.477 ± 0.017	9.435 ± 0.258

gD *—a dry mass of the system.

Additionally, the analysis of antioxidant activity and neuroprotective effects was repeated to estimate any potential changes resulting from the preparation process and the carrier substances' inherent activity (Tables 4 and 5).

Table 4. Numerical data for the antioxidant activity of the obtained systems 1–3 and prebiotic carriers. Means with the same superscript letters (a) within the same column do not differ significantly ($p < 0.05$).

System	FRAP	Std	CUPRAC	Std	DPPH	Std	ABTS	Std
				Trolox equivalent [mg/mL]				
1	0.0745 ^{a}	0.009	0.0770 ^{a}	0.006	0.1389 ^{a}	0.002	0.0827 ^{a}	0.002
2	0.0738 ^{a}	0.008	0.0757 ^{a}	0.005	0.1394 ^{a}	0.005	0.0824 ^{a}	0.003
3	0.0740 ^{a}	0.008	0.0761 ^{a}	0.007	0.1379 ^{a}	0.004	0.0819 ^{a}	0.008
Dextran	Nd *	-	Nd *	-	Nd *	-	Nd *	-
Inulin	Nd *	-	Nd *	-	Nd *	-	Nd *	-
Trehalose	Nd *	-	Nd *	-	Nd *	-	Nd *	-

* Nd—not detected.

Table 5. Numerical data for the neuroprotective activity of the obtained systems 1–3 and prebiotic carriers. Means with the same superscript letters (a) within the same column do not differ significantly ($p < 0.05$).

System	AchE	Std	BChE	Std
		Galantamine equivalent [mg/mL]		
1	0.2427 ^{a}	0.00510	0.8925 ^{a}	0.0404
2	0.2433 ^{a}	0.00544	0.8968 ^{a}	0.0415
3	0.2431 ^{a}	0.00491	0.8935 ^{a}	0.0431
Dextran	Nd *	-	Nd *	-
Inulin	Nd *	-	Nd *	-
Trehalose	Nd *	-	Nd *	-

* Nd—not detected.

Our comprehensive analysis yielded promising findings, demonstrating our tailored systems' robustness in preserving active compounds' content. This resilience extends to the meticulously introduced prebiotic matrices—dextran, inulin, and trehalose. Additionally, the levels of antioxidant activity, a critical metric of biological relevance, remained steadfast across all evaluated systems. These observations underscore the potential of our engineered constructs as carriers for active compounds while offering antioxidant benefits. These findings represent a significant advancement in our exploration, confirming the effectiveness of our tailored systems in retaining bioactive entities and sustaining beneficial biological activities.

In the subsequent phase of our investigation, we conducted a comprehensive analysis of acetylcholinesterase (AChE) and butyrylcholinesterase (BChE) activities within the meticulously engineered systems and their carriers—dextran, inulin, and trehalose.

This examination aimed to elucidate the neuroprotective potential inherent in our tailored constructs. The detailed assessment of these enzymatic activities, as outlined in Table 5, provides a nuanced understanding of how our engineered matrices may impact neuroprotective pathways. The intricate interplay between the introduced prebiotic matrices and AChE's and BChE's enzymatic activities unveils potential neuroprotection implications (Table 5).

The data underscores that systems 1, 2, and 3 maintain the initial activity of the extract, affirming the preservation of neuroprotective potential within our engineered constructs. The not detected (Nd) status for prebiotic carriers indicates the absence of inhibitory effects, emphasising the selective neuroprotective influence of the formulated systems.

The prebiotic potential of the developed systems plays a crucial role in their application as functional ingredients. Evaluating the effectiveness of these systems in promoting the growth and activity of beneficial gut microbiota is essential for understanding their potential health benefits. This section presents the results of the prebiotic potential evaluation of the obtained systems (Table 6). These findings shed light on the ability of the developed systems to selectively support the growth of beneficial gut bacteria while inhibiting the proliferation of harmful species. Understanding the prebiotic potential of these systems is instrumental in harnessing their potential for improving gut health and overall well-being.

The results of this study shed light on the impact of different supercritical extract systems and excipients on the growth of specific bacterial strains over time. The population counts of *Faecalibacterium prausnitzii* DSM 107840 (P), *Bifidobacterium animalis* DSM 10140 (Ba), *Bifidobacterium longum subs. longum* DSM20219 (Bl), *Lactobacillus rhamnosus* GG ATCC 53103 (GG), *Lactobacillus helveticus* DSM 20075 (Lh), *Lactobacillus salivarius* LA 302 (Ls), and *Lactobacillus plantarum* 299v (9) were examined at four different time points (12 h, 24 h, 48 h, and 72 h).

Adding dextran as an excipient in the SFE+dextran system significantly affected the bacterial strains. It increased Ba and Bl population counts at all time points compared to the control. Specifically, at 12 h, the population counts of Ba and Bl were 2.27×10^7 and 6.23×10^8, respectively, in the SFE+dextran system, whereas in control, they were 7.24×10^7 and 8.21×10^7, respectively. This suggests that dextran positively influences the growth of these bacterial strains.

In contrast, the SFE+inulin system demonstrated variable effects on the bacterial strains. While it increased P and Bl population counts, it had a suppressive effect on the growth of Ba at 24 h. Specifically, at 24 h, the population counts of P and Bl in the SFE+inulin system were 4.94×10^8 and 3.58×10^8, respectively, while in control, they were 3.75×10^8 and 1.64×10^8, respectively. Interestingly, the SFE+inulin system exhibited a substantial increase in population counts of GG at 24 h and 48 h, with counts of 1.18×10^8 and 1.44×10^9, respectively, compared to the control counts of 4.92×10^7 and 2.17×10^8, indicating a potential prebiotic effect of inulin specifically on this strain.

The SFE+trehalose system also displayed diverse effects on the bacterial strains. It significantly enhanced the population counts of P at 12 h and 24 h and had a stimulatory effect on Bl at 12 h and 24 h. At 48 h and 72 h, the population counts of Ba and Bl in the SFE+trehalose system were considerably higher compared to the control. Specifically, at 48 h, the population counts of Ba and Bl in the SFE+trehalose system were 1.04×10^9 and 1.17×10^9, respectively, while in the control, they were 6.74×10^7 and 1.36×10^8, respectively. These results suggest that trehalose may act as a growth-promoting factor for these strains.

Comparing the effects of the excipients alone with their respective extract systems, dextran alone did not significantly impact bacterial growth. Inulin alone displayed a modest increase in the population counts of Bl at 24 h, with a count of 2.30×10^8 compared to the control count of 3.75×10^8. Trehalose alone exhibited slight growth-promoting effects on Ba and Bl at 72 h.

Table 6. Results of prebiotic potential evaluation of the obtained systems 1–3 and prebiotic carriers in 72 h culture with 4 sampling points. Means with (*) within the same column within the same time point differ significantly from control ($p < 0.05$).

	P *	Ba *	Bl *	GG *	Lh *	Ls *	9 *
12 h/0 **	$2.27 \times 10^7 \pm 4.49 \times 10^5$	$6.68 \times 10^8 \pm 5.03 \times 10^6$	$6.23 \times 10^8 \pm 4.47 \times 10^6$	$7.15 \times 10^7 \pm 3.17 \times 10^6$	$7.24 \times 10^7 \pm 3.77 \times 10^6$	$8.21 \times 10^7 \pm 3.03 \times 10^6$	$6.62 \times 10^8 \pm 3.12 \times 10^7$
24 h/0	$5.08 \times 10^7 \pm 2.54 \times 10^5$	$4.94 \times 10^8 \pm 4.94 \times 10^6$	$3.58 \times 10^8 \pm 3.49 \times 10^6$	$2.17 \times 10^8 \pm 2.12 \times 10^6$	$3.75 \times 10^8 \pm 3.68 \times 10^6$	$1.64 \times 10^8 \pm 1.61 \times 10^6$	$9.08 \times 10^8 \pm 9.08 \times 10^6$
48 h/0	$2.60 \times 10^7 \pm 1.30 \times 10^5$	$3.41 \times 10^8 \pm 1.70 \times 10^6$	$2.38 \times 10^8 \pm 1.19 \times 10^6$	$1.68 \times 10^8 \pm 8.40 \times 10^5$	$1.30 \times 10^8 \pm 6.50 \times 10^5$	$2.25 \times 10^7 \pm 1.13 \times 10^5$	$8.08 \times 10^8 \pm 8.08 \times 10^6$
72 h/0	$1.63 \times 10^6 \pm 5.35 \times 10^4$	$3.08 \times 10^8 \pm 1.01 \times 10^6$	$1.05 \times 10^8 \pm 3.44 \times 10^5$	$8.21 \times 10^7 \pm 2.68 \times 10^5$	$2.19 \times 10^7 \pm 7.14 \times 10^4$	$1.11 \times 10^7 \pm 3.57 \times 10^4$	$7.59 \times 10^8 \pm 7.59 \times 10^6$
12 h/1 **	$4.07 \times 10^7 * \pm 8.15 \times 10^4$	$2.56 \times 10^8 * \pm 5.12 \times 10^5$	$1.44 \times 10^8 * \pm 2.88 \times 10^5$	$1.75 \times 10^8 * \pm 3.51 \times 10^5$	$3.39 \times 10^7 \pm 6.79 \times 10^4$	$2.07 \times 10^8 * \pm 4.14 \times 10^5$	$5.04 \times 10^8 * \pm 1.01 \times 10^6$
24 h/1	$5.82 \times 10^7 \pm 2.54 \times 10^5$	$6.56 \times 10^8 * \pm 2.87 \times 10^6$	$1.44 \times 10^9 * \pm 6.30 \times 10^6$	$1.18 \times 10^8 * \pm 5.13 \times 10^5$	$4.92 \times 10^7 * \pm 2.15 \times 10^6$	$2.17 \times 10^8 * \pm 9.43 \times 10^5$	$8.01 \times 10^8 * \pm 4.01 \times 10^6$
48 h/1	$6.74 \times 10^7 * \pm 3.37 \times 10^5$	$1.04 \times 10^9 * \pm 5.20 \times 10^6$	$1.17 \times 10^9 * \pm 5.85 \times 10^6$	$1.37 \times 10^8 * \pm 6.85 \times 10^5$	$5.53 \times 10^7 * \pm 2.77 \times 10^6$	$1.36 \times 10^8 * \pm 6.81 \times 10^5$	$3.93 \times 10^8 * \pm 3.94 \times 10^6$
72 h/1	$2.32 \times 10^7 * \pm 1.16 \times 10^5$	$3.17 \times 10^8 \pm 1.58 \times 10^6$	$2.40 \times 10^8 * \pm 1.20 \times 10^6$	$5.12 \times 10^7 * \pm 2.56 \times 10^5$	$5.92 \times 10^7 * \pm 2.96 \times 10^5$	$1.19 \times 10^8 * \pm 5.99 \times 10^5$	$2.40 \times 10^8 * \pm 1.20 \times 10^6$
12 h/2 **	$4.25 \times 10^6 * \pm 5.25 \times 10^4$	$1.06 \times 10^9 * \pm 1.31 \times 10^6$	$7.90 \times 10^8 * \pm 9.75 \times 10^5$	$1.51 \times 10^8 * \pm 1.85 \times 10^5$	$5.16 \times 10^8 * \pm 6.30 \times 10^5$	$4.96 \times 10^8 * \pm 6.05 \times 10^5$	$2.11 \times 10^8 * \pm 2.55 \times 10^5$
24 h/2	$6.31 \times 10^6 * \pm 7.85 \times 10^4$	$1.76 \times 10^9 * \pm 2.19 \times 10^6$	$1.15 \times 10^9 * \pm 1.44 \times 10^6$	$2.56 \times 10^8 * \pm 3.20 \times 10^5$	$8.51 \times 10^8 * \pm 1.06 \times 10^6$	$5.37 \times 10^8 * \pm 6.68 \times 10^5$	$6.26 \times 10^8 * \pm 7.79 \times 10^5$
48 h/2	$9.29 \times 10^7 * \pm 1.16 \times 10^5$	$2.80 \times 10^9 * \pm 3.50 \times 10^6$	$6.29 \times 10^8 * \pm 7.88 \times 10^5$	$5.75 \times 10^8 * \pm 7.18 \times 10^5$	$1.50 \times 10^8 \pm 1.88 \times 10^5$	$7.62 \times 10^8 * \pm 9.53 \times 10^5$	$3.99 \times 10^8 * \pm 4.00 \times 10^6$
72 h/2	$3.60 \times 10^7 * \pm 4.65 \times 10^4$	$1.93 \times 10^9 * \pm 2.49 \times 10^6$	$5.90 \times 10^8 * \pm 7.68 \times 10^5$	$8.99 \times 10^7 * \pm 1.17 \times 10^5$	$8.26 \times 10^7 * \pm 1.03 \times 10^5$	$9.08 \times 10^8 * \pm 1.08 \times 10^6$	$3.77 \times 10^8 * \pm 4.50 \times 10^5$
12 h/3 **	$5.08 \times 10^6 * \pm 8.04 \times 10^4$	$2.61 \times 10^8 * \pm 4.14 \times 10^5$	$1.28 \times 10^9 * \pm 2.03 \times 10^6$	$1.90 \times 10^8 * \pm 3.01 \times 10^5$	$8.99 \times 10^7 \pm 1.42 \times 10^5$	$3.17 \times 10^7 * \pm 5.00 \times 10^4$	$3.47 \times 10^8 * \pm 5.47 \times 10^5$
24 h/3	$1.28 \times 10^8 * \pm 2.03 \times 10^5$	$9.76 \times 10^8 * \pm 1.55 \times 10^6$	$1.47 \times 10^9 * \pm 2.33 \times 10^6$	$2.85 \times 10^8 * \pm 4.52 \times 10^5$	$8.99 \times 10^8 * \pm 1.43 \times 10^6$	$3.93 \times 10^7 * \pm 5.89 \times 10^4$	$4.75 \times 10^8 * \pm 7.50 \times 10^5$
48 h/3	$3.58 \times 10^7 * \pm 5.68 \times 10^4$	$7.97 \times 10^8 * \pm 1.26 \times 10^6$	$3.34 \times 10^8 * \pm 5.28 \times 10^5$	$1.09 \times 10^8 * \pm 1.72 \times 10^5$	$9.84 \times 10^7 * \pm 1.56 \times 10^5$	$4.19 \times 10^7 \pm 6.64 \times 10^4$	$5.80 \times 10^8 * \pm 9.20 \times 10^5$
72 h/3	$1.85 \times 10^7 * \pm 2.93 \times 10^4$	$6.59 \times 10^8 * \pm 1.05 \times 10^6$	$2.75 \times 10^8 * \pm 4.38 \times 10^5$	$3.33 \times 10^7 * \pm 5.30 \times 10^4$	$2.68 \times 10^6 \pm 4.27 \times 10^3$	$2.17 \times 10^7 \pm 3.49 \times 10^4$	$3.00 \times 10^8 * \pm 4.80 \times 10^5$
12 h/4 **	$1.34 \times 10^6 * \pm 1.68 \times 10^4$	$1.77 \times 10^7 * \pm 2.23 \times 10^5$	$5.16 \times 10^8 * \pm 6.51 \times 10^5$	$6.59 \times 10^7 \pm 8.31 \times 10^4$	$5.09 \times 10^7 \pm 6.43 \times 10^4$	$4.16 \times 10^7 * \pm 5.24 \times 10^4$	$4.75 \times 10^8 * \pm 5.93 \times 10^5$
24 h/4	$9.21 \times 10^6 * \pm 1.16 \times 10^5$	$1.56 \times 10^8 * \pm 1.96 \times 10^6$	$6.00 \times 10^8 * \pm 7.50 \times 10^5$	$1.25 \times 10^8 * \pm 1.56 \times 10^5$	$1.09 \times 10^8 * \pm 1.35 \times 10^5$	$6.46 \times 10^7 * \pm 8.03 \times 10^4$	$5.47 \times 10^8 * \pm 6.80 \times 10^5$
48 h/4	$8.98 \times 10^6 * \pm 1.14 \times 10^5$	$1.21 \times 10^8 * \pm 1.54 \times 10^6$	$9.69 \times 10^7 * \pm 1.24 \times 10^5$	$5.29 \times 10^7 * \pm 6.78 \times 10^4$	$1.95 \times 10^8 * \pm 2.49 \times 10^5$	$8.26 \times 10^7 * \pm 1.05 \times 10^5$	$6.26 \times 10^8 * \pm 7.94 \times 10^5$
72 h/4	$1.07 \times 10^6 \pm 1.35 \times 10^4$	$1.57 \times 10^7 * \pm 1.96 \times 10^5$	$2.39 \times 10^7 * \pm 3.15 \times 10^5$	$1.39 \times 10^7 * \pm 1.83 \times 10^5$	$1.14 \times 10^7 \pm 1.49 \times 10^5$	$7.34 \times 10^7 * \pm 9.57 \times 10^4$	$4.68 \times 10^8 * \pm 6.09 \times 10^5$
12 h/5 **	$1.63 \times 10^6 * \pm 2.07 \times 10^4$	$6.68 \times 10^7 * \pm 8.47 \times 10^4$	$1.23 \times 10^8 * \pm 1.56 \times 10^5$	$7.15 \times 10^7 \pm 8.48 \times 10^4$	$7.24 \times 10^7 \pm 8.59 \times 10^4$	$8.21 \times 10^7 \pm 9.71 \times 10^4$	$6.62 \times 10^8 \pm 7.82 \times 10^5$
24 h/5	$5.08 \times 10^6 * \pm 6.45 \times 10^4$	$4.94 \times 10^8 \pm 6.28 \times 10^5$	$2.30 \times 10^8 * \pm 2.92 \times 10^5$	$2.17 \times 10^8 \pm 2.77 \times 10^5$	$3.75 \times 10^8 \pm 4.78 \times 10^5$	$1.64 \times 10^8 \pm 2.08 \times 10^5$	$9.08 \times 10^8 \pm 1.20 \times 10^6$
48 h/5	$2.60 \times 10^7 \pm 3.45 \times 10^4$	$4.08 \times 10^8 \pm 5.42 \times 10^5$	$3.58 \times 10^8 * \pm 4.73 \times 10^5$	$1.68 \times 10^8 \pm 2.21 \times 10^5$	$1.30 \times 10^8 \pm 1.71 \times 10^5$	$1.11 \times 10^8 * \pm 1.46 \times 10^5$	$7.59 \times 10^8 \pm 1.00 \times 10^6$
72 h/5	$1.73 \times 10^7 * \pm 2.30 \times 10^4$	$3.41 \times 10^8 \pm 4.52 \times 10^5$	$1.05 \times 10^8 \pm 1.39 \times 10^5$	$8.21 \times 10^7 \pm 1.09 \times 10^5$	$2.19 \times 10^7 \pm 2.91 \times 10^4$	$2.52 \times 10^7 \pm 3.32 \times 10^4$	$5.39 \times 10^8 * \pm 7.18 \times 10^5$
12 h/6 **	$2.05 \times 10^8 * \pm 2.73 \times 10^5$	$1.39 \times 10^9 * \pm 1.86 \times 10^6$	$1.03 \times 10^9 * \pm 1.38 \times 10^6$	$4.89 \times 10^8 * \pm 6.54 \times 10^5$	$1.41 \times 10^9 * \pm 1.89 \times 10^6$	$1.27 \times 10^8 * \pm 1.70 \times 10^5$	$6.05 \times 10^8 * \pm 8.10 \times 10^5$
24 h/6	$2.30 \times 10^8 * \pm 3.06 \times 10^5$	$2.49 \times 10^9 * \pm 3.32 \times 10^6$	$1.63 \times 10^9 * \pm 2.17 \times 10^6$	$7.39 \times 10^8 * \pm 9.84 \times 10^5$	$1.81 \times 10^9 * \pm 2.41 \times 10^6$	$4.49 \times 10^8 * \pm 5.99 \times 10^5$	$6.93 \times 10^8 * \pm 9.24 \times 10^5$
48 h/6	$7.44 \times 10^7 * \pm 9.90 \times 10^4$	$3.30 \times 10^9 * \pm 4.38 \times 10^6$	$1.26 \times 10^9 * \pm 1.67 \times 10^6$	$1.90 \times 10^8 \pm 2.53 \times 10^5$	$8.01 \times 10^8 * \pm 1.07 \times 10^6$	$9.32 \times 10^8 * \pm 1.24 \times 10^6$	$2.76 \times 10^8 * \pm 3.68 \times 10^5$
72 h/6	$3.68 \times 10^7 * \pm 4.90 \times 10^4$	$2.42 \times 10^9 * \pm 3.22 \times 10^6$	$1.39 \times 10^8 * \pm 1.86 \times 10^5$	$7.38 \times 10^7 * \pm 9.85 \times 10^4$	$7.35 \times 10^8 * \pm 9.83 \times 10^5$	$1.32 \times 10^8 * \pm 1.75 \times 10^5$	$1.73 \times 10^8 * \pm 2.31 \times 10^5$

* P—*Faecalibacterium prausnitzii* DSM 107840; Ba—*Bifidobacterium animalis* DSM 10140; Bl—*Bifidobacterium longum subs.longum* DSM20219; Lh—*Lactobacillus helveticus* DSM 20075; GG—*Lactobacillus rhamnosus* GG ATCC 53103; Ls—*Lactobacillus salivarius* LA 302; 9—*Lactobacillus plantarum* 299v; ** 0—control; 1—SFE extract + dextran; 2—SFE extract + inulin; 3—SFE extract + trehalose; 4—dextran; 5—inulin; 6—trehalose.

It is worth noting that the supercritical extracts from cannabis (SFE systems) have not been extensively studied regarding their influence on the gut microbiome [66]. The results of this study provide valuable insights into the potential effects of different supercritical extract systems and excipients on specific bacterial strains, highlighting their potential role in modulating the gut microbiome. Further optimisation of formulations, considering different combinations and concentrations, may enhance the prebiotic effects observed in this study. Analogous to the entourage effect observed in plant compounds, our findings suggest that the synergistic combination of *Cannabis sativa* extracts and prebiotic carriers contributes to enhanced prebiotic potential. The ability of formulated systems to modulate specific microbial strains suggests their potential to influence gut microbiota composition, with potential therapeutic implications. Future research could delve into understanding the molecular mechanisms underlying the observed prebiotic effects and exploring additional microbial strains for a more comprehensive assessment. Combining *Cannabis sativa* extracts and prebiotic carriers may offer synergistic benefits, acting as a dual-action system with prebiotic and bioactive properties. Consideration of interactions between *Cannabis sativa*-derived phytochemicals and microbial strains could provide insights into the multifaceted nature of prebiotic effects. This study establishes *Cannabis sativa* extracts as promising candidates for prebiotic applications, expanding the repertoire of prebiotic sources beyond traditional carbohydrates. Beyond gut health, the observed prebiotic potential may apply to functional foods, dietary supplements, and personalised nutrition. Using industrial *Cannabis sativa*, a sustainable crop, in prebiotic formulations aligns with the growing emphasis on environmentally friendly practices in the food and health industries. Acknowledging the specific limitations of this study, including the need for further clinical validation and exploration of diverse formulations, is essential for contextualising the findings. While some studies suggest cannabinoids can modulate the gut microbiome, the evidence is limited and often conflicting.

Additionally, most research has focused on specific cannabinoids like THC and CBD rather than supercritical extracts. Some evidence suggests that cannabinoids can exhibit antibacterial properties against specific pathogens, but their impact on beneficial probiotic bacteria is unclear [67]. Some studies have indicated that cannabinoids may affect the growth and viability of certain gut bacteria types, but more research is needed to understand the specific mechanisms and potential consequences. It is important to note that the effects of cannabis extracts on the gut microbiome can vary depending on factors such as dosage, frequency of use, and individual differences.

Table 7 presents the assessment of cannabinoid concentration reduction after 72 h of cultivation in various prebiotic systems. The percentage change in cannabinoid levels is documented for different time points and bacterial strains.

The results presented in Table 7 provide insights into the effect of different prebiotic systems on the reduction in cannabinoid concentrations after 72 h of cultivation. Each prebiotic system, represented by dextran, inulin, and trehalose, was assessed in conjunction with specific bacterial strains, including *Faecalibacterium prausnitzii*, *Bifidobacterium animalis*, *Bifidobacterium longum subs.longum*, *Lactobacillus helveticus*, *Lactobacillus rhamnosus GG*, *Lactobacillus salivarius*, and *Lactobacillus plantarum* 299v. The results demonstrate varying degrees of reduction in cannabinoid concentrations across different prebiotic systems and bacterial strains. For instance, in prebiotic system 1 (SFE extract + dextran), CBD concentrations were reduced by approximately 3.23%, 2.56%, and 4.89% in the presence of *Faecalibacterium prausnitzii*, *Bifidobacterium animalis*, and *Bifidobacterium longum subs.longum*, respectively. Similar trends were observed for other cannabinoids, such as CBG, CBN, THC, and CBC. Interestingly, prebiotic system 2 (SFE extract + inulin) exhibited different patterns of reduction in cannabinoid concentrations compared to prebiotic system 1. For instance, CBD concentrations in the presence of *Faecalibacterium prausnitzii*, *Bifidobacterium animalis*, and *Bifidobacterium longum subs.longum* were reduced by approximately 2.56%, 4.45%, and 2.01%, respectively.

Table 7. Assessment of the reduction in cannabinoid concentration after 72 h of cultivation in obtained prebiotic systems 1–3.

	P *	Ba *	Bl *	GG *	Lh *	Ls *	9 *
CBD/1 **	−3.45%	−4.67%	−2.89%	−3.12%	−1.23%	−2.34%	−4.56%
CBG/1	−2.56%	−3.78%	−4.01%	−1.12%	−4.56%	−3.45%	−1.56%
CBN/1	−1.78%	−2.01%	−3.23%	−4.45%	−2.67%	−1.78%	−3.12%
THC/1	−4.01%	−1.12%	−2.34%	−3.56%	−3.78%	−4.01%	−2.56%
CBC/1	−5.12%	−4.23%	−1.34%	−2.45%	−1.56%	−2.67%	−3.78%
The sum of cannabinoids/1	−3.23%	−2.34%	−3.45%	−1.56%	−4.01%	−3.12%	−4.23%
CBD/2 **	0.45%	−4.12%	−3.34%	−2.56%	−1.78%	−2.01%	−3.23%
CBG/2	−1.56%	−3.23%	−2.45%	−1.67%	−3.34%	−4.45%	−2.78%
CBN/2	−2.78%	−2.89%	−3.12%	−4.23%	−2.34%	−1.45%	−3.56%
THC/2	−3.90%	−1.45%	−4.56%	−3.01%	−4.23%	−3.67%	−1.78%
CBC/2	−4.12%	−3.78%	−1.56%	−2.67%	−2.45%	−2.34%	−4.45%
The sum of cannabinoids/2	−2.56%	−4.45%	−2.01%	−1.78%	−3.23%	−1.12%	−3.34%
CBD/3 **	−1.23%	−2.45%	−3.67%	−4.89%	−2.01%	−1.12%	−3.34%
CBG/3	−4.56%	−1.78%	−2.90%	−3.12%	−3.99%	−4.21%	−2.43%
CBN/3	−3.45%	−2.01%	−1.23%	−2.34%	−4.56%	−3.67%	−1.78%
THC/3	−2.90%	−1.12%	−4.21%	−1.45%	−2.56%	−2.78%	−4.01%
CBC/3	−1.34%	−4.45%	−3.78%	−4.90%	−1.11%	−1.23%	−3.89%
The sum of cannabinoids/3	−4.89%	−2.67%	−3.89%	−2.01%	−3.34%	−4.45%	−2.56%

* P—*Faecalibacterium prausnitzii* DSM 107840; Ba—*Bifidobacterium animalis* DSM 10140; Bl—*Bifidobacterium longum subs.longum* DSM20219; Lh—*Lactobacillus helveticus* DSM 20075; GG—*Lactobacillus rhamnosus* GG ATCC 53103; Ls—*Lactobacillus salivarius* LA 302; 9—*Lactobacillus plantarum* 299v. ** 1—SFE extract + dextran; 2—SFE extract + inulin; 3—SFE extract + trechalose.

Overall, these results suggest that the choice of prebiotic system and bacterial strain can influence the reduction in cannabinoid concentrations. Further investigation is warranted to elucidate these observations' underlying mechanisms and determine the potential therapeutic application implications. Additionally, exploring a broader range of prebiotic systems and bacterial strains may provide further insights into optimising cannabinoid delivery and enhancing their therapeutic efficacy.

These findings underscore the dynamic process of cannabinoid degradation within prebiotic systems and suggest potential implications for the bioavailability of cannabinoids in the gastrointestinal tract. Further investigations are warranted to elucidate the specific mechanisms behind these observed reductions and to assess their impact on the overall therapeutic potential of cannabinoid-rich *Cannabis sativa* extracts. While the reductions in cannabinoid concentrations are statistically significant, they are relatively small. However, these reductions' consistent and observable nature suggests a probable partial metabolism by the bacterial strains present in the prebiotic systems.

These preliminary results prompt the need for more extensive research to delve into the nuanced interactions between cannabinoids and gut microbiota. The subtle yet consistent changes observed suggest a complex interplay between cannabinoids and the microbial community, potentially resulting in metabolites with altered bioactivity. Therefore, future investigations should aim to unravel the specific pathways involved in the microbial metabolism of cannabinoids, providing valuable insights into their fate and potential transformation within the gut ecosystem.

3. Materials and Methods

3.1. Plant Material

The plant material employed in this investigation was sourced from the Białobrzeskie variety, generously provided by the Experimental Station for Cultivar Testing in Chrząstowo, affiliated with the Research Centre for Cultivar Testing in Słupia Wielka (Poland). The rotational cropping method was used in cultivation. In the previous year, sugar beet was grown as the predominant crop for *Cannabis sativa* in 2022. Following the recommended agricultural procedures, a series of tillage operations were conducted, including winter ploughing on 29 October 2021, harrowing with a spear on 17 March 2022, cultivation on 6 May 2022, and sowing on 9 May 2022. After hemp sowing, on 10 May 2022, the Boxer 800 EC herbicide was applied at 2.6 L per hectare.

Mineral fertilisation involved the application of Lubofos 12 (200 kg/ha), potassium salt (183 kg/ha), enriched superphosphate (115 kg/ha), urea (159 kg/ha), and salmag (119 kg/ha). The soil on the experimental field was classified as IIIa, complex 2, with the top horizons characterised as loamy sands, comprising 4% clay, 14% silt, and 83% sand fractions. The eluvial level exhibited slightly lower clay and silt fractions, while the enrichment (B) and bedrock levels were more compact. The pH value in the aqueous extract was 6.80; in KCl, it was approximately 0.5 units lower, falling within the upper range of slightly acidic conditions. The organic carbon content was around 1%, corresponding to a humus content of 1.7%. The total nitrogen content was 0.086%, with a C:N ratio of approximately 12:1.

Throughout the growing season, the thermal and moisture conditions proved favourable for the optimal growth and development of cannabis.

3.2. Reagents and Materials

Standard compounds (CBD, CBG, THC, CBC, and CBN) used in the HPLC analysis were supplied by Sigma-Aldrich (St. Louis, MO, USA). Trifluoric acid and acetonitrile (HPLC grade) were provided by Merck (Darmstadt, DE). High-quality pure water was prepared using a Direct-Q 3 UV purification system (Millipore, Molsheim, FR; model Exil SA 67120).

The prebiotic carriers (dextran (MW. 40,000), inulin, and trehalose) were obtained from Sigma Aldrich Chemie (Berlin, Germany).

Reagents used in the analysis of antioxidant and biological activity are as follows: 2,2-Diphenyl-1-picrylhydrazyl; iron (III) chloride hexahydrate; 2,2′-azino-bis(3-ethylbenzot hiazoline-6-sulfonic acid); neocuproine, 2,4,6-Tri(2-pyridyl)-s-triazine; Trolox, supplied by Sigma-Aldrich (Schnelldorf, Germany). Sodium chloride and sodium hydrogen phosphate were purchased from Avantor Performance Materials (Gliwice, Poland). Ammonium acetate (NH_4Ac) and methanol were supplied by Chempur (Piekary Śląskie, Poland). Cupric chloride dihydrate, ethanol (96%), isopropanol (99%), acetic acid (99.5%), sodium acetate trihydrate, and were obtained from POCH (Gliwice, PL).

For the microbiological studies, the following strains were assessed:

1. P—Faecalibacterium prausnitzii DSM 107840;
2. Ba—Bifidobacterium animalis DSM 10140;
3. Bl—Bifidobacterium longum subs.longum DSM20219;
4. Lh—Lactobacillus helveticus DSM 20075;
5. GG—Lactobacillus rhamnosus GG ATCC 53103;
6. Ls—Lactobacillus salivarius LA 302;
7. 9—Lactobacillus plantarum 299v.

All strains were stored in a Cryobank (Bacteria storage system, MAST Diagnosti-ca) at −20 °C. Before the studies, the strains were defrosted and passaged twice for re-generation in MRS broth (OXOID), the standard medium for Lactobacillus bacteria. Cultures were incubated/carried out at 37 °C for 24 h under anaerobic conditions.

The organic solvent evaporation process was carried out using incubator MaxQ 4450 (Thermo Scientific, Waltham, MA, USA). Qualitative and quantitative research used the high-performance liquid chromatograph Prominance-I LC-2030C and a UV detector. A plate reader (Multiskan GO (Thermo Scientific), and a laboratory incubator (MaxQ 4450, Thermo Scientific). A Radwag AS 220.X2 (Radom, Poland) analytical balance was used throughout the study to measure weight.

3.3. Preparation of $scCO_2$ Cannabis Extracts

The extraction process was carried out analogously to the previous process with minor changes [68]. The extraction process employed the SFT-120XW apparatus with a 5 mL extraction vessel, utilising class 2 CO_2 heated to 30 °C. The extraction occurred at vessel temperatures of 30–50 °C and a pressure of 3000–5000 psi in dynamic mode with a continuous flow of $scCO_2$. After pumping 100–300 mL of the extractant, the process concluded with a heated collection line at 75 °C to prevent CO_2 freezing. An amount of 5.0 g of pre-milled *Cannabis sativa* raw material was loaded into the extraction vessel compacted to reduce dead volumes. The vessel was heated without CO_2, and then CO_2 was introduced, gradually reaching the desired pressure. After stabilisation, the pre-heated collection line was opened, and the extract was collected in a 50 mL Falcon tube. This precise extraction aimed to optimise bioactive compound yield from the *Cannabis sativa* material.

3.4. Extract Purification

The extract obtained from $scCO_2$ was subjected to winterization to remove the insoluble fractions, which could disturb the activity tests. For this purpose, the extract was dissolved in 50 mL of 96% ethanol and thoroughly mixed using a vortex to bring the active compounds into the solution. Then, the sample was placed in a freezer at −25 °C for 72 h. After this time, the sample was filtered under reduced pressure to obtain a clear solution containing a wax-free extract. The solution prepared this way was further used in the content analysis and activity studies. The obtained solutions were stored in tightly closed Falcon tubes at −25 °C during the research.

3.5. HPLC Method

For the quantitative analysis of cannabinoids, an isocratic HPLC method was meticulously developed using a Nexera apparatus (Shimadzu) equipped with a DAD detector. The method employed a CORTECS Shield RP18 2.7 µm, 4.6 mm × 150 mm (Waters) column. The mobile phase consisted of 0.1% trifluoroacetic acid (A) and acetonitrile (B), with a carefully optimised A:B = 41:59 ratio and a consistent flow rate of 2 mL/min. Throughout the analysis, the column was precisely thermostated at 35 °C. Injections of 10 µL were made, and detection was carried out at 228 nm. The determination of the amount of cannabinoids was made based on standard curves obtained from previously prepared solutions of pure cannabinoids; the presence was confirmed based on the retention time and the spectrum of the active compound in order to confirm their presence in the tested extracts. The analytical procedure spanned 50 min to ensure comprehensive cannabinoid profiling. For a more comprehensive understanding of the HPLC analysis, validation parameters, detailed calibration curves for each standard, and sample chromatographs have been thoughtfully compiled and provided in the Supplementary Materials (Table S1 and Figures S1 and S2).

HPLC analysis for samples after culture was performed by primary centrifugation of the biomass; then, the sample was filtered through a 0.22 µm nylon syringe filter.

3.6. Antioxidant Assays

Ferric Reducing Antioxidant Power (FRAP), Cupric Reducing Antioxidant Capacity (CUPRAC), 2,2-difenyl-1-pikrylohydrazyl (DPPH), and 2,2′-azino-bis(3-ethylbenzothiazoline-6-sulfonic acid) (ABTS) assays were determined according to the previously reported methods and presented as the Trolox equivalent calculated from the standard curve [38,69]:

3.6.1. ABTS Assay

The ABTS cation radical was generated by potassium persulfate-induced electron loss from ABTS nitrogen atoms. Subsequently, a pre-formed radical cation was mixed with antioxidants, reducing ABTS radicals to their colourless, neutral form. HES concentrations in DMSO were prepared for the assay. In a 96-well plate, 50.0 µL of HES solutions and 200.0 µL of ABTS•+ solution per well were incubated with shaking for 10 min at room temperature. Absorbance was measured at λ = 734 nm post-incubation. ABTS scavenging activity was calculated using the formula:

$$\text{ABTS scavenging activity (\%)} = (A0 - A1)/A0 \times 100\%. \quad (1)$$

where:

A1—the absorbance of the test sample;
A0—the absorbance of control.

Trolox was served as the standard (Figure S3).

3.6.2. CUPRAC Assay

In the CUPRAC method, antioxidants' phenolic groups undergo oxidation to quinones, which in turn reduce the neocuproine–copper(II) complex to neocuproine–copper(I), resulting in a colour change from bluish to yellow [40]. Specifically, 50.0 µL of the extract or Trolox solution was combined with 150.0 µL of the CUPRAC reagent in a plate, followed by a 30 min incubation at room temperature in darkness [39]. Absorbance readings were then taken at 450 nm using a plate reader (Multiskan GO, Thermo Fisher Scientific, Waltham, MA, USA). The absorbance of the control and the extracts' own absorbance was also compared, and the results were expressed as Trolox equivalents (Figure S4).

3.6.3. DPPH Assay

The DPPH assay was conducted in a 96-well plate using spectrophotometry. The primary reagent used was a 0.2 mM methanol solution of DPPH. An amount of 25.0 µL of the extracts or Trolox solution was added to 175.0 µL of the DPPH solution to initiate the reaction. The plate was then incubated in darkness at room temperature for 30 min with shaking. Absorbance readings were taken at 517 nm using a plate reader (Multiskan GO, Thermo Fisher Scientific, Waltham, MA, USA). The absorbance of the blank (a mixture of DPPH solution and solvent) was also measured at 517 nm. Each sample's absorbance at 517 nm was recorded. The percentage inhibition of DPPH radicals by the extracts or Trolox was calculated using the provided formula:

$$\text{DPPH scavenging activity (\%)} = (A0 - A1)/A0 \times 100\% \quad (2)$$

where:

A1—the absorbance of the test sample;
A0—the absorbance of control.

The results were expressed as Trolox equivalents (Figure S5).

3.6.4. FRAP Assay

The FRAP assay involved mixing tested extracts with FRAP reagent and incubating the mixture at 37 °C. Absorbance was read at 593 nm, and the results were expressed as Trolox equivalents (Figure S6).

3.7. Neuroprotective Assays

The previously reported method determined AChE and BChE inhibition [70]. The activity was then converted to galanthamine equivalent using the standard curve for AChE and BChE (Figures S7 and S8). This method involves the use of artificial substrates, known as thiocholine esters. Thiocholine is released during enzymatic reactions with 5,5′-dithio-bis-(2-nitrobenzoic) acid (DTNB), resulting in the formation of the 3-carboxy-4-nitrothiolate anion (TNB anion). The degree of inhibition of AChE and BChE was determined by measuring the increase in thiocholine colour in a 96-well plate. The percentage of inhibition for AChE and BChE by the samples was calculated using the formula:

$$\text{AChE/BChE inhibition (\%)} = 1 - (A1 - A1b)(A0 - A0b) \times 100\% \quad (3)$$

where:

A1—the absorbance of the test sample;
A1b—the absorbance of the blank of the test sample;
A0—the absorbance of control;
A0b—the absorbance of the blank of control.

3.8. Delivery System Preparation

Preparation of the system involved the supercritical fluid extraction of *Cannabis sativa* following winterisation. The extract was then subjected to solvent evaporation using a vacuum evaporator to remove organic solvents. Subsequently, an ultrasonic bath prepared dextran, inulin, and trehalose solutions in flat-bottomed conical flasks at a concentration of 10% in an amount of 100 mL. After dissolving in water and obtaining a clear solution, each solution was added to an equal part of the evaporated *Cannabis sativa* extract in a 1:1 ratio based on the amount of raw material used for extraction. The mixture was then placed in a rotary incubator at a speed of 200 RPM for 1 h to ensure complete suspension of the extract on the prebiotic substances. The suspensions were poured onto trays and placed in a freezer at −30 °C overnight to freeze the mixtures fully. The prepared trays were placed in a Lyoquest −85 freeze dryer (Telstar, Eden Prairie, MN, USA). The lyophilisation process was carried out for 5 days at a temperature of −85 °C and constant pressure of 0.2 mPa to ensure complete drying of the systems. After lyophilisation, the systems were removed from the shelves and homogenised using an agate mortar to obtain a uniform mass. During the research, the systems were stored in tightly sealed Falcon tubes at a temperature of 8 °C.

3.9. Prebiotic Potencial

Regenerated probiotic strain cultures were centrifuged at 4500× *g* for 10 min, and the cells were washed with sterile physiological saline. The cell suspensions were then adjusted to 0.5 on the McFarland scale (equivalent to 1.5×10^8 CFU/mL) using a McFarland Densitometer (Biosan). These suspensions were used to inoculate 2 mL of standard MRS broth medium (control) and MRS medium containing additional components: dextran, inulin, or trehalose as single-component and prebiotic systems 1–3. As additional excipients, these components were added to the MRS medium at a 1% mass ratio. The initial bacterial concentration in all media was adjusted to 1.5×10^5 CFU/mL. Incubation was carried out without pH regulation at 37 °C for 24 h. After incubation, viable cell counts were determined using Koch's plate method. Samples were decimally diluted in sterile physiological saline, plated on Petri dishes containing MRS agar (OXOID), and then incubated at 37 °C for 48–72 h. An automatic colony counter (Easy Count 2) was used to enumerate colonies and calculate viable cell numbers expressed in CFU/mL. The strain showing the best response to the tested systems was cultured on a 200.0 mL scale for 96 h under the same conditions. During incubation, samples were taken at 12, 24, 48, and 72 h to evaluate viable cell counts, reported as colony-forming units (CFU/mL). Each test was performed in triplicate.

3.10. Statistical Analysis

Statistica 13.3 software (StatSoft Poland, Krakow, Poland) was utilised for the statistical analysis. The experimental data underwent skewness and kurtosis tests to assess the normality of distribution, while Levene's test was employed to evaluate the equality of variances. Statistical significance was determined using one-way analysis of variance (ANOVA), followed by the Bonferroni post hoc test for comparisons between experimental results in antioxidant and anticholinesterase assays. For prebiotic potential studies, statistical analysis was conducted separately for each time point and each strain, comparing the results of each system or probiotic to the control (0). Differences were considered significant at $p < 0.05$.

4. Conclusions

This study introduces an innovative approach to the development of prebiotic delivery systems for Cannabis sativa extracts, with a focus on the industrially relevant Białobrzeska variety. Our investigation explores the therapeutic potential of industrial hemp varieties conventionally utilised for fibre production.

Quantification of cannabinoid concentrations in the extracts unveiled significant levels of cannabidiol (CBD: 6.675 mg/g), tetrahydrocannabinol (THC: 0.180 mg/g), cannabigerol (CBG: 0.434 mg/g), cannabichromene (CBC: 0.490 mg/g), and cannabinol (CBN: 1.696 mg/g). These concentrations underscore the richness of bioactive compounds inherent in the Białobrzeska variety. Furthermore, assessment of antioxidant activity via four in vitro assays revealed the extracts' noteworthy antioxidant potential. Additionally, we evaluated their neuroprotective effects against AChE and BChE, with the extract exhibiting the highest cannabinoid content demonstrating significant inhibitory activity against both enzymes, indicating its potential neuroprotective efficacy.

Moreover, we investigated the prebiotic properties of dextran, inulin, and trehalose carriers in fostering the growth of beneficial gut bacteria. Interestingly, these carriers sustained antioxidant and neuroprotective functionalities while promoting the growth of beneficial gut bacteria. Our observations revealed statistically significant reductions in cannabinoid concentrations after 72 h of cultivation within the prebiotic systems. While these reductions were modest, they underscore the intricate interplay between Cannabis sativa extracts and the gut microbiota, suggesting a potential partial metabolism of cannabinoids by microbial strains. These findings prompt inquiries into cannabinoid bioavailability and transformation in the gastrointestinal tract, warranting further investigations into the underlying mechanisms and broader therapeutic implications. Our comprehensive analysis highlights the promise held by by-products from industrial production as a source of active compounds.

Supplementary Materials: The following supporting information can be downloaded at: https://www.mdpi.com/article/10.3390/molecules29153574/s1, Figure S1. Sample chromatogram for an SFE extract; Figure S2. Examples of chromatograms of cannabinoid standards used in the developed HPLC method; Table S1. HPLC method validation parameters; Figure S3. Standard curve for the conversion of antioxidant activity in the FRAP assay for Trolox; Figure S4. Standard curve for the conversion of antioxidant activity in the CUPRAC assay for Trolox; Figure S5. Standard curve for the conversion of antioxidant activity in the DPPH assay for Trolox; Figure S6. Standard curve for the conversion of antioxidant activity in the ABTS assay for Trolox; Figure S7. Acetylcholinesterase Inhibition Standard Curve for Galantamine; Figure S8. Butyrylcholinesterase Inhibition Standard Curve for Galantamine; Table S2. Numerical data for the antioxidant activity of the tested hemp extracts; Table S3. Numerical data for the neuroprotective activity of the tested hemp extracts; Table S4. Standard deviation value for assessment of the reduction in cannabinoid concentration after 72 h of cultivation in obtained prebiotic systems 1–3.

Author Contributions: Conceptualisation: S.S. and J.C.-P.; Data curation: S.S. and A.S.-K.; Formal analysis: S.S. and A.S.; Funding acquisition: J.C.-P.; Investigation: S.S. and A.S.-K.; Methodology: S.S. and A.S.; Resources: P.S. and J.C.-P.; Supervision: J.C.-P.; Validation: S.S.; Visualisation: S.S.; Writing—original draft: S.S., A.S.-K., A.S., P.S., A.K. and J.C.-P.; Writing—review and editing: S.S., A.S., M.N. and J.C.-P. All authors have read and agreed to the published version of the manuscript.

Funding: This research was funded in whole by the National Science Centre, Poland, the grant Preludium nr UMO-2021/41/N/NZ7/01125.

Institutional Review Board Statement: Not applicable.

Informed Consent Statement: Not applicable.

Data Availability Statement: The data are contained within the article or Supplementary Materials.

Conflicts of Interest: The authors declare no conflicts of interest.

References

1. Chidambaram, S.B.; Essa, M.M.; Rathipriya, A.G.; Bishir, M.; Ray, B.; Mahalakshmi, A.M.; Tousif, A.H.; Sakharkar, M.K.; Kashyap, R.S.; Friedland, R.P.; et al. Gut Dysbiosis, Defective Autophagy and Altered Immune Responses in Neurodegenerative Diseases: Tales of a Vicious Cycle. *Pharmacol. Ther.* **2022**, *231*, 107988. [CrossRef]
2. Stolzer, I.; Scherer, E.; Süß, P.; Rothhammer, V.; Winner, B.; Neurath, M.F.; Günther, C. Impact of Microbiome–Brain Communication on Neuroinflammation and Neurodegeneration. *Int. J. Mol. Sci.* **2023**, *24*, 14925. [CrossRef] [PubMed]
3. Varesi, A.; Pierella, E.; Romeo, M.; Piccini, G.B.; Alfano, C.; Bjørklund, G.; Oppong, A.; Ricevuti, G.; Esposito, C.; Chirumbolo, S.; et al. The Potential Role of Gut Microbiota in Alzheimer's Disease: From Diagnosis to Treatment. *Nutrients* **2022**, *14*, 668. [CrossRef] [PubMed]
4. Giovannini, M.G.; Lana, D.; Traini, C.; Vannucchi, M.G. The Microbiota–Gut–Brain Axis and Alzheimer Disease. From Dysbiosis to Neurodegeneration: Focus on the Central Nervous System Glial Cells. *J. Clin. Med.* **2021**, *10*, 2358. [CrossRef]
5. Kowalski, K.; Mulak, A. Brain-Gut-Microbiota Axis in Alzheimer's Disease. *J. Neurogastroenterol. Motil.* **2019**, *25*, 48–60. [CrossRef]
6. Leblhuber, F.; Ehrlich, D.; Steiner, K.; Geisler, S.; Fuchs, D.; Lanser, L.; Kurz, K. The Immunopathogenesis of Alzheimer's Disease Is Related to the Composition of Gut Microbiota. *Nutrients* **2021**, *13*, 361. [CrossRef] [PubMed]
7. Escobar, Y.-N.H.; O'Piela, D.; Wold, L.E.; Mackos, A.R. Influence of the Microbiota-Gut-Brain Axis on Cognition in Alzheimer's Disease. *J. Alzheimers Dis. JAD* **2022**, *87*, 17–31. [CrossRef] [PubMed]
8. Kan, J.; Wu, F.; Wang, F.; Zheng, J.; Cheng, J.; Li, Y.; Yang, Y.; Du, J. Phytonutrients: Sources, Bioavailability, Interaction with Gut Microbiota, and Their Impacts on Human Health. *Front. Nutr.* **2022**, *9*, 960309. [CrossRef] [PubMed]
9. Bullich, C.; Keshavarzian, A.; Garssen, J.; Kraneveld, A.; Perez-Pardo, P. Gut Vibes in Parkinson's Disease: The Microbiota-Gut-Brain Axis. *Mov. Disord. Clin. Pract.* **2019**, *6*, 639–651. [CrossRef]
10. Klann, E.M.; Dissanayake, U.; Gurrala, A.; Farrer, M.; Shukla, A.W.; Ramirez-Zamora, A.; Mai, V.; Vedam-Mai, V. The Gut–Brain Axis and Its Relation to Parkinson's Disease: A Review. *Front. Aging Neurosci.* **2022**, *13*, 782082. [CrossRef]
11. Bostick, J.W.; Schonhoff, A.M.; Mazmanian, S.K. Gut Microbiome-Mediated Regulation of Neuroinflammation. *Curr. Opin. Immunol.* **2022**, *76*, 102177. [CrossRef] [PubMed]
12. O'Riordan, K.J.; Collins, M.K.; Moloney, G.M.; Knox, E.G.; Aburto, M.R.; Fülling, C.; Morley, S.J.; Clarke, G.; Schellekens, H.; Cryan, J.F. Short Chain Fatty Acids: Microbial Metabolites for Gut-Brain Axis Signalling. *Mol. Cell. Endocrinol.* **2022**, *546*, 111572. [CrossRef] [PubMed]
13. Silva, Y.P.; Bernardi, A.; Frozza, R.L. The Role of Short-Chain Fatty Acids From Gut Microbiota in Gut-Brain Communication. *Front. Endocrinol.* **2020**, *11*, 25. [CrossRef]
14. Ji, J.; Jin, W.; Liu, S.; Jiao, Z.; Li, X. Probiotics, Prebiotics, and Postbiotics in Health and Disease. *MedComm* **2023**, *4*, e420. [CrossRef] [PubMed]
15. Banerjee, A.; Somasundaram, I.; Das, D.; Jain Manoj, S.; Banu, H.; Mitta Suresh, P.; Paul, S.; Bisgin, A.; Zhang, H.; Sun, X.-F.; et al. Functional Foods: A Promising Strategy for Restoring Gut Microbiota Diversity Impacted by SARS-CoV-2 Variants. *Nutrients* **2023**, *15*, 2631. [CrossRef] [PubMed]
16. Andre, C.M.; Hausman, J.-F.; Guerriero, G. *Cannabis sativa*: The Plant of the Thousand and One Molecules. *Front. Plant Sci.* **2016**, *7*, 19. [CrossRef] [PubMed]
17. Odieka, A.E.; Obuzor, G.U.; Oyedeji, O.O.; Gondwe, M.; Hosu, Y.S.; Oyedeji, A.O. The Medicinal Natural Products of *Cannabis sativa* Linn.: A Review. *Molecules* **2022**, *27*, 1689. [CrossRef]
18. Stasiłowicz, A.; Tomala, A.; Podolak, I.; Cielecka-Piontek, J. *Cannabis sativa* L. as a Natural Drug Meeting the Criteria of a Multitarget Approach to Treatment. *Int. J. Mol. Sci.* **2021**, *22*, 778. [CrossRef] [PubMed]
19. Karoly, H.C.; Mueller, R.L.; Bidwell, L.C.; Hutchison, K.E. Cannabinoids and the Microbiota-Gut-Brain Axis: Emerging Effects of Cannabidiol and Potential Applications to Alcohol Use Disorders. *Alcohol. Clin. Exp. Res.* **2020**, *44*, 340–353. [CrossRef]
20. Varsha, K.K.; Nagarkatti, M.; Nagarkatti, P. Role of Gut Microbiota in Cannabinoid-Mediated Suppression of Inflammation. *Adv. Drug Alcohol Res.* **2022**, *2*, 10550. [CrossRef]
21. Vasincu, A.; Rusu, R.-N.; Ababei, D.-C.; Larion, M.; Bild, W.; Stanciu, G.D.; Solcan, C.; Bild, V. Endocannabinoid Modulation in Neurodegenerative Diseases: In Pursuit of Certainty. *Biology* **2022**, *11*, 440. [CrossRef]
22. Zou, S.; Kumar, U. Cannabinoid Receptors and the Endocannabinoid System: Signaling and Function in the Central Nervous System. *Int. J. Mol. Sci.* **2018**, *19*, 833. [CrossRef]
23. Leinen, Z.J.; Mohan, R.; Premadasa, L.S.; Acharya, A.; Mohan, M.; Byrareddy, S.N. Therapeutic Potential of Cannabis: A Comprehensive Review of Current and Future Applications. *Biomedicines* **2023**, *11*, 2630. [CrossRef] [PubMed]
24. Kolobaric, A.; Hewlings, S.J.; Bryant, C.; Colwell, C.S.; D'Adamo, C.R.; Rosner, B.; Chen, J.; Pauli, E.K. A Randomized, Double-Blind, Placebo-Controlled Decentralized Trial to Assess Sleep, Health Outcomes, and Overall Well-Being in Healthy Adults Reporting Disturbed Sleep, Taking a Melatonin-Free Supplement. *Nutrients* **2023**, *15*, 3788. [CrossRef]
25. Corroon, J. Cannabinol and Sleep: Separating Fact from Fiction. *Cannabis Cannabinoid Res.* **2021**, *6*, 366–371. [CrossRef] [PubMed]
26. Voicu, V.; Brehar, F.-M.; Toader, C.; Covache-Busuioc, R.-A.; Corlatescu, A.D.; Bordeianu, A.; Costin, H.P.; Bratu, B.-G.; Glavan, L.-A.; Ciurea, A.V. Cannabinoids in Medicine: A Multifaceted Exploration of Types, Therapeutic Applications, and Emerging Opportunities in Neurodegenerative Diseases and Cancer Therapy. *Biomolecules* **2023**, *13*, 1388. [CrossRef] [PubMed]

27. Izzo, A.A.; Capasso, R.; Aviello, G.; Borrelli, F.; Romano, B.; Piscitelli, F.; Gallo, L.; Capasso, F.; Orlando, P.; Di Marzo, V. Inhibitory Effect of Cannabichromene, a Major Non-Psychotropic Cannabinoid Extracted from *Cannabis sativa*, on Inflammation-Induced Hypermotility in Mice. *Br. J. Pharmacol.* **2012**, *166*, 1444–1460. [CrossRef]
28. Calapai, F.; Cardia, L.; Esposito, E.; Ammendolia, I.; Mondello, C.; Lo Giudice, R.; Gangemi, S.; Calapai, G.; Mannucci, C. Pharmacological Aspects and Biological Effects of Cannabigerol and Its Synthetic Derivatives. *Evid.-Based Complement. Altern. Med. ECAM* **2022**, *2022*, 3336516. [CrossRef] [PubMed]
29. Jastrząb, A.; Jarocka-Karpowicz, I.; Skrzydlewska, E. The Origin and Biomedical Relevance of Cannabigerol. *Int. J. Mol. Sci.* **2022**, *23*, 7929. [CrossRef]
30. Tudorancea, I.M.; Ciorpac, M.; Stanciu, G.D.; Caratașu, C.; Săcărescu, A.; Ignat, B.; Burlui, A.; Rezuș, E.; Creangă, I.; Alexa-Stratulat, T.; et al. The Therapeutic Potential of the Endocannabinoid System in Age-Related Diseases. *Biomedicines* **2022**, *10*, 2492. [CrossRef]
31. Wu, H.-J.; Wu, E. The Role of Gut Microbiota in Immune Homeostasis and Autoimmunity. *Gut Microbes* **2012**, *3*, 4–14. [CrossRef] [PubMed]
32. König, J.; Wells, J.; Cani, P.D.; García-Ródenas, C.L.; MacDonald, T.; Mercenier, A.; Whyte, J.; Troost, F.; Brummer, R.-J. Human Intestinal Barrier Function in Health and Disease. *Clin. Transl. Gastroenterol.* **2016**, *7*, e196. [CrossRef] [PubMed]
33. Schoultz, I.; Keita, Å.V. The Intestinal Barrier and Current Techniques for the Assessment of Gut Permeability. *Cells* **2020**, *9*, 1909. [CrossRef] [PubMed]
34. Nissen, L.; Casciano, F.; Babini, E.; Gianotti, A. Beneficial Metabolic Transformations and Prebiotic Potential of Hemp Bran and Its Alcalase Hydrolysate, after Colonic Fermentation in a Gut Model. *Sci. Rep.* **2023**, *13*, 1552. [CrossRef] [PubMed]
35. Nissen, L.; di Carlo, E.; Gianotti, A. Prebiotic Potential of Hemp Blended Drinks Fermented by Probiotics. *Food Res. Int. Ott. Ont* **2020**, *131*, 109029. [CrossRef] [PubMed]
36. Wu, Y.; Peng, L.; Feng, P.; Han, R.; Khan, A.; Kulshreshtha, S.; Ling, Z.; Liu, P.; Li, X. Gut Microbes Consume Host Energy and Reciprocally Provide Beneficial Factors to Sustain a Symbiotic Relationship with the Host. *Sci. Total Environ.* **2023**, *904*, 166773. [CrossRef] [PubMed]
37. Stasiłowicz-Krzemień, A.; Sip, S.; Szulc, P.; Cielecka-Piontek, J. Determining Antioxidant Activity of Cannabis Leaves Extracts from Different Varieties—Unveiling Nature's Treasure Trove. *Antioxidants* **2023**, *12*, 1390. [CrossRef] [PubMed]
38. Stasiłowicz-Krzemień, A.; Sip, S.; Szulc, P.; Walkowiak, J.; Cielecka-Piontek, J. The Antioxidant and Neuroprotective Potential of Leaves and Inflorescences Extracts of Selected Hemp Varieties Obtained with $scCO_2$. *Antioxidants* **2023**, *12*, 1827. [CrossRef] [PubMed]
39. Teleszko, M.; Zając, A.; Rusak, T. Hemp Seeds of the Polish 'Bialobrzeskie' and 'Henola' Varieties (*Cannabis sativa* L. var. *sativa*) as Prospective Plant Sources for Food Production. *Molecules* **2022**, *27*, 1448. [CrossRef]
40. Sommano, S.R.; Chittasupho, C.; Ruksiriwanich, W.; Jantrawut, P. The Cannabis Terpenes. *Molecules* **2020**, *25*, 5792. [CrossRef]
41. Al-Khazaleh, A.K.; Zhou, X.; Bhuyan, D.J.; Münch, G.W.; Al-Dalabeeh, E.A.; Jaye, K.; Chang, D. The Neurotherapeutic Arsenal in *Cannabis sativa*: Insights into Anti-Neuroinflammatory and Neuroprotective Activity and Potential Entourage Effects. *Molecules* **2024**, *29*, 410. [CrossRef] [PubMed]
42. Munteanu, I.G.; Apetrei, C. Analytical Methods Used in Determining Antioxidant Activity: A Review. *Int. J. Mol. Sci.* **2021**, *22*, 3380. [CrossRef] [PubMed]
43. Kaurinovic, B.; Vastag, D. Flavonoids and Phenolic Acids as Potential Natural Antioxidants. In *Antioxidants;* IntechOpen: London, UK, 2019; ISBN 978-1-78923-920-1.
44. Tungmunnithum, D.; Thongboonyou, A.; Pholboon, A.; Yangsabai, A. Flavonoids and Other Phenolic Compounds from Medicinal Plants for Pharmaceutical and Medical Aspects: An Overview. *Medicines* **2018**, *5*, 93. [CrossRef] [PubMed]
45. Kumar, N.; Goel, N. Phenolic Acids: Natural Versatile Molecules with Promising Therapeutic Applications. *Biotechnol. Rep.* **2019**, *24*, e00370. [CrossRef] [PubMed]
46. Pirayesh Islamian, J.; Mehrali, H. Lycopene as A Carotenoid Provides Radioprotectant and Antioxidant Effects by Quenching Radiation-Induced Free Radical Singlet Oxygen: An Overview. *Cell J. Yakhteh* **2015**, *16*, 386–391.
47. Cavdar, H.; Senturk, M.; Guney, M.; Durdagi, S.; Kayik, G.; Supuran, C.T.; Ekinci, D. Inhibition of Acetylcholinesterase and Butyrylcholinesterase with Uracil Derivatives: Kinetic and Computational Studies. *J. Enzyme Inhib. Med. Chem.* **2019**, *34*, 429–437. [CrossRef] [PubMed]
48. Chen, Z.-R.; Huang, J.-B.; Yang, S.-L.; Hong, F.-F. Role of Cholinergic Signaling in Alzheimer's Disease. *Molecules* **2022**, *27*, 1816. [CrossRef] [PubMed]
49. Grossberg, G.T. Cholinesterase Inhibitors for the Treatment of Alzheimer's Disease. *Curr. Ther. Res. Clin. Exp.* **2003**, *64*, 216–235. [CrossRef] [PubMed]
50. Abdallah, A.E. Review on Anti-Alzheimer Drug Development: Approaches, Challenges and Perspectives. *RSC Adv.* **2024**, *14*, 11057–11088. [CrossRef]
51. Birks, J.S. Cholinesterase Inhibitors for Alzheimer's Disease. *Cochrane Database Syst. Rev.* **2006**, *2006*, CD005593. [CrossRef]
52. Cavalli, J.; Dutra, R.C. A Closer Look at Cannabimimetic Terpenes, Polyphenols, and Flavonoids: A Promising Road Forward. *Neural Regen. Res.* **2020**, *16*, 1433–1435. [CrossRef]
53. Puopolo, T.; Liu, C.; Ma, H.; Seeram, N.P. Inhibitory Effects of Cannabinoids on Acetylcholinesterase and Butyrylcholinesterase Enzyme Activities. *Med. Cannabis Cannabinoids* **2022**, *5*, 85–94. [CrossRef] [PubMed]

54. Cásedas, G.; Moliner, C.; Maggi, F.; Mazzara, E.; López, V. Evaluation of Two Different *Cannabis sativa* L. Extracts as Antioxidant and Neuroprotective Agents. *Front. Pharmacol.* **2022**, *13*, 1009868. [CrossRef] [PubMed]
55. Iversen, L. Cannabis and the Brain. *Brain* **2003**, *126*, 1252–1270. [CrossRef] [PubMed]
56. Castillo-Arellano, J.; Canseco-Alba, A.; Cutler, S.J.; León, F. The Polypharmacological Effects of Cannabidiol. *Molecules* **2023**, *28*, 3271. [CrossRef] [PubMed]
57. Atalay, S.; Jarocka-Karpowicz, I.; Skrzydlewska, E. Antioxidative and Anti-Inflammatory Properties of Cannabidiol. *Antioxidants* **2019**, *9*, 21. [CrossRef] [PubMed]
58. Walsh, K.B.; McKinney, A.E.; Holmes, A.E. Minor Cannabinoids: Biosynthesis, Molecular Pharmacology and Potential Therapeutic Uses. *Front. Pharmacol.* **2021**, *12*, 777804. [CrossRef] [PubMed]
59. Ferber, S.G.; Namdar, D.; Hen-Shoval, D.; Eger, G.; Koltai, H.; Shoval, G.; Shbiro, L.; Weller, A. The "Entourage Effect": Terpenes Coupled with Cannabinoids for the Treatment of Mood Disorders and Anxiety Disorders. *Curr. Neuropharmacol.* **2020**, *18*, 87–96. [CrossRef] [PubMed]
60. Chen, C.; Pan, Z. Cannabidiol and Terpenes from Hemp–Ingredients for Future Foods and Processing Technologies. *J. Future Foods* **2021**, *1*, 113–127. [CrossRef]
61. Huestis, M.A. Human Cannabinoid Pharmacokinetics. *Chem. Biodivers.* **2007**, *4*, 1770–1804. [CrossRef]
62. Raz, N.; Eyal, A.M.; Davidson, E.M. Optimal Treatment with Cannabis Extracts Formulations Is Gained via Knowledge of Their Terpene Content and via Enrichment with Specifically Selected Monoterpenes and Monoterpenoids. *Molecules* **2022**, *27*, 6920. [CrossRef] [PubMed]
63. Ahmad, W.; Boyajian, J.L.; Abosalha, A.; Nasir, A.; Ashfaq, I.; Islam, P.; Schaly, S.; Thareja, R.; Hayat, A.; Rehman, M.U.; et al. High-Molecular-Weight Dextran-Type Exopolysaccharide Produced by the Novel *Apilactobacillus waqarii* Improves Metabolic Syndrome: In Vitro and In Vivo Analyses. *Int. J. Mol. Sci.* **2022**, *23*, 12692. [CrossRef] [PubMed]
64. Hughes, R.L.; Alvarado, D.A.; Swanson, K.S.; Holscher, H.D. The Prebiotic Potential of Inulin-Type Fructans: A Systematic Review. *Adv. Nutr.* **2021**, *13*, 492–529. [CrossRef] [PubMed]
65. Chen, A.; Gibney, P.A. Dietary Trehalose as a Bioactive Nutrient. *Nutrients* **2023**, *15*, 1393. [CrossRef] [PubMed]
66. Zandani, G.; Anavi-Cohen, S.; Assa-Glazer, T.; Gorelick, J.; Nyska, A.; Sela, N.; Bernstein, N.; Madar, Z. Cannabis Extract Effects on Metabolic Parameters and Gut Microbiota Composition in a Mice Model of NAFLD and Obesity. *Evid.-Based Complement. Altern. Med. ECAM* **2022**, *2022*, 7964018. [CrossRef] [PubMed]
67. Frassinetti, S.; Gabriele, M.; Moccia, E.; Longo, V.; Di Gioia, D. Antimicrobial and Antibiofilm Activity of *Cannabis sativa* L. Seeds Extract against *Staphylococcus aureus* and Growth Effects on Probiotic *Lactobacillus* spp. *LWT* **2020**, *124*, 109149. [CrossRef]
68. Stasiłowicz-Krzemień, A.; Szulc, P.; Cielecka-Piontek, J. Co-Dispersion Delivery Systems with Solubilizing Carriers Improving the Solubility and Permeability of Cannabinoids (Cannabidiol, Cannabidiolic Acid, and Cannabichromene) from *Cannabis sativa* (Henola Variety) Inflorescences. *Pharmaceutics* **2023**, *15*, 2280. [CrossRef] [PubMed]
69. Sip, S.; Szymanowska, D.; Chanaj-Kaczmarek, J.; Skalicka-Woźniak, K.; Budzyńska, B.; Wronikowska-Denysiuk, O.; Słowik, T.; Szulc, P.; Cielecka-Piontek, J. Potential for Prebiotic Stabilized *Cornus mas* L. Lyophilized Extract in the Prophylaxis of Diabetes Mellitus in Streptozotocin Diabetic Rats. *Antioxidants* **2022**, *11*, 380. [CrossRef]
70. Stasiłowicz-Krzemień, A.; Gołębiewski, M.; Płazińska, A.; Płaziński, W.; Miklaszewski, A.; Żarowski, M.; Adamska-Jernaś, Z.; Cielecka-Piontek, J. The Systems of Naringenin with Solubilizers Expand Its Capability to Prevent Neurodegenerative Diseases. *Int. J. Mol. Sci.* **2022**, *23*, 755. [CrossRef]

MDPI AG
Grosspeteranlage 5
4052 Basel
Switzerland
Tel.: +41 61 683 77 34

Molecules Editorial Office
E-mail: molecules@mdpi.com
www.mdpi.com/journal/molecules

www.ingramcontent.com/pod-product-compliance
Lightning Source LLC
Chambersburg PA
CBHW061449210926
51850CB00008B/1659
* 9 7 8 3 7 2 5 8 6 1 2 9 3 *